Oligonu

The Practical Approach Series

Affinity Chromatography
Anaerobic Microbiology
Animal Cell Culture
Animal Virus Pathogenesis
Antibodies I and II
Biochemical Toxicology
Biological Membranes
Biosensors
Carbohydrate Analysis
Cell Growth and Division
Cellular Calcium
Cellular Neurobiology
Centrifugation (2nd edition)
Clinical Immunology
Computers in Microbiology
Crystallization of Proteins and Nucleic Acids
Cytokines
The Cytoskeleton
Directed Mutagenesis
DNA Cloning I, II, and III
Drosophila
Electron Microscopy in Biology
Electron Microscopy in Molecular Biology
Enzyme Assays
Essential Molecular Biology I and II
Fermentation
Flow Cytometry
Gel Electrophoresis of Nucleic Acids (2nd edition)
Gel Electrophoresis of Proteins (2nd edition)
Genome Analysis
HPLC of Small Molecules
HPLC of Macromolecules
Human Cytogenetics
Human Genetic Diseases
Immobilised Cells and Enzymes
Iodinated Density Gradient Media
Light Microscopy in Biology
Liposomes
Lymphocytes
Lymphokines and Interferons

Mammalian Cell Biotechnology
Mammalian Development
Medical Bacteriology
Medical Mycology
Microcomputers in Biology
Microcomputers in Physiology
Mitochondria
Molecular Neurobiology
Mutagenicity Testing
Neurochemistry
Nucleic Acid and Protein Sequence Analysis
Nucleic Acids Hybridisation
Nucleic Acids Sequencing
Oligonucleotide Synthesis
Oligonucleotides and Analogues
PCR
Peptide Hormone Action
Peptide Hormone Secretion
Photosynthesis: Energy Transduction
Plant Cell Culture
Plant Molecular Biology
Plasmids
Post-implantation Mammalian Embryos
Prostaglandins and Related Substances
Protein Architecture
Protein Function
Protein Purification Applications
Protein Purification Methods
Protein Sequencing
Protein Structure
Proteolytic Enzymes
Radioisotopes in Biology
Receptor Biochemistry
Receptor–Effector Coupling
Receptor–Ligand Interactions
Ribosomes and Protein Synthesis
Solid Phase Peptide Synthesis
Spectrophotometry and Spectrofluorimetry
Steroid Hormones
Teratocarcinomas and Embryonic Stem Cells
Transcription and Translation
Virology
Yeast

Oligonucleotides and Analogues

A Practical Approach

Edited by

F. ECKSTEIN

Max-Planck-Institut für Experimentelle Medizin, Göttingen, Germany

OXFORD UNIVERSITY PRESS

Oxford New York Tokyo

This book has been printed digitally and produced in a standard specification in order to ensure its continuing availability

OXFORD
UNIVERSITY PRESS

Great Clarendon Street, Oxford OX2 6DP

Oxford University Press is a department of the University of Oxford.
It furthers the University's objective of excellence in research, scholarship,
and education by publishing worldwide in

Oxford New York

Auckland Cape Town Dar es Salaam Hong Kong Karachi
Kuala Lumpur Madrid Melbourne Mexico City Nairobi
New Delhi Shanghai Taipei Toronto

With offices in

Argentina Austria Brazil Chile Czech Republic France Greece
Guatemala Hungary Italy Japan South Korea Poland Portugal
Singapore Switzerland Thailand Turkey Ukraine Vietnam

Oxford is a registered trade mark of Oxford University Press
in the UK and in certain other countries

Published in the United States
by Oxford University Press Inc., New York

Reprinted 2011

ISBN 978-0-19-963279-4

Preface

THE chemical synthesis of oligodeoxynucleotides has come a long way since the last volume on this subject, *Oligonucleotide synthesis: a practical approach*, edited by M. Gait, was published in 1984. Still being somewhat of an art for the specialist in those days, the methodology has since undergone tremendous development. Synthesis of unmodified oligodeoxynucleotides has now become routine in most instances and certainly does not require a particular expertise any longer. This, of course, is entirely due to the automation of this process and the ongoing improvements by manufacturers in the design of synthesizers and the development of new reagents. In addition to the ease of synthesis, the realization that these compounds have a much wider application than originally anticipated makes the chemical synthesis of oligodeoxynucleotides such a fast growing area. Moreover, the discovery that certain RNAs can have catalytic activities and that RNA–protein interaction plays an important role in the control of gene expression has led to an enormous interest in the automated chemical synthesis of oligoribonucleotides as well. All these developments justify the publication of a new Practical Approach book on the chemical synthesis of oligonucleotides, including examples of their applications.

The reader will find that most of the areas included are those which have undergone considerable changes since the appearance of M. Gait's book. These include two chapters on the state of the art in automated synthesis of oligonucleotides and oligoribonucleotides, although the latter has yet to reach the level of perfection of the former. Several chapters deal with the synthesis of modified oligonucleotides. There are three chapters which describe the modification of the phosphate backbone to phosphorothioates, phosphorodithioates, and the methyl phosphonates. The importance of these lie to a considerable degree in their potential application as therapeutics. Other chapters describe the synthesis of sugar-modified oligonucleotides such as the 3′-O-methyl derivatives, the introduction of base modifications, and the attachment of reporter groups at various positions in the oligonucleotides. These latter modifications are of considerable interest to those looking for non-radioactive probes in hybridization and for suitable reporter groups for studying DNA–DNA or DNA–protein interactions.

The authors of the various chapters are all experts in their field and very often are the persons who have developed the particular area. They all submitted manuscripts of a very high standard which greatly simplified my task as editor, a job which I had accepted with some trepidation. Their enthusiastic co-operation guaranteed the delivery of manuscripts and facilitated rapid publication. I thank them all for their active support in preparing this book, which, I hope, will meet the needs of the many who have an active interest in exploring the scope of applications of oligonucleotides.

F.E.

Göttingen
January 1991

Contents

List of contributors xix

Abbreviations xxiii

1. Modern machine-aided methods of oligodeoxyribonucleotide synthesis 1

Tom Brown and Dorcas J. S. Brown

1. Introduction 1
    The history of the solid phase method 1
2. Solid phase deoxyoligonucleotide synthesis 2
    Principles of solid phase deoxyoligonucleotide synthesis 2
    Solid phase phosphoramidite oligonucleotide synthesis 4
3. Automation of solid phase oligonucleotide synthesis 9
    Advantages of automation 9
    A practical guide to automated DNA synthesis 10
    Oligonucleotide cleavage and deprotection 18
    Limitations of solid phase deoxyoligonucleotide synthesis 19
4. Conclusions 23

References 23

2. Oligoribonucleotide synthesis 25

Michael J. Gait, Clare Pritchard, and George Slim

1. Introduction 25
2. Basic chemistry of oligoribonucleotide synthesis 26
3. Materials and reagents 32
4. Assembly of oligoribonucleotide chains and deprotection 35
    Machine and reagent preparation 35
    Column packing 35
    Assembly cycle 35
    Deprotection 36
5. Desalting and purification of oligoribonucleotides 36
    Precautions to avoid ribonuclease contamination 36
    Desalting 37

HPLC purification of oligoribonucleotides 39
Polyacrylamide gel electrophoresis 42

6. Enzymatic analysis of oligoribonucleotides 45

7. Special applications 45
Mixed RNA–DNA oligonucleotides 45
Phosphorothioate oligoribonucleotides 46

8. Future developments 47

References 48

3. 2′-O-Methyloligoribonucleotides: synthesis and applications 49

B. S. Sproat and A. I. Lamond

1. Introduction 49

2. Synthesis of protected 2′-O-methylribonucleoside-3′-O-phosphoramidites, coupling agent, succinates, supports, and biotinylation reagent 50
Preparation of 2′-O-methylribonucleoside-3′-O-phosphoramidites 50
Synthesis of the condensing agent 5-(4-nitrophenyl)-1H-tetrazole 70
Preparation of protected 2′-O-methylribonucleoside-3′-O-succinates and loading of controlled pore glass 71
Synthesis of a building block for direct biotinylation during solid phase synthesis 72

3. Synthesis of 2′-O-methyloligoribonucleotides 78

4. Antisense affinity selection and affinity depletion of RNA–protein complexes 79
General points 79
Affinity depletion 79
Affinity selection 83

References 85

4. Phosphorothioate oligonucleotides 87

Gerald Zon and Wojciech J. Stec

1. Introduction 87

2. Applications of phosphorothioate analogues of DNA and RNA 88

Issues in antisense applications of phosphorothioate oligonucleotides 91
3. Synthetic methods 96
Overview 96
4. Product isolation 100
5. ^{35}S-Labelling 102
References 103

5. Synthesis of oligonucleotide phosphorodithioates 109

Graham Beaton, Douglas Dellinger, William S. Marshall, and Marvin H. Caruthers

1. Introduction 109
2. Outline of the chemistry 110
3. Synthesis of protected deoxynucleoside-3′-O-pyrrolidino-S-(2,4-dichlorobenzyl)-phosphorothioamidites 113
Preparation of *tris*(pyrrolidino)phosphine 113
Preparation of deoxyribonucleoside-3′-O-pyrrolidino-S-(2,4-dichlorobenzyl)phosphorothioamidites 114
4. Synthesis of DNA containing phosphorodithioate linkages by solid phase methods 119
Common synthesis reagents 120
The sulphur solution 121
Preparation of deoxyribonucleoside-3′-O-phosphoramidite and deoxyribonucleoside-3′-O-phosphorothioamidite solutions 121
The synthesis cycle 122
5. Deprotection and isolation of oligodeoxynucleotides containing the phosphorodithioate linkage 124
6. Analysis and purification of oligodeoxyribonucleotide products 125
Analysis and purification using PAGE 126
HPLC methods 129
Acknowledgements 132
References 133
Appendix: Chemical suppliers 134

6. Synthesis of oligo-2′-deoxyribonucleoside methylphosphonates 137

Paul S. Miller, Cynthia D. Cushman, and Joel T. Levis

1. Introduction 137
2. Synthesis 137
   Chemicals required 139
   Synthesis of oligonucleoside methylphosphonates in a DNA synthesizer 139
   Deprotection of oligonucleoside methylphosphonates 140
   Purification of oligonucleoside methylphosphonates 142
   Desalting oligonucleoside methylphosphonates 143
3. Characterization of oligonucleoside methylphosphonates 143
   Phosphorylation and gel electrophoresis of oligonucleoside methylphosphonates 144
   Hydrolysis of oligonucleoside methylphosphonates with piperidine 147
   Determining the location of pyrimidines and purines in oligonucleoside methylphosphonates 147
4. Use of antisense oligonucleoside methylphosphonates in cell culture experiments 149
   Inhibition of virus protein synthesis in vesicular stomatitis virus infected cells 150

   References 153

7. Oligodeoxynucleotides containing modified bases 155

Bernard A. Connolly

1. Introduction 155
2. General methods 155
   Safety precautions 155
   Analytical and preparative techniques 155
   Necessity for anhydrous conditions 156
3. Synthesis of modified bases 157
   Preparation of a purine-9-ß-D-2′-deoxyribofuranoside derivative suitable for oligodeoxynucleotide synthesis 157
   Preparation of 2-aminopurine-9-ß-O-2′-deoxyribofuranoside and 6-thiodeoxyguanosine derivatives suitable for oligonucleotide synthesis 162

Preparation of 4-thiothymidine and 5-methyl-2-pyrimidinone -1-ß-D-2′-deoxyribofuranoside derivatives suitable for oligonucleotide synthesis 168
Preparation of 3′-phosphoramidites 174
Commercially available modified bases suitable for oligonucleotide synthesis 176

4. Synthesis, purification, and characterization of oligonucleotides containing modified bases 176
Synthesis 176
Purification 177
Characterization 179

5. Applications of oligodeoxynucleotides containing modified bases 181
As probes of protein–DNA interactions 181
As spectral probes 181
As reactive probes 182

Acknowledgements 182

References 183

8. Oligonucleotides with reporter groups attached to the 5′-terminus 185

Nanda D. Sinha and Steve Striepeke

1. Introduction 185
2. Selection, synthesis, and application of non-nucleosidic reagents 186
3. Synthesis and purification of linkers 189
4. Synthesis of the 5′-terminal modifying agents 191
Synthesis of phosphoramidite derivatives as modifying agents 191
Synthesis of H-phosphonate linker 194
5. Incorporation of linkers on to oligonucleotide chains and purification 195
Synthesis of 5′-end modified oligonucleotides using phosphoramidite chemistry 195
Synthesis of 5′-end modified oligonucleotides using H-phosphonate chemistry 197
Deprotection and isolation of 5′-modified oligonucleotides from solid supports 198
6. Incorporation of non-radioactive marker molecules on to free primary amine or sulphydryl-linked oligonucleotides 200

7. Applications of oligonucleotides carrying non-radioactive reporter molecules 205
DNA sequencing 205
Diagnostic probes 205

Acknowledgements 207

References 207

Appendix: Chemical suppliers 208

9. Site specific attachment of labels to the DNA backbone 211

Nancy E. Conway, Jacqueline A. Fidanza, Maryanne J. O'Donnell, Nicole D. Narekian, Hiroaki Ozaki, and Larry W. McLaughlin

1. Introduction 211
Labelling techniques directed towards the DNA backbone 212

2. The synthesis of oligonucleotides containing phosphorothioate diesters 213
Synthesis of an oligodeoxynucleotide containing a single internucleotidic phosphorothioate diester 214
Synthesis of stereochemically pure oligodeoxynucleotides containing phosphorothioate diesters 216

3. Reactions of phosphorothioate diesters 219
Analysis and purification of the labelled oligodeoxynucleotides 222
Variations in reactivity 226
Stability of the product phosphorothioate triesters 227
Thermal stability of labelled phosphorothioate triester-containing DNA 228

4. Applications for site-specifically modified DNA fragments 228
Detection of sequence-specific protein–DNA binding 229
Detection of binding between the tryptophan repressor and operator 229

5. DNA containing multiple phosphorothioate diesters 231
Synthesis of DNA with multiple phosphorothioate diesters 232

6. Fluorescent labelling of phosphorothioate diesters for the detection of DNA 233
Introduction of multiple fluorophores into DNA containing phosphorothioates 234
Post-assay labelling for high detection sensitivity 234
Detection sensitivity of DNA labelled with multiple fluorophores 236

7. Fluorescent detection of DNA 237
DNA sequencing 237
DNA hybridization 237
8. Conclusions 237
Acknowledgements 238
References 238

10. Oligodeoxynucleotides for affinity chromatography 241
Robert Blanks and Larry W. McLaughlin
1. Introduction 241
2. Matrix-bound oligodeoxynucleotide ligands 241
Design of the oligodeoxynucleotide ligand 242
Choice of chromatographic matrix 243
3. Preparation of oligodeoxynucleotide matrices for affinity chromatography 244
Preparation of the oligodeoxynucleotide ligand 245
Attachment of the oligodeoxynucleotide ligand to the Sepharose matrix 247
4. Protein isolation with oligodeoxynucleotide affinity columns 248
Isolation of the *Eco*RI restriction endonuclease 249
5. Conclusions 253
Acknowledgements 253
References 253

11. Oligodeoxynucleotides with reporter groups attached to the base 255
Jerry L. Ruth
1. Introduction 255
2. Synthesis of 'linker arm' oligodeoxynucleotides 256
Synthesis of pyrimidine nucleosides with linker arms at C-5 257
Pyrimidine nucleosides with linker arms 261
Synthesis of purine nucleosides with linker arms 263
Conversion of the blocked nucleoside to 3'-O-ß-cyanoethyl N,N-diisopropylphosphoramidites for oligonucleotide synthesis 264

Oligodeoxynucleotide synthesis using linker arm nucleoside amidites 265

3. Attaching reporter groups to linker arm oligomers 267
Attaching biotin to amine linker arm oligonucleotides 267
Attaching fluorophores and lumiphores to amine linker arm oligomers 268
Conjugation of enzymes to linker arm oligodeoxynucleotides 270

4. Use and behaviour of oligodeoxynucleotides with non-isotopic reporter groups 275
Hybridization of oligomer probes 275
Effect of modification and reporter group on hybridization 276

5. Concluding remarks 279

References 281

12. Oligonucleotides attached to intercalators, photoreactive and cleavage agents 283

N. T. Thuong and U. Asseline

1. Introduction 283

2. Synthesis of intercalating and active derivatives with a hydroxyl or halogenoalkyl group via a linker 284
Synthesis of the acridine derivatives (**1b**) and (**1c**) 286
Preparation of the methylpyrroporphyrin XXI derivative (**2b**) 287
Synthesis of orthophenanthroline intermediates (**3**) 288
Preparation of EDTA intermediate (**4**) 289
Preparation of the psoralen intermediates (**5**) 290

3. Attachment of intercalators and cleavage reagents to oligomers 291
Synthesis by the phosphotriester method in solution 291
Synthesis on a solid support of oligodeoxyribonucleotides covalently linked to an acridine derivative 294

4. Synthesis via a phosphorothioate group 295
Preparation of oligodeoxyribonucleotides involving a phosphorothioate group at their 5′-end (**15**) 295
Preparation of oligodeoxyribonucleotides carrying a phosphorothioate group at their 3′-end 297
Coupling reaction between oligodeoxyribonucleotides containing a phosphorothioate group (**20**) and the substituent group involving a halogenoalkyl linker to give (**21**) 299

5. Purification, analysis, and identification[a,b] 300
General methods 300
Systems commonly used for analysis and purification 302
6. Properties and potential applications of oligodeoxyribonucleotides covalently linked to intercalating and reactive groups 303
Interactions between the oligodeoxyribonucleotide—intercalator conjugates and their complementary sequences 303
Biological effects of oligodeoxyribonucleotides covalently linked to intercalating agents 304
Irreversible modifications induced by modified oligonucleotides in the target sequences 304
Conclusions 305
References 305
Appendix: Chemical suppliers 306
Acknowledgements 308

Appendix

Suppliers of specialist items 309

Index 311

Contributors

U. ASSELINE
Centre de Biophysique Moléculaire, CNRS, 1A Avenue de la Recherche Scientifique, F-45071 Orléans Cedex 2, France.

GRAHAM BEATON
Department of Chemistry and Biochemistry, University of Colorado, Boulder, Colorado 80309, USA.

ROBERT BLANKS
Department of Chemistry, Boston College, Chestnut Hill, Massachusetts 02167, USA.

TOM BROWN
University of Edinburgh, Department of Chemistry, West Mains Road, Edinburgh EH9 3JJ, UK.

DORCAS J. S. BROWN
University of Edinburgh, Department of Chemistry, West Mains Road, Edinburgh EH9 3JJ, UK.

MARVIN H. CARUTHERS
Department of Chemistry and Biochemistry, University of Colorado, Boulder, Colorado 80309, USA.

BERNARD A. CONNOLLY
Department of Biochemistry, Medical School, University of Southampton, Bassett Crescent, Southampton SO9 3TU, UK.

NANCY E. CONWAY
Department of Chemistry, Boston College, Chestnut Hill, Massachusetts 02167, USA.

CYNTHIA D. CUSHMAN
Department of Biochemistry, School of Hygiene and Public Health, The Johns Hopkins University, Baltimore, Maryland 21205, USA.

DOUGLAS DELLINGER
Department of Chemistry and Biochemistry, University of Colorado, Boulder, Colorado 80309, USA.

JACQUELINE A. FIDANZA
Department of Chemistry, Boston College, Chestnut Hill, Massachusetts 02167, USA.

MICHAEL J. GAIT
Medical Research Council, Laboratory of Molecular Biology, Hills Road, Cambridge CB2 2QH, UK.

A. I. LAMOND
EMBL, Meyerhofstrasse 1, Postfach 102209, D-6900 Heidelberg, Germany.

JOEL T. LEVIS
Department of Biochemistry, School of Hygiene and Public Health, The Johns Hopkins University, Baltimore, Maryland 21205, USA.

WILLIAM S. MARSHALL
Department of Chemistry and Biochemistry, University of Colorado, Boulder, Colorado 80309, USA.

LARRY W. McLAUGHLIN
Department of Chemistry, Boston College, Chestnut Hill, Massachusetts 02167, USA.

PAUL S. MILLER
Department of Biochemistry, School of Hygiene and Public Health, The Johns Hopkins University, Baltimore, Maryland 21205, USA.

NICOLE D. NAREKIAN
Department of Chemistry, Boston College, Chestnut Hill, Massachusetts 02167, USA.

MARYANNE J. O'DONNELL
Department of Chemistry, Boston College, Chestnut Hill, Massachusetts 02167, USA.

HIROAKI OZAKI
Department of Chemistry, Boston College, Chestnut Hill, Massachusetts 02167, USA.

CLARE PRITCHARD
Medical Research Council, Laboratory of Molecular Biology, Hills Road, Cambridge CB2 2QH, UK.

JERRY L. RUTH
Molecular Biosystems, 10030 Barnes Canyon Road, San Diego, California 92121, USA.

NANDA D. SINHA
Millipore Corporation, 89 Ashby Road, Bedford, Massachusetts 01730, USA.

GEORGE SLIM
Medical Research Council, Laboratory of Molecular Biology, Hills Road, Cambridge CB2 2QH, UK.

B. S. SPROAT
EMBL, Meyerhofstrasse 1, Postfach 102209, D-6900 Heidelberg, Germany.

WOJCIECH J. STEC
Polish Academy of Sciences, Centre of Molecular and Macromolecular Studies, Department of Bioorganic Chemistry, 90–363 Lódz, Poland.

STEVE STRIEPEKE
Millipore Corporation, 80 Ashby Road, Bedford, Massachusetts 01730, USA.

N. T. THUONG
Centre de Biophysique Moléculaire, CNRS, 1A Avenue de la Recherche Scientifique, F-45071 Orléans Cedex 2, France.

GERALD ZON
Applied Biosystems, 850 Lincoln Centre Drive, Foster City, California 94404, USA.

Abbreviations

AdCOCl	adamantane carbonylchloride
AP	alkaline phosphatase
BCIP	5-bromo-4-chloro-3-indoyl phosphate
BDDDP	2-tert-butylimino-2-diethylamino-1,3-dimethylperhydro-1,3, 2-diazaphosphorin
BPB	bromophenol blue
BSA	bovine serum albumin
Bz	benzoyl-
CPG	controlled pore glass
DAB	diaminobenzidine
DABCO	1,4-diazabicyclo[2.2.2]octan
DBU	1,8-diazabicyclo[5.4.0]undec-7-ene
DCA	dichloroacetic acid
DMEM	Dulbecco's modified Eagle's medium
DMF	N,N-dimethylformamide
DMSO	dimethylsulphoxide
DMT(r)	dimethoxytrityl-
DSS	di-succinimidyl suberate
DTT	dithiothreitol
EDTA	ethylenediamine tetraacetic acid
EMEM	Eagle's minimal essential medium
EtOAc	ethylacetate
FITC	fluorescein isothiocyanate
FPMP	1-(2-fluorophenyl)-4-methoxypiperidin-4-yl
HPLC	high performance liquid chromatography
HRP	horseradish peroxidase
Ibu	isobutyryl-
IPP	triethylammonium isopropylphosphonate
MMT	monomethoxytrityl
MUBP	4-methylumbelliferone phosphate
NBD	N-(7-nitrobenz-2-oxa-1,3-diazol-4-yl-)
NBT	nitro blue tetrazolium
NHS	N-hydroxysuccinimide
NMI	N-methylimidazole
OD	optical density
PAGE	polyacrylamide gel electrophoresis
PCR	polymerase chain reaction
PMSF	phenylmethane sulphonylfluoride
SAX	strong ion exchanger
SDS	sodium dodecylsulphate

t-BOC	tert-butyloxycarbomyl
TBAF	tetrabutylammonium fluoride
TBDMS	tert-butyldimethylsilyl-
TBE	tris-borate EDTA buffer
TCA	trichloroacetic acid
TEA	triethylamine
TEAA	tetraethylammonium acetate
TEAB	triethylammonium bicarbonate
TETD	tetraethylthiuram disulphide
TFA	trifluoroacetyl
THF	tetrahydrofuran
TLC	thin layer chromatography
UV	ultraviolet light
XC	xylene cyanol

1

Modern machine-aided methods of oligodeoxyribonucleotide synthesis

TOM BROWN and DORCAS J. S. BROWN

1. Introduction

During the 1980s the development of efficient automated methods of solid phase oligonucleotide synthesis has had a major impact in the fields of molecular biology, biotechnology, and biological chemistry (1). Advances in nucleic acid chemistry have made possible the synthesis of oligonucleotides with modified backbones, non-standard bases, or with non-radioactive labels attached to the 3′- or 5′-termini (2). The synthesis of ribo-oligonucleotides has recently become possible and this will undoubtedly have far reaching implications in molecular biology (3, 4). New techniques of DNA sequencing have now been fully automated and the polymerase chain reaction (PCR) has facilitated the detection and amplification of very small amounts of natural DNA (5). Novel and imaginative applications and modifications of oligonucleotides emerge at a remarkable rate and this is clearly a very exciting era for nucleic acid chemists. Some of the most significant recent advances in the field of synthetic oligonucleotides are referred to in subsequent chapters.

1.1 The history of the solid phase method

The principle of solid phase synthesis was developed by Bruce Merrifield at the Rockefeller Institute (now Rockefeller University) in the 1950s and 1960s (6). This simple and ingenious technique was initially applied to the synthesis of polypeptides. Merrifield realized that the key to successful peptide synthesis was the anchoring of the first amino acid to an insoluble polymeric support. Other amino acids could then be joined, one by one, to the fixed terminus. At the end of the sequence, the completed chain could be detached from the insoluble polymer and purified. The process proved to be highly efficient and of great significance and fundamental importance for research on hormones, enzymes, and many commercial peptide-based drugs. Indeed, the impact and importance of Merrifield's work was such that he was awarded the Nobel Prize for Chemistry in 1984. Soon after the solid phase method had been shown to be valuable for peptide synthesis, the technique was applied to the synthesis of oligonucleotides (2). In the past 20 years there have been several important developments (3, 8) and the synthesis of deoxyoligonucleotides is now a relatively simple exercise.

2. Solid phase deoxyoligonucleotide synthesis

2.1 Principles of solid phase deoxyoligonucleotide synthesis

Today almost all synthetic oligonucleotides are prepared by solid phase phosphoramidite techniques (9–13). The method is illustrated schematically in *Figure 1*, which shows all stages of solid phase synthesis of a dimer. The nucleoside at the 3′-terminus is attached by means of a linker arm to a solid support, usually a bead of borosilicate glass. In the first step in the synthesis the support-bound nucleoside is deprotected to provide a free 5′-hydroxyl group for the attachment of the second nucleotide. An excess of the second nucleotide, protected at the 5′-hydroxyl position to prevent self-polymerization and activated at the 3′-phosphate position to facilitate condensation, is then added. This results in the for
must be oxidized. A capping
inert to further monomer additi
the whole several times until
Finally, the fully assembled olig
deprotected, and purified by
advantages of solid phase synth

- An oligonucleotide can be
  nucleotide addition). When c
  protecting groups is taken int
  oligonucleotide within one d
- As the growing oligonucleo
  excess reagents are simply w
  avoiding laborious purificati
- All chemical reactions can
  excesses of solution reagents r
- The process is amenable to aut

The basic chemical step that results in oligonucleotide chain extension involves reaction of an activated mononucleotide in solution with the free 5′-hydroxyl group of the support-bound oligonucleotide. As the activated species is in large excess in solution, the deleterious effect of any side reactions, such as hydrolysis or oxidation, which might deactivate the mononucleotide, are minimized.

In theory it would be possible to devise a method of oligonucleotide

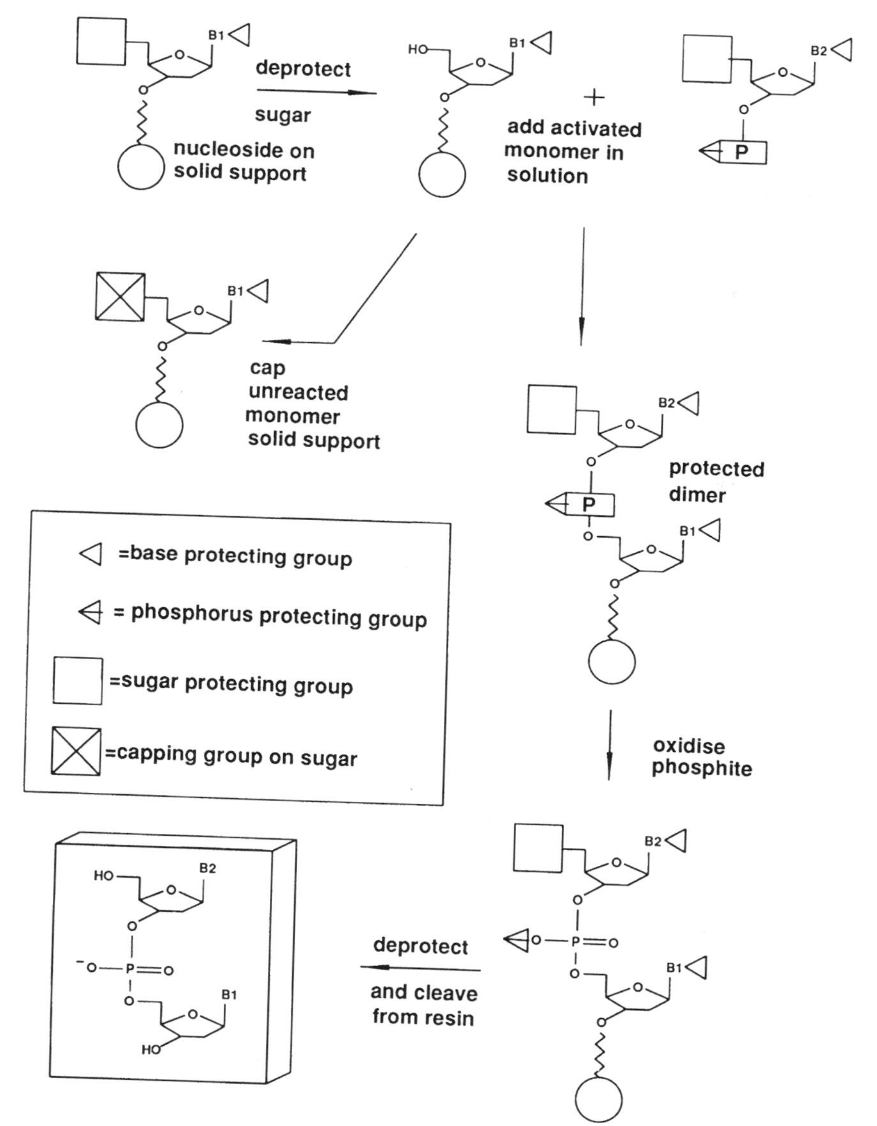

Figure 1. A schematic diagram of solid phase oligodeoxynucleotide synthesis.

synthesis that would involve activation of a phosphate group on the support-bound oligonucleotide. However, such a method would probably be unsuccessful, as the occasional deactivation of the phosphate group due to inevitable side reactions would lead to premature chain termination and the accumulation of failure sequences. The chemical synthesis of DNA is normally carried out in the 3′- to 5′-direction to take advantage of the high chemical reactivity of the primary 5′-hydroxyl group. This is in contrast to the biosynthesis of nucleic acids, which proceeds in the 5′- to 3′-direction. The above points illustrate the importance of strategy in solid phase synthesis.

2.2 Solid phase phosphoramidite oligonucleotide synthesis

The chemistry of the phosphoramidite technique of deoxyoligonucleotide synthesis (9–13) is illustrated in *Figures 2–5*. These figures show the

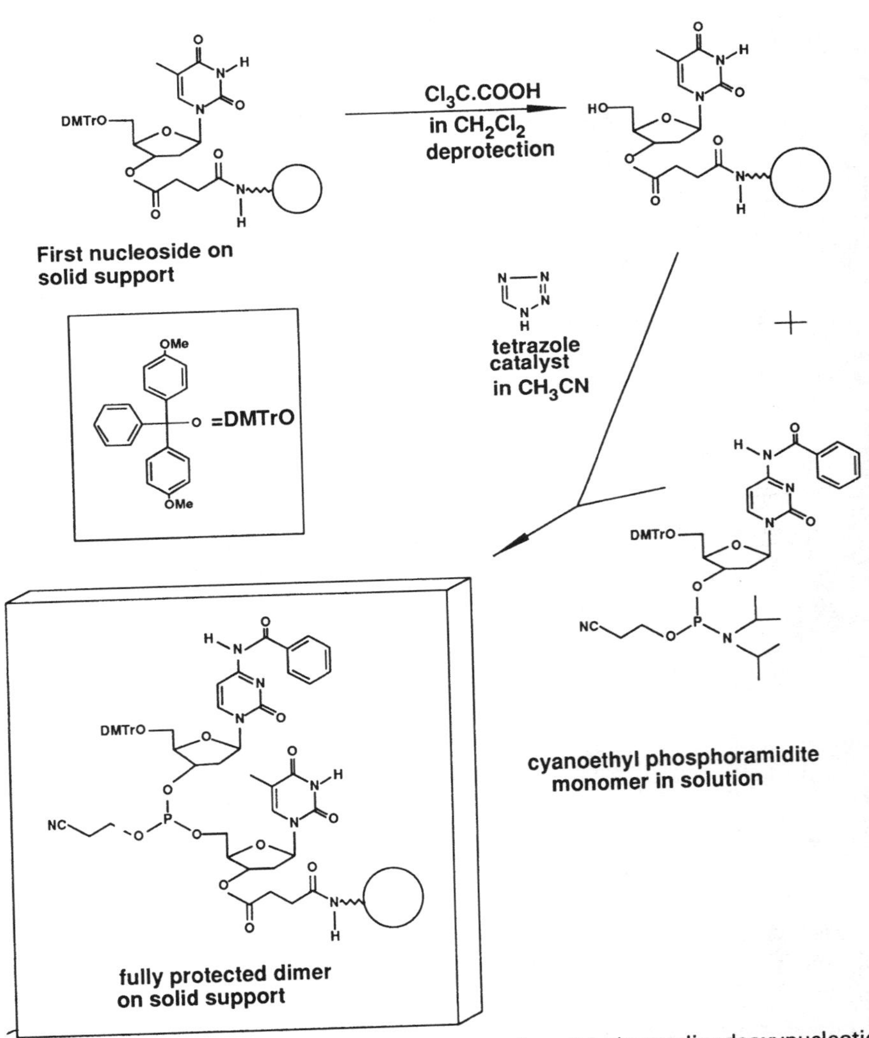

Figure 2. The 5′-deprotection and coupling steps in solid phase oligodeoxynucleotide synthesis.

unreacted support bound monomer

capped support bound monomer

Figure 3. The capping step in solid phase oligodeoxynucleotide synthesis.

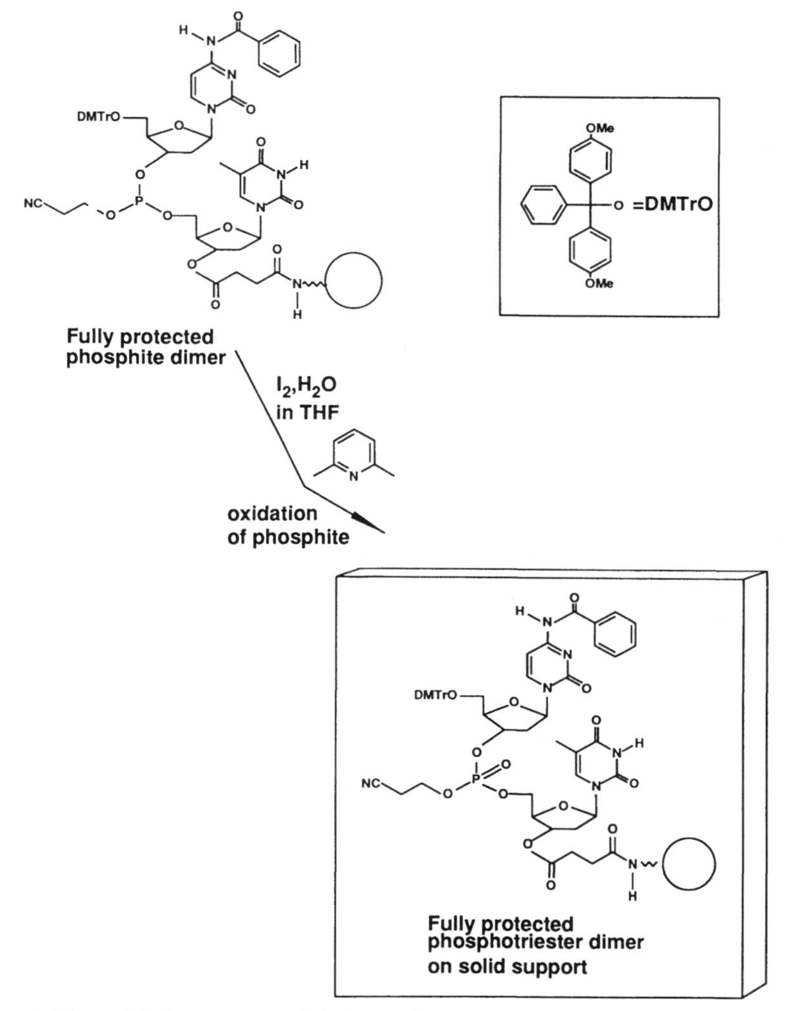

Figure 4. The oxidation step in solid phase oligodeoxynucleotide synthesis.

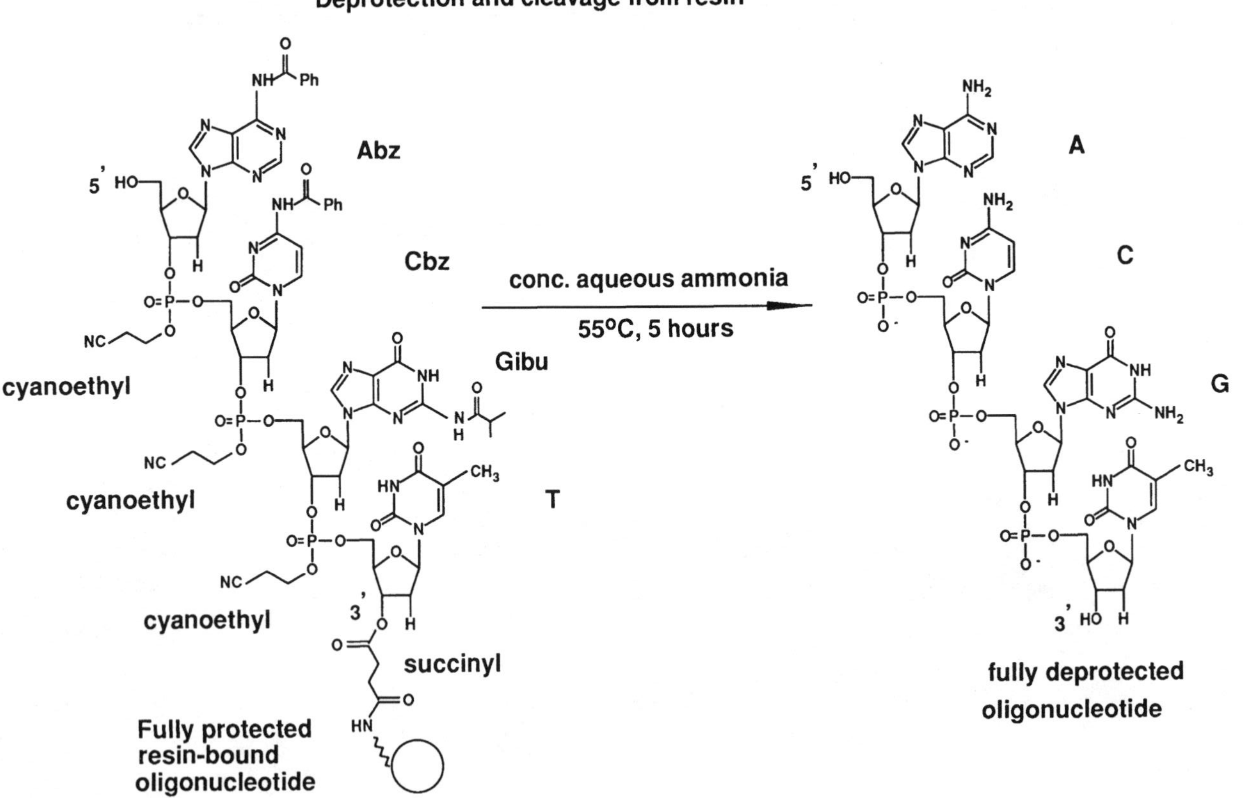

Figure 5. The base and phosphate deprotection and cleavage steps in solid phase oligodeoxynucleotide synthesis.

condensation, capping, oxidation, and cleavage/deprotection respectively. The coupling reaction is catalysed by tetrazole, which protonates the N,N-diisopropyl phosphoramidite, and converts the diisopropylamino moiety into a good leaving group. The protonated amino group is displaced by the 5′-hydroxyl group of the support-bound mononucleotide and the dimer is formed. After a capping step, the dimer is oxidized with aqueous iodine to convert the phosphite triester into a more stable phosphate triester. The essentials of the synthesis cycle are listed in *Table 1*. This is discussed in more detail in Section 3.1. The average coupling efficiency in the phosphoramidite method of DNA synthesis, when carried out on an automatic machine is *c*.98.5% and sequences in excess of 100 bases can be prepared. However, in each cycle around 1.5% of the oligonucleotide chains on the glass beads fail to react with the activated monomer and if this situation were ignored, a complex mixture of truncated sequences would accumulate, the majority of which would be only one nucleotide shorter than the correct product. These impurities would obviously have very similar properties to the desired oligonucleotide and purification would be impossible. Hence, the capping step is extremely important. In this step (*Figure 3*) any unreacted 5′-hydroxyl groups are acetylated and thereby rendered inert to subsequent monomer additions. Assuming the capping reaction is efficient, the crude product will be a mixture of the desired oligomer and an even distribution of n−1, n−2, n−3, etc. failure sequences. Such a mixture can be easily purified by ion exchange HPLC or by polyacrylamide gel electrophoresis.

Table 1. Essentials of the phosphoramidite oligonucleotide synthesis cycle

Operation	Reagent/solvent	Time (seconds)
1. Wash	acetonitrile	30
2. Detritylate	3% trichloroacetic acid in dichloromethane	50
3. Wash	acetonitrile	30
4. Couple	0.1 M phosphoramidite monomer in acetonitrile 0.5 M tetrazole in acetonitrile	40
5. Wash	acetonitrile	30
6. Cap	acetic anhydride/lutidine/THF 1/1/8 17.6% w/v N-methyl imidazole in THF	30
7. Wash	acetonitrile	30
8. Oxidise	0.1 M iodine in water/pyridine/THF 2/20/80	50

2.2.1 Protecting and linker groups in solid phase phosphoramidite oligonucleotide synthesis

The 5′-hydroxyl group of the deoxyribose sugar of each nucleoside phosphoramidite must be protected during monomer addition to prevent self-polymerization. However, this protecting group must be removed when the monomer has been added to the growing oligonucleotide chain so that the 5′-OH group can participate in the next monomer addition. As the 5′-OH protecting group on the support-bound oligonucleotide is removed prior to each monomer addition, it is known as temporary protection. The protecting group of choice is the 4′,4″-dimethoxytrityl group (14), (*Figures 2* and *4*), which is removed by acid (TCA). Cleavage of the DMTr group produces an orange colour which is extremely useful for monitoring the progress of the synthesis.

Side reactions in DNA synthesis are minimized by the use of protecting groups which mask potentially reactive sites on the base and phosphate moieties throughout oligonucleotide synthesis. These 'semi-permanent' protecting groups are removed after the oligonucleotide is cleaved from the solid support. The groups in common use to protect the heterocyclic bases are illustrated in *Figure 6*. Thus, the exocyclic amino groups of A, G, and C are protected by acylation (15–17), whereas the thymine base does not require protection. As the slowest single step in the synthesis of deoxyoligonucleotides by the solid phase phosphoramidite method is the removal of these protecting groups from the heterocyclic bases (5 hours at 55 °C in concentrated aqueous ammonia), it is not surprising that a significant amount of research has been directed at designing more labile base protecting groups. Some such groups (18, 19), which are now commercially available, are illustrated in *Figure 7*.

For a long time the protecting group of choice for phosphorus in phosphoramidite oligonucleotide synthesis was the methyl group (20, 21), which can be removed by treatment with thiophenol. However, thiophenol is an unpleasant reagent and for this reason methyl phosphoramidites have been superseded by 2-cyanoethyl phosphoramidites (*Figures 2, 4, 5*). The 2-

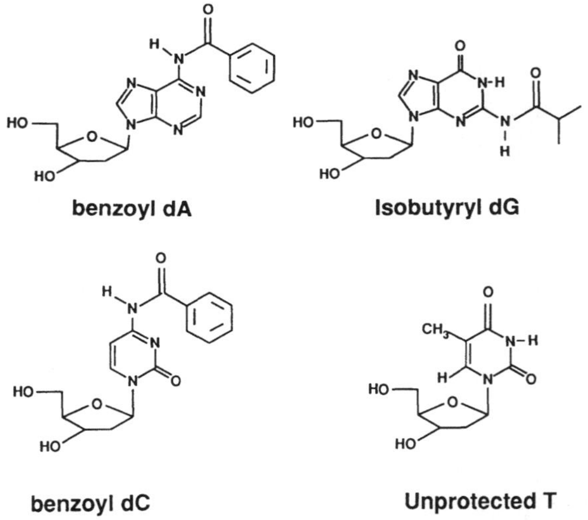

benzoyl dA **Isobutyryl dG**

benzoyl dC **Unprotected T**

Figure 6. Standard base-protecting groups used in solid phase oligodeoxynucleotide synthesis.

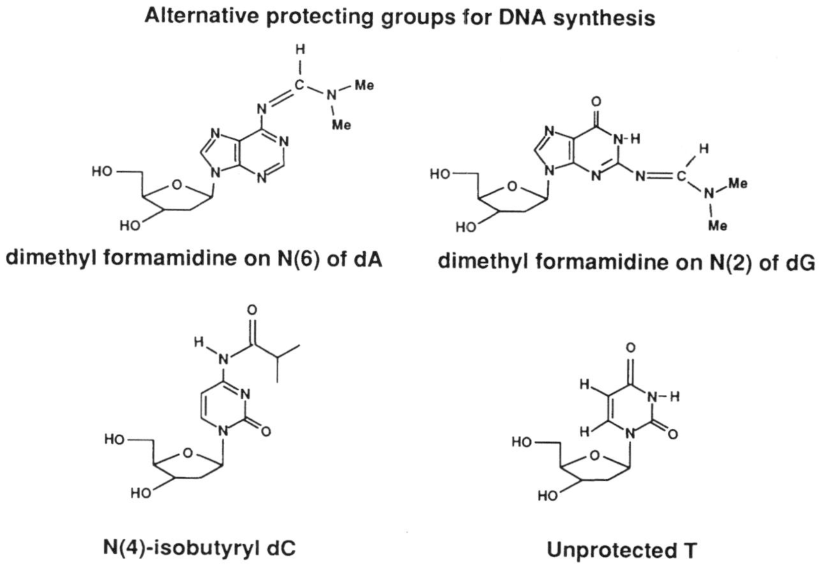

Alternative protecting groups for DNA synthesis

dimethyl formamidine on N(6) of dA

dimethyl formamidine on N(2) of dG

N(4)-isobutyryl dC

Unprotected T

Figure 7. Base-protecting groups for oligodeoxynucleotide synthesis that are designed to be removed rapidly in aqueous ammonia.

cyanoethyl group (10, 21, 22) is reasonably stable in the phosphoramidite monomer, but can be rapidly removed from the oligonucleotide phosphotriester by treatment with base (by a beta-elimination mechanism). 2-Cyanoethyl N,N-diisopropyl phosphoramidites (22) are currently the most widely used monomers for deoxyoligonucleotide synthesis.

Post-synthetic cleavage of the oligonucleotide from the solid support involves hydrolysis of the succinyl ester linkage. This requires *c.* one hour at ambient temperature in aqueous ammonia, the reagent used for phosphate and base deprotection (*Figure 5*).

3. Automation of solid phase oligonucleotide synthesis

3.1 Advantages of automation

The assembly of an oligonucleotide by manual methods has always been a laborious and demanding task as the multitude of chemical reactions and washes involved are extremely sensitive to moisture and are not well suited to manual techniques. Fortunately, the tedium of solid phase oligonucleotide synthesis has now been relieved due to the commercial availability of automatic DNA synthesizers. The synthesis of oligonucleotides on automatic machines is a classic example of advances in several areas of science being brought together at a critical time, namely:

- The solid phase principle of Merrifield.
- Important developments in synthetic chemistry.
- The use of modern computer technology.
- Advances in engineering and valve design.

The driving force for the intense activity that culminated in the production of commercial DNA synthesizers came from exciting discoveries in the field of molecular biology. Many critical experiments could not be carried out without access to highly pure oligonucleotides of defined base sequence. The success of automated DNA synthesis further increased the demand for synthetic DNA in a manner that could not be predicted as recently as five years ago. This has led to a chain reaction in supply and demand of synthetic DNA. Some modern DNA synthesizers can produce up to four sequences simultaneously and the output from a single machine may be more than 1000 sequences per year. The ability of such machines to deliver exact volumes of reagents for precise periods of time has made it possible to assemble oligonucleotides with an efficiency which was hitherto considered impossible. *Table 1* lists the reagents and essential steps in the phosphoramidite oligonucleotide synthesis cycle. The full phosphoramidite cycle on an automatic synthesizer may consist of up to 100 separate steps, most of which consist of solvent washes and column purges. The synthesis column contains the solid support to which the first nucleoside is attached. The support, normally consisting of a few milligrammes of tiny glass beads, is trapped between two filters in the column. The glass beads, which have great mechanical strength, are repeatedly soaked in reagents and washed with an appropriate solvent. An air gap above the beads enables them to be displaced as the solvent passes through them. This ensures efficient mixing of solid phase and solution reagents and is known as the raining bed method. On some automatic machines the solutions, after being delivered to the synthesis column for the requisite time, are rapidly blown out of the column to waste by a stream of inert gas. This minimizes mixing of reagents and permits an extremely rapid and efficient process.

3.2 A practical guide to automated DNA synthesis

Automated DNA synthesis may appear deceptively simple but even the most sophisticated automatic DNA synthesizer will not correct mistakes made by the operator. Hence, in order to avoid costly errors it is very important to take great care in setting up the synthesis of an oligonucleotide. The protocol we supply to new users of the Applied Biosystems ABI 380B DNA synthesizer in our laboratory is outlined below. It can easily be adapted for any other type of DNA synthesizer. Do not rely on the manual supplied with the instrument, as this may lack important details.

Protocol 1. General procedure for phosphoramidite oligonucleotide synthesis on an ABI 380B DNA synthesizer

Materials

Reagents for phosphoramidite DNA synthesis: base-protected 5′-dimethoxytrityl, A,G,C,T phosphoramidite monomers, tetrazole coupling catalyst, acetic anhydride and N-methyl imidazole capping reagents, trichloroacetic acid deprotection solution, iodine oxidation mixture, acetonitrile wash solvent, aqueous ammonia cleavage solution. All obtained from Applied Biosystems, Cruachem Ltd, or alternative suppliers.

Method

1. Carefully check all reagent levels on the synthesizer and change bottles as necessary. Pay particular attention to the solvent level in reservoir 18 (acetonitrile).
2. Screw three clean collection vials onto the synthesizer.
3. Type the sequences to be prepared into the synthesizer.
4. Switch to the column monitor, clear the set-up for each individual synthesis column, and load in the new sequences to be synthesized.
5. Check that the choice of synthesis cycle is correct. This will normally be **ssce103A** for small scale (0.2 μmol) synthesis and **ce103A** for 1 μmol synthesis.
6. Select the ending method.[a] **Trityl Off Auto** will normally be the system default. Switch over to the **Trityl On Auto** ending method if you intend to purify the oligonucleotide with the 5′-trityl group attached.
7. Select **deprce03** as ending method. This will normally be the system default.
8. Type in your initials under **User Name** on each column monitor.
9. Label each collection vial with a code number for the relevant DNA sequence.
10. Print out the details of each column set-up.
11. Carefully check that all details outlined on the print out are correct.
12. Run through the **Start Column** procedure for each column position, being sure to put the correct synthesis columns on the synthesizer. *Remember*, 1 μmol columns have a continuous coloured band and 0.2 μmol columns have a broken coloured band.
13. Check that the solvent residue (waste) bottle connected to the synthesizer is not full.

Protocol 1. *Continued*

14. If a fraction collector is to be used to collect the effluent from the detritylation step at each deprotection, ensure that sufficient clean dry test-tubes are present and that the fraction collector is switched on. Ensure that the trityl waste line leads to the fraction collector.

15. Start the cycle, always running a begin procedure to purge the reagent lines and remove old reagents.

[a] **Trityl On Auto** signifies that after synthesis the 5′-terminal nucleotide of the oligonucleotide will bear a dimethoxytrityl group, and the oligonucleotide will be cleaved from the solid support and deposited in the collection vial.
Trityl Off Auto signifies that after synthesis the 5′-terminal nucleotide of the oligonucleotide will be unprotected, and the oligonucleotide will be cleaved from the solid support and deposited in the collection vial.
Trityl On Manual signifies that the 5′-terminal nucleotide of the oligonucleotide will bear a dimethoxytrityl group, and that the oligonucleotide will not be cleaved from the solid support, but will be left on the synthesis column.
Trityl Off Manual signifies that the 5′-terminal nucleotide of the oligonucleotide will be unprotected, and that the oligonucleotide will not be cleaved from the solid support, but will be left on the synthesis column.

3.2.1 Points to check after commencement of synthesis

When the synthesis has started, examine the orange colour produced by detritylation of the second nucleotide to ensure all is well. The first trityl colour is not very informative as it is a measure of the loading of the first nucleoside on the resin. Check that the synthesis columns are not leaking. Column leakage may be due to a faulty column or a column which has not been pushed on to the Luer fittings on the synthesizer tightly enough. If the DNA synthesizer has been switched off for any reason the software system defaults may have changed. Therefore, it is essential to carefully check the details on the column monitors and if necessary to restore the desired defaults. This is done on the ABI 380B DNA synthesizer using the **Config/Std/End** option on the column monitor. After doing this double check that all the options displayed on the column monitors are those desired.

No matter how well conceived and powerful the purification procedure, it is not possible to obtain a pure oligonucleotide from a bad synthesis. The coupling efficiency of the synthesis can be accurately calculated by measuring spectrometrically at 495 nm the intensity of the orange colour produced by the deprotection of the 5′-hydroxyl group of the terminal nucleotide during each cycle. This colour results from release of the dimethoxytrityl cation and it is extremely useful in determining stepwise coupling efficiency. In a successful synthesis the intensity of the colour falls very slightly from one cycle to the next, due to incomplete coupling and capping of unreacted chains.

Table 2. The relationship between coupling efficiency of oligonucleotide synthesis and percentage yield for a series of oligodeoxynucleotides of varying length. The normal coupling efficiency of phosphoramidite DNA synthesis is 98.5%. (yields in italics)

			Coupling efficiency		
Oligonucleotide length	**90%**	**95%**	**97%**	**98.5%**	**99.5%**
10mer	*38.7*	*63.0*	*76.0*	***87.3***	*95.6*
20mer	*13.5*	*37.7*	*56.1*	***75.0***	*90.9*
50mer	–	*8.10*	*22.5*	***47.7***	*78.2*
100mer	–	–	*4.90*	***22.4***	*60.88*
150mer	–	–	*1.07*	***10.52***	*47.38*
200mer	–	–	–	***4.94***	*36.9*

The importance of a high average stepwise yield is illustrated in *Table 2*, which represents the effect of coupling efficiency on the overall percentage yield of oligonucleotides of various lengths. It is clear from *Table 2* that coupling efficiency has a dramatic effect on the overall yield. This effect is much greater for longer sequences and an average stepwise yield of less than 98% is totally unacceptable for routine oligonucleotide synthesis. The cumulative effect of a series of low couplings is two-fold;

- a poor overall yield of the desired oligonucleotide,
- a product that is extremely difficult to purify.

It should be possible to achieve an average coupling efficiency of 98.5% on a regular basis on well-designed automatic DNA synthesizer using good quality reagents. *Table 2* illustrates that there is still scope for increasing coupling efficiency in oligonucleotide synthesis. If an average coupling efficiency in excess of 99.5% could be routinely achieved, the synthesis of DNA strands in excess of 200 nucleotides would be feasible. Such very high coupling efficiencies have been achieved in peptide synthesis but this has not yet been possible in oligonucleotide synthesis. Assuming that the synthesis and purification have been successful, the overall yield of HPLC purified product in milligrammes or micromoles can be determined from the UV absorbance of an aqueous solution of the oligonucleotide. Note that when the pure oligonucleotide is isolated as a lyophilized solid, it is highly hydrated. Because of this, 1 mg of lyophilized DNA is usually equivalent to between 20 and 25 OD units at 254 nm.

Protocol 2. Calculating coupling efficiency and overall yield of oligonucleotide synthesis

Materials

p-Toluene sulphonic acid monohydrate and HPLC grade acetonitrile.
A 3 litre beaker, volumetric flasks: 1000 ml, 25 ml, 5 ml.

Protocol 2. *Continued*

Method

1. Prepare a solution of 0.1 M *p*-toluene sulphonic acid (tosic acid) in acetonitrile by dissolving tosic acid monohydrate (19.02 g) in acetonitrile (1000 ml) in a 3 litre beaker. **Care, these reagents are toxic.**
2. For a normal 0.2 μmol synthesis, dilute the contents of the test-tubes containing the effluent from each detritylation to 25 ml with the tosic acid solution prepared above. Stopper the flask and shake well. Use clean dry volumetric flasks for the dilution.
   For a 1 μmol synthesis, dilute the contents of the tubes first to 25 ml then dilute 1 ml of the 25 ml solution to 5 ml. Remember to shake the flasks to mix the solutions. The overall dilution factor for the 1 μmol detritylation is 125.
3. Measure the absorbance of the diluted trityl colour in a 1 cm cuvette in a UV-visible spectrometer at 495 nm. Carry out this operation soon after diluting the detritylation solutions to ensure accurate results. The trityl colour may fade with time.

Calculations

Let the absorbance of the fully diluted detritylation from the first oligonucleotide coupling step (second tube) be x. Let the absorbance of the fully diluted detritylation from the final coupling (last tube) be y.

Then: overall yield = y/x.
Overall percentage yield = overall yield × 100.
Stepwise yield = (overall yield)$^{(1/y - x)}$.
Percentage stepwise yield = stepwise yield × 100.

Example

For a 0.2 μmol synthesis of a 52mer, the absorbance obtained from the last detritylation, using *Protocol 2* was found to be 0.11. Absorbance of second detritylation, using the same protocol was 0.23.

Overall yield, y/x = 0.11/0.23 = 0.48.
Hence *overall percentage yield = 48%.*

Stepwise yield = (overall yield)$^{(1/52 - 2)}$ = (0.48)$^{(1/50)}$ = 0.985.
Hence *stepwise percentage yield = 98.5%.*

Stepwise percentage yield is referred to as coupling efficiency.

3.2.2 Stability of reagents on an automatic DNA synthesizer

The reagents on all automatic DNA synthesizers are maintained under an inert gas atmosphere for protection from air. This is essential as phosphoramidite chemistry is extremely sensitive to moisture and oxygen. The inert gas also serves to pressurize the reagent bottles so that the reagents can be delivered to the synthesis columns. On some synthesizers the gas is helium and on others argon is used; argon has the advantage of being heavier than air and therefore protects the reagents if a slight leak occurs in the machine. The reagents are less well protected when they are in the reagent lines (the flexible tubes that carry the chemicals from their bottles to the synthesis columns), as this tubing is not totally impermeable. However, whenever the synthesis is in progress, the monomers in the reagent lines will be replenished every few minutes and no significant decomposition of the phosphoramidites occurs. If, however, the machine has been unused for more than one hour, the reagent lines must be purged before a synthesis is started to remove decomposed reagents.

A particular problem can arise if for some reason a synthesis is interrupted for more than one hour. This can occur if the machine's automatic interrupt facility has caused the synthesis to stop in the middle of the night. The phosphoramidite monomers will have been stationary in the reagent lines for a long period of time and will have slowly hydrolysed and oxidized due to the diffusion of air through the tubing. Hence, when the synthesis is recommenced the next day, failure will result. This can be avoided by carrying out a manual purge of the phosphoramidite, tetrazole, and acetonitrile lines before recommencing the synthesis after the interruption.

Even under the most favourable conditions reagents have a finite lifetime. Therefore, always discard reagents which have been on the synthesizer for more than two weeks regardless of the manufacturers' advice. For synthesis of oligonucleotides of length in excess of 30 residues, use only phosphoramidites and tetrazole that have been on the synthesizer for less than one week. For 50mers or longer, use fresh phosphoramidites, tetrazole, and acetonitrile.

3.2.3 Changing the chemistry on an automatic DNA synthesizer

It is very important to purge all lines on an automatic DNA synthesizer when changing from one type of chemistry to another. If this is not done, a failed synthesis will result. This is particularly crucial when changing from H-phosphonate chemistry to phosphoramidite chemistry or vice versa, as the coupling step in the former is base-catalysed and in the latter is acid-catalysed. A reliable method of thoroughly purging lines is to carry out a dummy synthesis as shown in *Protocol 3*.

Protocol 3. Purging reagent lines on an automatic DNA synthesizer after changing the chemistry

Method

1. Change all necessary reagents in the normal manner (Section 3.24, *Protocol 5*).

Protocol 3. *Continued*

2. Put a used synthesis column on each column position on the synthesizer.
3. Load the sequence AGCTT onto each column position.[a]
4. Carry out a synthesis with ending method **Trityl On Manual** according to *Protocol 2*.
5. At the end of the dummy synthesis, discard the old synthesis columns and commence normal oligonucleotide synthesis.

[a] Synthesizing AGCTT ensures that each monomer line is thoroughly purged as well as every other line that is used in a standard synthesis. There is obviously no need to carry out an end procedure (ammonia cleavage).

3.2.4 Changing reagents on an automatic DNA synthesizer

If any reagent runs out during a synthesis this will obviously result in a failed synthesis, so the levels in all bottles should be inspected prior to synthesis. The reagents used on a typical DNA synthesizer for phosphoramidite deoxyoligonucleotide synthesis are shown in *Figures 2–5* and in *Table 1*. Bottle positions quoted below are those used for the Applied Biosystems ABI 380B synthesizer. Usage of iodine (bottle 15) and TCA (bottle 14) is particularly high and as not all automatic machines have alarms, the levels of these reagents should be carefully monitored. Capping reagents and tetrazole need to be changed much less frequently as only small amounts of these chemicals are consumed during a synthesis cycle. Tetrazole tends to crystallize if the temperature in the laboratory falls below *c*.18 °C. If this problem arises, warm the bottle with a hair dryer to redissolve the crystals before commencing the bottle change procedure. Be particularly careful to check the level of acetonitrile (bottle 18) and the level of ammonia (bottle 10) before the synthesis.

If a position is fitted with a removable reagent bottle seal, examine this and ensure that it is replaced when the new bottle is put on the synthesizer. Any split seal must be discarded and a new one fitted. Permanent seals should also be checked. In addition, examine the rim of the mouth of the reagent bottles for chipping. Chipped bottles will not seal on the synthesizer and the reagent will not be delivered.

Protocol 4. Changing phosphoramidite monomer bottles

Materials

4′,4″-Dimethoxytrityl 2-cyanoethyl deoxynucleoside phosphoramidites and anhydrous acetonitrile. Syringes (10.0 ml) and syringe needles. Source of inert gas. Septa to fit acetonitrile bottles.

Protocol 4. *Continued*

Method

Caution—wear safety glasses and gloves throughout this protocol

1. Wash a 10 ml glass syringe by filling with and expelling acetonitrile. Dry the syringe thoroughly in a hot oven.
2. Fit a septum on a bottle of anhydrous acetonitrile and pressurize the bottle by delivering argon into it. This can be done on the ABI 380B synthesizer by fitting a syringe needle onto column 1 and activating function 2 and column 1. (i.e. deliver argon to column 1). The acetonitrile bottle is pushed into the syringe for a few seconds. The bottle is then withdrawn and function 2 is deactivated.
3. Take a 1.0 g 2-cyanoethyl deoxynucleoside phosphoramidite monomer bottle and expose the septum by lifting the aluminium cap. Insert a syringe needle so it just protrudes through the septum into the bottle. This needle serves as a gas outlet when the bottle is filled with acetonitrile.
4. Take the pressurized anhydrous acetonitrile bottle and a 10.0 ml syringe fitted with a needle. Insert the syringe needle through the septum into the acetonitrile. Hold your thumb over the end of the syringe barrel and allow the syringe to slowly fill to a volume of 10.0 ml. Withdraw the syringe from the acetonitrile and into the argon gas just above it. Allow a small volume of argon to enter the syringe.
5. Withdraw the syringe needle from the acetonitrile bottle holding the plunger tightly. Insert the syringe needle into the monomer bottle until it protrudes slightly through the septum into the bottle. Expel the acetonitrile from the syringe into the monomer bottle and withdraw the syringe needle.
6. Repeat steps 4 and 5 but instead of delivering 10.0 ml of acetonitrile, into the monomer bottle, deliver the following volumes:[a]

   For **A** monomer 1.0 ml **G** 2.0 ml **C** 2.5 ml **T** 3.5 ml
7. Withdraw the syringe and syringe needle and gently agitate the monomer bottle to dissolve and mix the monomer.
8. Carry out the relevant bottle change procedure on the synthesizer. At the appropriate time remove the septum from the freshly dissolved monomer bottle and quickly fit the bottle into position on the machine.

Never pour the remains of one monomer bottle into a new bottle.

[a] The *total* volume of acetonitrile used to dissolve 1.0 g of a 2-cyanoethyl deoxynucleotside phosphoramidite monomer (steps 4–6) is as follows:

For **A** monomer 11.0 ml **G** 12.0 ml **C** 12.5 ml **T** 13.5 ml

On most DNA synthesizers, the **A** monomer fits on position **1**, **G** on position **2**, **C** on position **3**, and **T** on position **4**. Ensure that the appropriate bottles are inserted in the correct positions. The worst mistake that can be made during oligonucleotide synthesis is to insert the monomers in the wrong positions. The mistake may not be detected and the resultant syntheses will obviously appear fine on HPLC and on the printout of the sequence. However, the wrong sequence will have been synthesized! Ensure that you are using *cyanoethyl* amidites and that the bottle contains *1.0 g*. Occasionally a 0.5 g bottle will be encountered in which case the following volumes of anhydrous acetonitrile should be used:

For **A** monomer 5.5 ml **G** 6.0 ml **C** 6.2 ml **T** 6.7 ml

Always have a spare monomer bottle and septum available when carrying out a bottle change, as loose, badly fitting bottles which will not seal on the synthesizer are occasionally encountered. If a monomer bottle feels very loose when pushed into position on the synthesizer, quickly pour the contents of the bottle into a spare clean dry monomer bottle and continue the bottle change. If the faulty bottle is used, the monomer will not be delivered during the synthesis cycle and this will result in failed syntheses.

3.2.5 Reagent bottle changes during a synthesis.

This can be done according to *Protocol 4* with the following modifications: wait until an acetonitrile wash (18 to column) is indicated *after monomer addition, capping, and oxidation* on a particular cycle and interrupt the cycle, column 1, column 2, and column 3. The bottle can then be changed in the normal way. Cancel all interrupts and recommence the synthesis. Never allow a synthesis to be interrupted for more than a few minutes as this will result in a low coupling efficiency or even a failure.

3.3 Oligonucleotide cleavage and deprotection

Cleavage of ester linkage between the 3′-OH group of the oligonucleotide and the succinyl linker is carried out automatically on some synthesizers, but it may also be performed manually according to *Protocol 5*.

Protocol 5. Cleavage of the oligonucleotide from the resin

Materials

- Concentrated aqueous ammonia
- Screw-capped vials (5 ml)
- 1.0 ml gas-tight syringe
- Rotary evaporator or SpeedVac

Caution—carry out all operations in a fume cupboard.

Protocol 5. *Continued*

Method

1. After the synthesis, dry the synthesis column by blowing argon through it. This can be done by activating function 2 along with the relevant column on the ABI 380B, 381A, or 391A synthesizers.
2. After drying the synthesis column, remove it from the instrument. Fill a syringe with 1 ml of concentrated aqueous ammonia and attach it to one end of the column. Fit the other end of the synthesis column with a Luer adaptor and a needle. Inject the ammonia solution into the column until the first drop emerges through the needle. The glass beads in the column should now be totally immersed in the ammonia solution.
3. Seal the syringe by pushing the needle into a rubber stopper and set aside for one hour at room temperature.
4. Remove the syringe needle from the rubber stopper and pass the ammonia solution through the column by syringe pressure into a screw-capped vial. Temporarily screw the cap on to the via and repeat the entire procedure (steps 1–4) twice. Finally, tightly screw the cap onto the via which now contains all three ammonia washings (3 ml).
5. To complete base deprotection, the sample is heated overnight at 55 °C. The vial is cooled to room temperature, the screw cap removed, and the solution of the fully deprotected oligonucleotide is transferred to a 50 ml round-bottomed flask and evaporated to dryness in preparation for purification.

3.4 Limitations of solid phase deoxyoligonucleotide synthesis

Despite the efficiency of DNA synthesis, impurities inevitably accumulate. The synthesis of a 20mer involves more than 100 separate chemical steps, each with a finite probability of producing an undesirable side reaction. The accumulation of failure sequences due to incomplete coupling has already been discussed, but truncated sequences can also arise from depurination followed by oligonucleotide chain cleavage. Chemical modifications to the heterocyclic bases sometimes occur and the resultant modified oligonucleotides, which are potentially mutagenic, are liable to cause serious problems in biological experiments. Examples of mutagenic modifications are the conversion of adenine to hypoxanthine and guanine to 2,6-diaminopurine. Because of the occurrence of side reactions it is clear that oligonucleotides of very high purity can only be obtained by employing two orthogonal purification methods such as reversed phase and ion-exchange HPLC. For the majority of biological applications either one of these purification methods alone will probably be sufficient, but for more stringent applications such as X-ray crystallography, both methods *must* be used.

Protocol 6. Calculation of the molecular weight and extinction coefficients of synthetic oligonucleotides

Method

1. Calculate the molecular weight of DNA as follows:

Mol. wt =
$\{(251 \times nA) + (245 \times nT) + (267 \times nG) + (230 \times nC) + (61 \times n - 1) + (54 \times n) + (17 \times n - 1) + 2\}$
where:

(i) nA = number of adenine bases in the DNA sequence etc. and n = total number of bases.
(ii) $(61 \times n - 1)$ accounts for the molecular weight of the phosphate groups. For phosphorothioates this is $(78 \times n - 1)$.
(iii) $(54 \times n)$ accounts for the hydration of the DNA, approximately three water molecules per nucleotide as a rule of thumb.
(iv) $(17 \times n - 1)$ accounts for the ammonium cations associated with the phosphate groups. If the DNA is a sodium salt this is $(23 \times n - 1)$.

Example

The 20mer d(AGCTCTGAACGTAGCTCTGA)
Mol. wt = $\{(5 \times 251) + (5 \times 245) + (5 \times 267) + (5 \times 230) + (61 \times 19) + (54 \times 20) + (17 \times 19) + 2\}$
Mol. wt = 7529 i.e. 1 μmol of the above DNA sequence = 7.529 mg.

2. Calculation of micromolar extinction coefficient, (E_{254})
$E_{254} = \{(8.8 \times nT) + (7.3 \times nC) + (11.7 \times nG) + (15.4 \times nA)\} \times 0.9$[a] for the 20mer d(AGCTCTGAACGTAGCTCTGA):
$E = \{(8.8 \times 5) + (7.3 \times 5) + (11.7 \times 5) + (15.4 \times 5)\} \times 0.9$
Micromolar extinction coefficient E = 194.4

3. To convert OD units to milligrammes
From the above calculations, 7.5 mg of the oligonucleotide d(AGCTCTGAA CGTAGCTCTGA) (lyophilized solid) will have an absorbance of approximately 200 OD units at 254 nm.
7.5 mg = 200 OD units. Therefore, 1 mg = 200/7.5 OD units.

1 mg = 27 OD units

Generally, 1 mg of a lyophilized oligonucleotide corresponds to *c.* 25 OD units at 254 nm. Similar figures will be obtained at 264 nm.

[a] Note that it is advisable to multiply the sum of the extinction coefficients of the individual bases by a factor of 0.9. This is because base stacking interactions in the single strand suppress the absorbance of DNA relative to the value calculated from the extinction coefficients of the individual nucleosides. This effect is greater for a duplex and the multiplication factor for a self-complementary sequence is *c.* 0.8. These figures are estimates for typical DNA sequences.

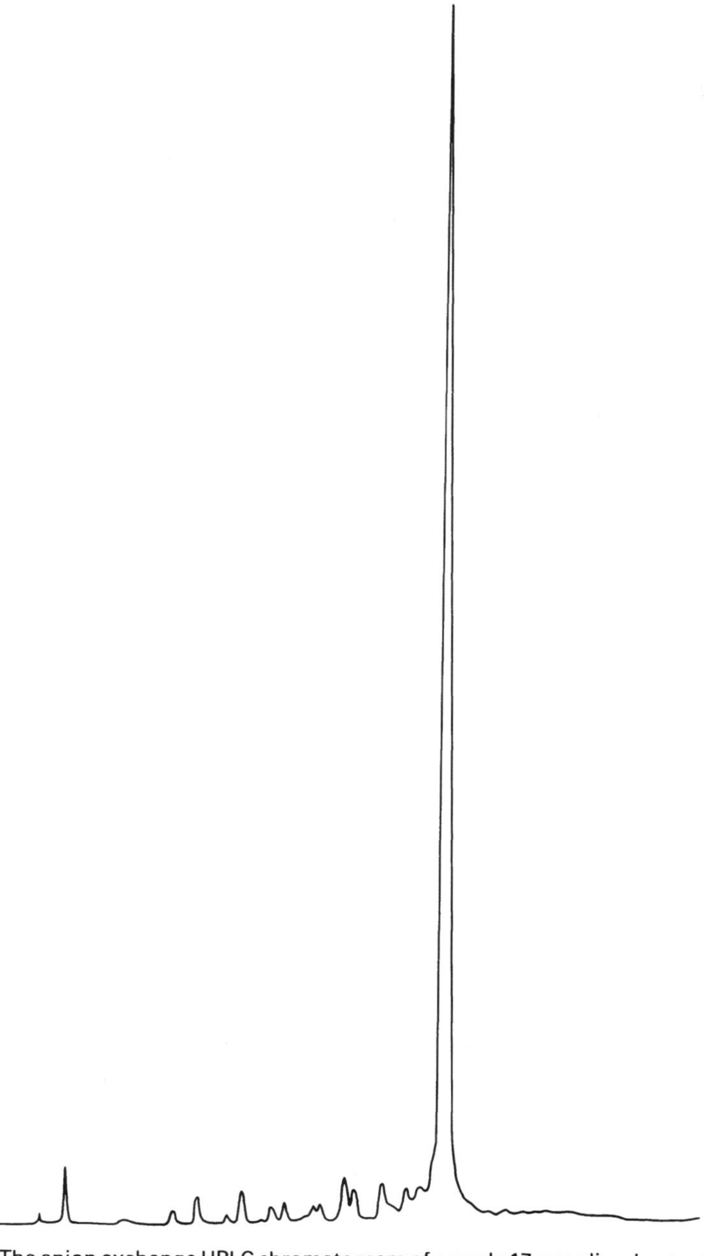

Figure 8. The anion exchange HPLC chromatogram of a crude 17mer oligodeoxynucleotide synthesized by the solid phase phosphoramidite method described in this chapter. A Partisil SAX column (0.8 cm × 25 cm) was used. The mobile phase was a linear gradient of 95% A, 5–100% B over a period of 30 min. Buffer A: 0.04 M potassium dihydrogen phosphate in water/acetonitrile (8/2), adjusted to pH 6.4 with sodium hydroxide. Buffer B: 0.67 M potassium dihydrogen phosphate in water/acetonitrile (8/2), adjusted to pH 6.4 with sodium hydroxide.

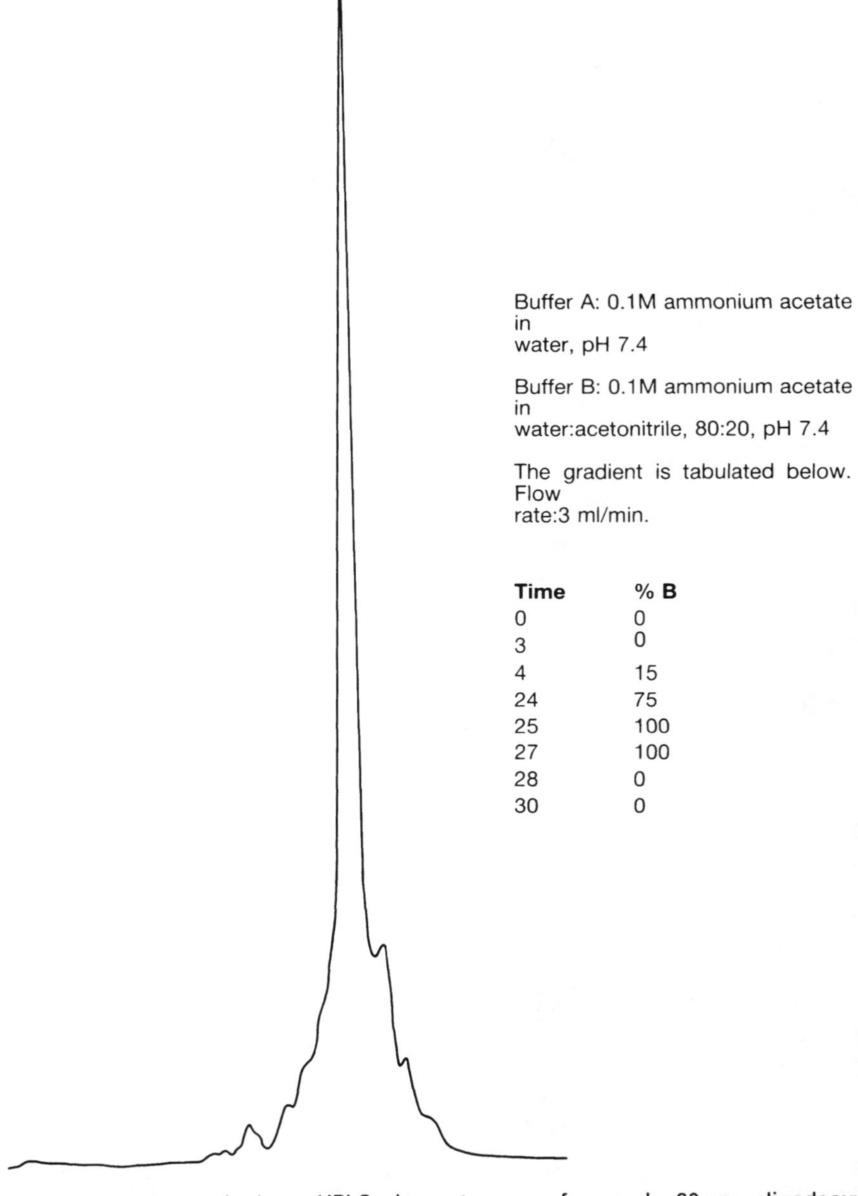

Time	% B
0	0
3	0
4	15
24	75
25	100
27	100
28	0
30	0

Figure 9. The reversed phase HPLC chromatogram of a crude 20mer oligodeoxynucleotide synthesized by the solid phase phosphoramidite method described in this chapter. A Brownlee aquapore RP200 column (octyl, 1 cm × 25 cm) was used. A binary solvent system was used to give a gradient of acetonitrile in aqueous ammonium acetate buffer.

The practical maximum length of oligonucleotide that can be synthesized depends on the efficiency of the automatic synthesizer. Most machines should be able to produce 100mers and some can produce sequences in excess of 150 nucleotides in length. The yield of the synthesis can be estimated by measuring the UV absorbance of the oligonucleotide at a convenient wavelength, e.g. 254 nm using the following protocol, which also illustrates the conversion of OD units to milligrammes of lyophilized solid.

4. Conclusions

The protocols described in this chapter are those that are used routinely in the OSWEL DNA Service to prepare several thousands of oligonucleotides per year. Examples of oligonucleotides prepared by automated solid phase methods are illustrated in *Figures 8* and *9*. The anion exchange HPLC chromatogram of a crude 17mer deoxyoligonucleotide is shown in *Figure 8*, and *Figure 9* shows the reversed phase (octyl) HPLC chromatogram of a crude 20mer. A description of oligonucleotide purification techniques is beyond the scope of this chapter but a number of articles have been written on the subject (23, 24).

References

1. Bannwarth, W. (1987). *Chimia*, **41**, 302.
2. Letsinger, R. L., Finnan, J. L., Heavner, G. A., and Lunsford, W. B. (1975). *J. Am. Chem. Soc.*, **97**, 3278.
3. Uhlmann, E. and Peyman, A. (1990). *Chem. Rev.*, **90**, 543.
4. Usman, N., Ogilvie, K. K., Jiang, M.-Y., and Cedergren, R. J. (1987). *J. Am. Chem. Soc.*, **109**, 7845.
5. Ogilvie, K. K., Usman, N., Nicoghosian, K., and Cedergren, R. J. (1988). *Proc. Natl. Acad. Sci. USA*, **85**, 5764.
6. Saiki, R. K., Scharf, S., Faloona, F., Mullis, K. B., Horn, G. T., Erlich, H. A., and Arnheim, N. (1985). *Science*, **37**, 170.
7. Merrifield, B. (1963). *J. Am. Chem. Soc.*, **85**, 2149.
8. Engels, J. W., and Uhlmann, E. (1989). *Angew. Chemie Int. Ed. Engl.*, **28**, 716.
9. Beaucage, S. L. and Caruthers, M. H. (1981). *Tetrahedron Lett.*, **22**, 1859.
10. Sinha, N. D., Biernat, J., McManus, J., and Koster, H. (1984). *Nucleic Acids Res.*, **12**, 4539.
11. McBride, L. J. and Caruthers, M. H., (1983). *Tetrahedron Lett.*, **24**, 245.
12. Daub, G. W. and van Tamelen, E. E. (1977). *J. Amer. Chem. Soc.*, **99**, 3526.
13. Adams, S. P. (1983). *J. Am. Chem. Soc.*, **105**, 661.
14. Smith, M., Rammler, D. H., Goldberg, I. H. and Khorana, H. G. (1963). *J. Am. Chem. Soc*, **84**, 430.
15. Lohrman, R., Soll., D., Hayatsu, H., Ohtsuka, E., and Khorana, H. G. (1966). *J. Am. Chem. Soc.*, **88**, 819.

16. Ti, G. S., Gaffney, B. L., and Jones, R. A. (1982). *J. Amer. Chem. Soc.*, **104**, 1316.
17. Agarwal, K. L., Yamazaki, A., Cashion, P. J., and Khorana, H. G. (1972). *Angew. Chemie Int. Ed. Engl.*, **11**, 451.
18. McBride, L. J., Kierzek, R., Beaucage, S. L., and Carruthers, M. H. (1986). *J. Am. Chem. Soc.*, **108**, 2040.
19. FOD monomers available from Applied Biosystems Inc.
20. Matteucci, M. D. and Caruthers, M. H. (1980). *Tetrahedron Lett.*, **21**, 719.
21. Ogilvie, K. K., Theriault, N. Y., Seifert, J., Pon, R. T., and Nemer, M. J. (1980). *Canad. J. Chem.*, **58**, 2686.
22. Sinha, N. D., Biernat, J., and Koster, H. (1983). *Tetrahedron Lett.*, **24**, 5843.
23. Zon, G. and Thompson, J. A. (1986). *BioChromatography*, **1**, 22.
24. Becker, C. R., Efcavitch, J. W., Heiner, C. R., and Kaiser, N. F. (1985). *J. Chromatography*, **326**, 293.

2

Oligoribonucleotide synthesis

MICHAEL J. GAIT, CLARE PRITCHARD, and GEORGE SLIM

1. Introduction

It is remarkable to note that, compared with a 2′-deoxyribonucleoside, the presence of a single extra hydroxyl group at the 2′-position of a ribonucleoside has given rise in recent years to so many headaches for chemists attempting to assemble oligoribonucleotide chains. For example, whereas machine-aided assembly of oligodeoxyribonucleotides has been established for some years (Chapter 1), only very recently has it been possible to assemble oligoribonucleotides satisfactorily using mechanized solid phase procedures. Part of the difficulty here has been the need to find a combination of compatible protecting groups for the ribonucleoside 2′- and 5′-hydroxyl groups. Thus in solid phase synthesis with chain assembly proceeding in a conventional 3′- to 5′-direction, 5′-terminal protecting groups must be selectively removed at every cycle of ribonucleotide addition, whereas 2′-protecting groups must remain intact throughout all steps of oligoribonucleotide assembly and must be removed specifically at the end of the synthesis without leading to chain migration or internucleotidic cleavage. Further problems stem from the less facile coupling reactions and hence slightly poorer coupling efficiences obtained hitherto in oligoribonucleotide synthesis. In addition, oligoribonucleotides are highly susceptible to degradation by ribonucleases, which are ubiquitous and difficult to remove, and therefore the utmost care must be taken during all purification steps.

Completely satisfactory solutions to all these problems are not yet available. Nevertheless, it has now become possible to synthesize oligoribonucleotides of moderate length in reasonable yield and purity by machine-aided methods and by use of commercially available reagents. We now present protocols that have worked reasonably well in our hands but we are mindful that further improvements are likely to follow soon in terms of better reagents and methods. We therefore concentrate where possible on basic techniques which should still be of relevance as new materials become available.

1.2 Basic chemistry of oligoribonucleotide synthesis

The solid phase strategy for synthesis of oligoribonucleotides is very similar to that employed in the preparation of oligodeoxyribonucleotides (see Chapter 1). Thus a suitably protected nucleoside derivative is attached to a controlled pore glass support via a succinate linker. Cycles of addition of nucleotide units are then carried out by removal of 5′-terminal protecting groups, coupling to the next nucleotide unit, capping and oxidation. The completed oligonucleotide is then cleaved from the support, and phosphate and heterocyclic base protecting groups are removed by treatment with ammonia solution. The final step is the complete removal of 2′-protecting groups, followed by purification of the oligonucleotide by gel electrophoresis or by HPLC.

Many alternative 2′-protecting groups have been explored for use in oligoribonucleotide synthesis. The most reliable groups in terms of their selectivity of introduction and cleanness of removal without formation of side products are acid-labile acetal groups, such as tetrahydropyran-1-yl (1) and 4-methoxytetrahydropyran-4-yl (2). Unfortunately their use is not totally compatible with conventional 5′-dimethoxytrityl groups because of partial loss during 5′-deprotection (3). One solution to this problem is the use of the modified acetal 1-(2-fluorophenyl)-4-methoxypiperidin-4-yl (Fpmp) as a 2′-protecting group (4). The Fpmp group is reported to be stable during the non-aqueous acidic treatment required to remove 5′-dimethoxytrityl groups, but is readily removed under aqueous conditions at pH 2. Hopefully, recently reported successes in oligoribonucleotide synthesis using this protecting group (5) will be translated into a commercially viable route (Cruachem, UK).

An alternative solution to the problem of 2′- and 5′-protecting group compatibility is the replacement of dimethoxytrityl as a 5′-protecting group with a group that is removed under non-acidic conditions. In this context, both the levulinyl group (6) and the 9-fluorenylmethoxycarbonyl group (7) have been successfully utilized in conjunction with an acid-labile 2′-protecting group for synthesis of oligoribonucleotides up to 25 residues in length. Neither of these routes has thus far attracted much commercial interest principally because such ribonucleotide materials cannot be used interchangeably with conventional 5′-dimethoxytrityl-containing deoxyribonucleotide materials to make mixed RNA–DNA sequences.

The method which has gained most favour amongst commercial suppliers is that which has been developed principally by Ogilvie and co-workers over many years and which utilizes the t-butyldimethylsilyl group for 2′-protection (8). Only more recently, however, has the use of this protecting group become reasonably accepted and *Figure 1* shows the currently preferred route for synthesis of the ribonucleoside 3′-phosphoramidite derivatives. First, it

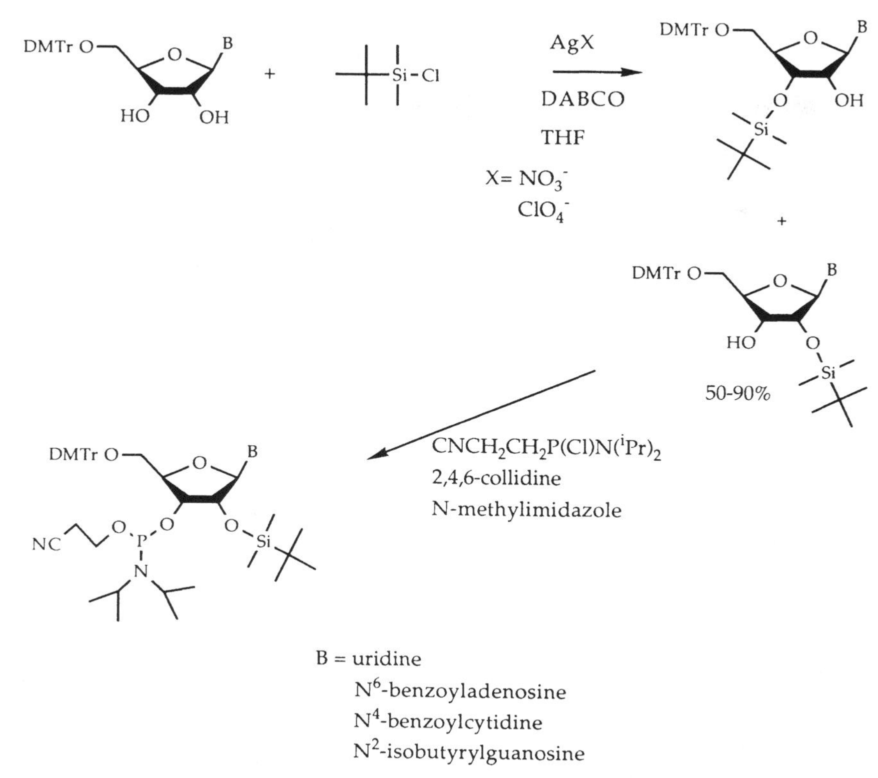

Figure 1. Steps in synthesis of the four ribo amidites.

was necessary for a selective procedure of 2′-silylation to be developed for each of the four 5′- and N-protected ribonucleosides (9). Secondly, the danger of migration of the 2′-silyl group to the 3′-position under mildly alkaline conditions (10) needed to be recognized and appropriate precautions taken. This is particularly important in the case of 3′-phosphitylation where under the standard conditions previously recommended (N,N-diisopropylamino, 2-cyanoethyl chlorophosphite in the presence of diisopropylethylamine and N,N-dimethylaminopyridine) (11) some migration may be expected. Very recently it has been shown that replacement of base and catalyst by 2,4,6-collidine and N-methylimidazole respectively leads to isomeric purity levels of the ribonucleoside phosphoramidites of $> 99.95\%$ (12). It should also be noted that the currently used heterocyclic protecting groups (benzoyl for A, benzoyl for C, and isobutyryl for G), developed many years ago by Khorana and co-workers to meet the needs of solution phase oligonucleotide synthesis, require the use of quite harsh ammoniacal conditions for their removal at the end of oligonucleotide assembly. Under these conditions, there is a danger of partial loss of *tert*-butyldimethylsilyl groups, which can lead eventually to reductions in synthetic yield due to chain cleavage reactions. Although adjustments to the conditions of ammonia treatment have alleviated this problem somewhat, the use of protecting groups removable under milder conditions may be preferable in future.

Another requirement for oligoribonucleotide synthesis by modern solid phase chemistry is a solid support to which is attached one of the four ribonucleosides to act as a 3′-end of an oligoribonucleotide chain (*Figure 2*). The linkage to the support is via a succinate group. This is introduced on to a ribonucleoside derivative by treatment with succinic anhydride largely as previously outlined for deoxyribonucleosides (11, 13). The nucleoside succinate derivative is then coupled to a controlled pore glass support via intermediate preparation of the corresponding pentachlorophenyl (11) or pentafluorophenyl (C. Pritchard, unpublished) ester. It should be noted that because the nucleoside attached to the support becomes the 3′-residue of an oligoribonucleotide, it does not matter if the succinate linkage is formed through the 2′- or 3′-position, since both positions become deprotected in the final oligonucleotide. Usman *et al.* (11) suggested that the unwanted 3′-*tert*-butyldimethylsilyl nucleoside derivatives prepared as by-products during the preparation of ribonucleotide monomers (*Figure 1*) could be effectively used for succinate derivation. Alternatively, succinylation of 2′(3′)-acetyl ribonucleosides is quite acceptable. In either case, any migration of protecting groups at this 3′-terminus is immaterial.

Assembly of oligoribonucleotide chains takes place in a very similar way to that already described for oligodeoxyribonucleotides (see Chapter 1). First, terminal 5′-dimethoxytrityl groups are removed by treatment with trichloroacetic acid solution in dichloromethane (*Figure 3*). Then, coupling of a ribonucleoside phosphoramidite derivative is carried out in acetonitrile

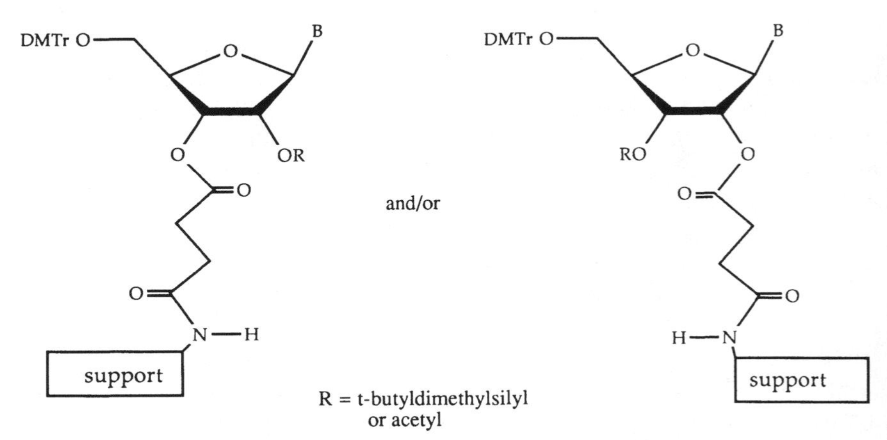

Figure 2. Structure of solid supports functionalized with ribonucleosides.

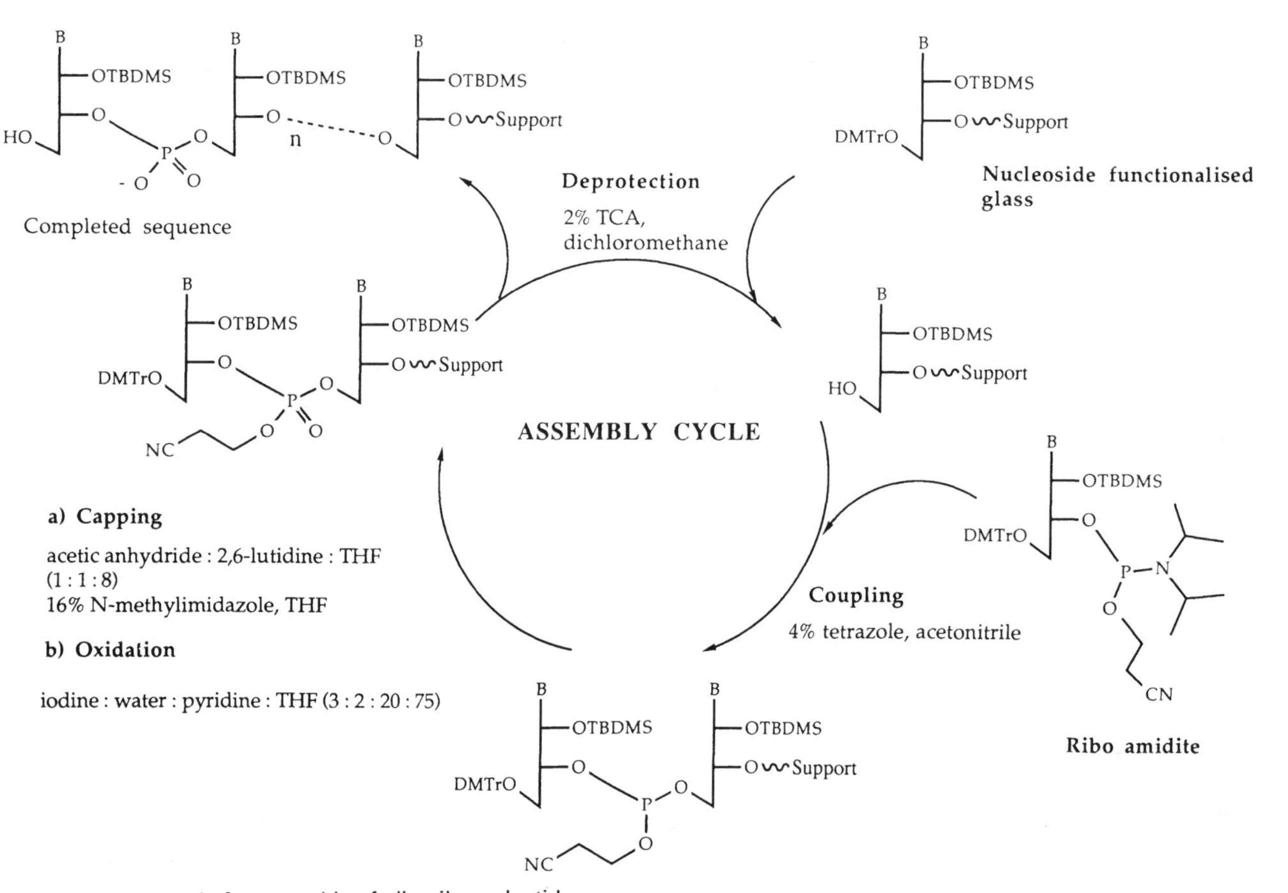

Figure 3. The cycle for assembly of oligoribonucleotides.

solution using tetrazole as catalyst. It should be noted that such coupling reactions are much slower than those between deoxyribonucleotides. The reason for this is not fully understood. It cannot be ascribed totally to the proximity of a bulky 2′-protecting group, since the much less hindered 2′-O-methyl derivatives give rise to similarly slow couplings (Chapter 3). The more activated 5-(4-nitrophenyl)tetrazole has been used as a catalyst to increase the speed of ribonucleotide coupling reactions (5, 7). However, it is our experience that the sparing solubility of this reagent in acetonitrile solution (*c*. 0.1 M) causes some problems in routine use on commercial DNA synthesizers due to lack of consistency in coupling reactions, perhaps because of partial precipitation of the reagent when forced through narrow tubing. At the time of writing, therefore, tetrazole is the preferred catalyst despite the slower internucleotide couplings.

The cycle of nucleotide addition is completed by a 'capping' reaction, involving treatment of the support with acetic anhydride and N-methylimidazole in the presence of 2,6-lutidine in tetrahydrofuran followed by oxidation of intermediate phosphite triesters to phosphate triesters using iodine in aqueous pyridine solution.

Further cycles of assembly involving terminal deprotection, coupling, capping, and oxidation are carried out to elaborate the desired sequence. All these cycles are usually carried out using a commercial DNA/RNA synthesis machine. Finally, 5′-terminal dimethoxytrityl groups are removed using a further trichloracetic acid treatment, since it is our experience that the conditions used later for 2′-deprotection of the oligoribonucleotide, treatment with tetrabutylammonium fluoride (TBAF), give rise to partial loss of 5′-dimethoxytrityl groups and thus it is better to remove these groups completely immediately after oligonucleotide assembly.

Removal of phosphate protecting groups (2-cyanoethyl), heterocyclic base protecting groups (benzoyls and isobutyryl), and cleavage of the oligonucleotide from the glass support is achieved by heating in ammonia solution (*Figure 4*). Here the use of totally aqueous ammonia (see Chapter 1) is not recommended since there is substantial loss of 2′-*tert*-butyldimethylsilyl groups. This leads to some cleavage of the internucleotide linkages by attack of the liberated 2′-hydroxyl groups on neighbouring phosphotriesters or phosphodiesters (14). Use of concentrated ammonia/ethanol (3:1) is far better in this respect (10,13) leading to only a small amount of oligonucleotide degradation. However, very recently the use of anhydrous ammonia in ethanol has been reported to result in no degradation at all (12). Unfortunately this reagent is both hygroscopic and rather volatile and must be prepared freshly, which makes it less attractive for automated use.

Finally, 2′-*tert*-butyldimethylsilyl protecting groups are removed using 1 M tetrabutylammonium fluoride (TBAF) in tetrahydrofuran. This process requires a full 24 h for completion but is highly specific (14). However, great care must be taken to ensure that the TBAF reagent is completely removed

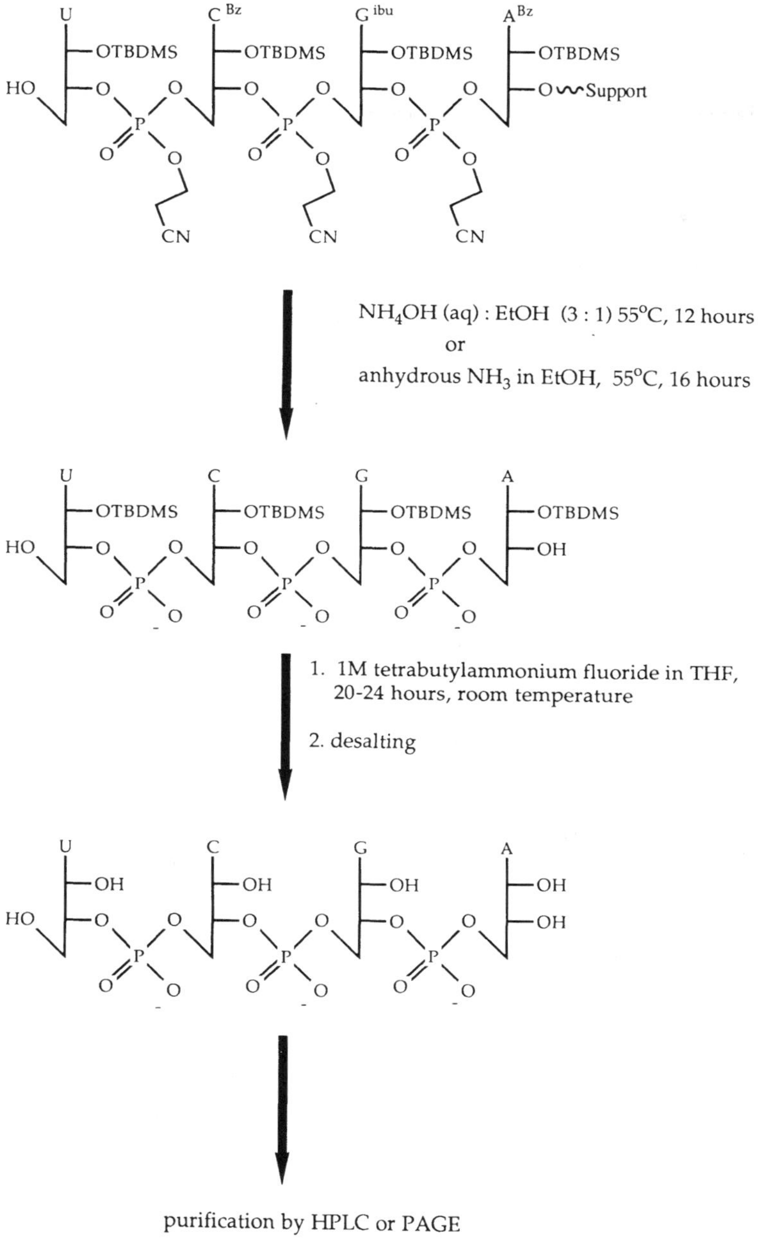

U, C^{Bz}, G^{ibu}, A^{Bz} — OTBDMS — HO — O — P — CN — O~Support

NH_4OH (aq) : EtOH (3 : 1) 55°C, 12 hours
or
anhydrous NH_3 in EtOH, 55°C, 16 hours

U, C, G, A — OTBDMS — HO — O — P — OH

1. 1M tetrabutylammonium fluoride in THF, 20-24 hours, room temperature
2. desalting

U, C, G, A — OH — HO — O — P

purification by HPLC or PAGE

Figure 4. Steps in deprotection and purification of oligoribonucleotides.

subsequently in order to preserve the oligonucleotide chain intact (see below). The deprotected oligonucleotide is now ready for purification and analysis.

3. Materials and reagents

Practically all reagents and materials for oligoribonucleotide synthesis are commercially available but at this early stage of development of oligoribonucleotide synthesis chemistry, not all suppliers' materials are of sufficient quality. We recommend the following:

Ribonucleoside phosphoramidites Milligen/Biosearch

5′-O-(4,4′-Dimethoxytrityl)-2′-O-*tert*-butyldimethylsilyl-6-N-benzoyl-adenosine-3′-O-(2-cyanoethyl)-N,N′-diisopropylphosphoramidite
0.5 g Ribo Amidite A GEN 067001

5′-O-(4,4′-Dimethoxytrityl)-2′-O-*tert*-butyldimethylsilyl-4-N-benzoyl-cytidine-3′-O-(2-cyanoethyl)-N,N′-diisopropylphosphoramidite
0.5 g Ribo Amidite C GEN 067011

5′-O-(4,4′-Dimethoxytrityl)-2′-O-*tert*-butyldimethylsilyl-2-N-isobutyryl guanosine-3′-O-(2-cyanoethyl)-N,N′-diisopropylphosphoramidite
0.5 g Ribo Amidite G GEN 067021

5′-O-(4,4′-Dimethoxytrityl)-2′-O-*tert*-butyldimethylsilyluridine-3′-O-(2-cyanoethyl)-N,N′-diisopropylphosphoramidite
0.5 g Ribo Amidite U GEN 067031

Nucleoside-functionalized controlled pore glass supports
(For self-packing of columns) Peninsula Labs
DMT-r-A(bz)-2′-tBuSiR-CPG NR2023S
DMT-rC(bz)-2′-tBuSi-CPG NR2024S
DMT-rG(ibu)-2′-tBuSi-CPG NR2027S
DMT-rU-2′-tBuSi-CPG NR2026S
(Prepacked columns for use on Milligen/Biosearch machines are also available from Milligen)

Other reagents for oligoribonucleotide assembly
These are identical to those normally used for DNA synthesis:
Activator: 4% tetrazole/acetonitrile
Terminal deprotection: 2% trichloracetic acid/dichloromethane
Capping: acetic anhydride/2,6-lutidine/THF (1:1:8)
16% 1-methylimidazole/THF
Oxidation: iodine/water/pyridine/THF (3:2:20:75)
Anhydrous acetonitrile (for dissolution of amidites)

We recommend that these reagents are purchased according to the specifications of the supplier of DNA/RNA synthesizer equipment. In our case we have used an Applied Biosystems 380B Synthesizer and Applied Biosystems reagents throughout. HPLC grade acetonitrile (e.g. from Rathburn, Romil, Fisons etc.) can be used for all intermediate washing steps.

Deprotection reagents

(i) *Ammonium hydroxide/ethanol (3:1)*

35% ammonia solution (Analar)	BDH, Poole UK
Absolute ethanol (Analar)	BDH, Poole UK

Mix the reagents just prior to use in the synthesizer and transfer to an appropriate bottle. We recommend that the reagent is made up freshly each day, since some loss of ammonia is observed upon standing at room temperature. We have very recently attempted to use anhydrous ammonia in ethanol for deprotection (12). This is prepared by bubbling ammonia gas through absolute ethanol which is cooled on ice. Care must be taken to exclude moisture in this procedure. Also the reagent is very unstable and should be stored at −20 °C for a maximum of 3–4 days. We also recommend that the reagent is not connected to the synthesis machine but used manually by syringe injection into the column after disconnection from the machine. Very preliminary results suggest that yields are at least as good as in the case of partially aqueous ethanolic ammonia and possibly better.

(ii) *2'-Hydroxyl deprotection*

1 M tetrabutylammonium fluoride in THF (less than 5% water)	Aldrich Chemical Co.

Volatile buffers

(i) *2 M triethylammonium bicarbonate (TEAB) solution*

Add triethylamine (28 ml, BDH Analar) to sterile water (50 ml) and bubble carbon dioxide gas through the mixture with occasional swirling until the triethylamine is completely dissolved. Continue bubbling until the pH of the solution drops below 8. Make up to 100 ml with sterile water and store in a sterile bottle at 4 °C. Dilute as necessary with sterile water to the required molarity.

(ii) *0.1 M triethylammonium acetate solution*

Add to sterile water (500 ml) triethylamine (6.7 ml) and glacial acetic acid (2.8 ml, analytical grade, Fisons). After thorough mixing, adjust the pH to 7.0 by addition of either a few drops of acetic acid or a few drops of triethylamine as required.

Reagents used in oligonucleotide purification

(i) *Sterile water*

A major source or ribonucleases derives from the use of insufficiently pure water. We recommend that double distilled and autoclaved water is used for all buffers and aqueous reagent that come into contact with oligoribonucleotides (see also precautions against ribonuclease cleavage, Section 5).

(ii) *HPLC*:

- Reagents

formamide (Analar)	BDH
water (HiPerSolv)	BDH
potassium dihydrogenorthophosphate (Analar)	BDH
potassium hydroxide (Analar)	BDH
acetonitrile (HPLC grade S)	Rathburn
ammonium acetate (Analar)	BDH

- Preparation of ion exchange buffers
  Make a stock solution of 1.0 M potassium dihydrogenorthophosphate (KH_2PO_4) with pH adjusted to 6.3 by careful addition of potassium hydroxide. Solvent A (15 mM phosphate): 15 ml KH_2PO_4 stock solution + 600 ml water + 900 ml formamide; Solvent B (300 mM phosphate): 300 ml KH_2PO_4 stock solution + 100 ml water + 600 ml formamide.
  Notes: (1) Higher phosphate concentrations cannot be used with this system because of the danger of precipitation of potassium phosphate; (2) it is best to use the same batch of formamide in both the buffers to ensure a good UV baseline in gradient elution; and (3) 40% acetonitrile may be used instead of formamide as disaggregant but care should be taken here to avoid precipitation of potassium phosphate in pump B by ensuring that it is always pumping, even if at low speed.

- Preparation of reversed phase buffers
  Solvent A: 0.1 M ammonium acetate solution (pH unadjusted).
  Solvent B: 0.1 M ammonium acetate solution/acetonitrile (6:4).

- Columns

Partisphere 5-SAX (0.46 × 12.5 cm) analytical cartridge	Whatman
Partisil 10-SAX (1.0 × 25 cm) semi-preparative column	Hichrom, Reading, UK or Whatman
μ-Bondapak C18 (0.39 × 30 cm) analytical column or (0.78 × 30 cm) semi-preparative column	Waters/Millipore

(iii) *Desalting*

Sephadex G25 or Sephadex NAP columns	Pharmacia
Qiagen pack 500 strong anion exchange cartridges	Hybaid Ltd, Middlesex, UK
Dialysis tubing (Visking)	Medicell International

4. Assembly of oligoribonucleotide chains and deprotection

4.1 Machine and reagent preparation

We have carried out all assemblies on an Applied Biosystems 380B DNA Synthesizer, although other manufacturers machines are also applicable in general. Solutions of ribo amidites are made up at 0.1 M concentration by direct injection of anhydrous acetonitrile into 0.5 g bottles as follows: A 5.2 ml, G 5.3 ml, C 5.4 ml, U 6.0 ml. The bottles are now connected to the machine (*Note that if the machine has previously been used for deoxynucleotide synthesis, very careful washing of the amidite delivery lines is necessary. A single bottle change procedure may not be sufficient in all cases.*)

4.2 Column packing

Currently, prepacked columns are not available for oligoribonucleotide synthesis using Applied Biosystems machines. This is not a difficulty since empty columns can be packed very simply (*Protocol 1*).

Protocol 1. Packing of columns

1. Weigh out the appropriate amount of support on a micro-balance. For 1 μmol scale this will be about 30 mg of 33–35 μmol/g nucleoside functionalized support. Do not fill an ABI standard column with more than 40 mg of support, since there is insufficient mixing during flow and synthesis results become poor.
2. Using a small funnel, transfer the support into an empty column and crimp or snap close.
3. Test the column for leaks by connecting to the synthesizer and flow in both directions using acetonitrile.

4.3 Assembly cycle

For oligoribonucleotide assembly, we use an identical cycle to that used for DNA synthesis on 1 μmole scale *except* that the wait step of the coupling reaction is increased to 600 s (as opposed to 30 s for DNA synthesis).

To check the efficiency of coupling, use exactly the same procedure as is recommended for assembly of DNA chains (ABI under bulletin 13).

- Dilute each TCA eluate (collected by fraction collector in glass tubes) to 50 or 100 ml with 0.1 M toluene-*p*-sulphonic acid in acetonitrile.
- Measure the absorbance at 498 nm (ε 71 700) and compare for each synthesis cycle as a percentage of the previous cycle result.

In our experience, the efficiences are about 98% for U or G couplings, 96% for A, and 95% for C couplings. Hopefully these values will improve as commercial materials become of higher quality.

4.4 Deprotection

The first part of the deprotection is pre-programmed on the 380B Synthesizer, (end procedure) but this step can also be carried out manually by connecting the column to a syringe having a Luer fitting (*Protocol 2*).

Protocol 2. Deprotection of assembled oligoribonucleotide chains

1. Treat the support with the ammonia/ethanol deprotection reagent in 5 batches for 1450 s each. Note that this is double the time normally allowed for aqueous ammonia treatment in oligodeoxyribonucleotide synthesis. Collect the five ammoniacal eluates together in a small screw-capped glass vial.
2. Seal the vial and heat at 55 °C (water bath, oven, or heating block) for 8–12 hours.
3. Freeze the sample in dry ice/acetone and then lyophilize. We recommend a Savant SpeedVac Concentrator for this purpose, since the vial can be placed directly in one of the rotor buckets. The freezing to low temperature prevents 'bumping' of the ammonia solution.
4. Resuspend the residue in 1 M TBAF solution (1 ml).
5. Seal the vial again and keep in the dark at room temperature for 20–24 h. Swirl occasionally to ensure that a homogeneous solution is obtained (NB. If total solubility is not achieved eventually, the silyl removal may not proceed to completion. Repeat treatment perhaps at lower TBAF concentration may be needed.)
6. To quench the reaction, add to the mixture 0.1 M triethylammonium acetate solution (5 ml, if dialysis is used for desalting or 1 ml if gel filtration is used, see below) or 0.1 M TEAB solution (10 ml) (if Qiagen cartridge desalting is used) and the sample is now ready for desalting.

---

5. Desalting and purification of oligoribonucleotides

5.1 Precautions to avoid ribonuclease contamination

The importance of good handling techniques and care in preparation of buffers, reagents, and apparatus cannot be overstressed. Ribonucleases are ubiquitous and the slightest trace can give rise to degradation of the oligoribonucleotide. All water used should be sterile (see Section 3) and reagents should be of the highest purity. Wherever possible, sterile disposable plastic tips and tubes should be used for storage and handling of solutions of oligoribonucleotides. If glassware must be used, wash well with chromic acid, rinse exhaustively in glass-distilled water, and bake in the hottest possible oven (or autoclave). Disposable plastic gloves should be used at all times when handling RNA. Separate HPLC columns should be used for RNA purification and for enzymatic analysis of RNA (Section 6).

5.2 Desalting

It is vital to remove all traces of TBAF before attempting evaporation of the deprotected oligoribonucleotide sample. In our experience, there is great danger of degradation of the oligoribonucleotide chain during evaporation and therefore we recommend an initial desalting step. Three methods have been successfully used in our laboratory.

5.2.1 Dialysis against water (*Protocol 4*)

The dialysis tubing must first be very carefully prepared (15) (*Protocol 3*). Gloves should be worn.

Protocol 3. Preparation of dialysis tubing

1. Boil appropriate length pieces of tubing in a large volume of 2% sodium bicarbonate and 1 mM ethylenediaminetetraacetic acid disodium salt (EDTA), for 10 min.
2. Rinse the tubing thoroughly with sterile water.
3. Boil the tubing for 10 min in sterile water. This step can be replaced by 'wet' autoclaving in a loosely capped jar of sterile water, which is recommended if possible.
4. Cool and store at 4 °C under sterile water.
5. Before use, wash inside and outside of the tubing with sterile water.

Protocol 4. Desalting by dialysis

1. Transfer the oligonucleotide solution into a section of dialysis tubing and seal the ends.
2. Place the dialysis tube in a flask containing 5 litres of gently stirred distilled water (it is not essential to use sterile water on the outside of the dialysis bag).
3. Allow to equilibrate for at least 2 h.
4. Change the external water three times. It is recommended that the minimum time be used for dialysis (usually 8–16 h) to avoid the possibility of RNA degradation.

5.2.2 Strong anion-exchange cartridge method (12) (*Protocol 5*)

An alternative to ion exchange cartridge desalting is the use of a reversed phase cartridge. We have currently insufficient experience in their routine use for desalting oligoribonucleotides to make a positive recommendation, but we have found both OPC cartridges (Applied Biosystems) and Poly-Pak cartridges (Glen Research) are useful for oligodeoxyribonucleotide desalting.

Protocol 5. Desalting by use of ion exchange cartridge

1. Using a disposable syringe or via gravity flow elution, apply the oligonucleotide sample, which must be less than 0.1 M in TBAF, to a Qiagen pack 500 cartridge which has been pre-washed with 15 ml 0.1 M TEAB containing 0.15% Triton X-100.
2. Collect the eluate and reapply this to the Qiagen cartridge to ensure complete absorption of the oligonucleotide.
3. Wash the column with 0.1 M TEAB solution (7 ml) and then elute the oligonucleotide from the cartridge using 2 M TEAB (12 ml).
4. Collect the eluate and lyophilize.

5.2.3 Sephadex gel filtration (11) (*Protocol 6*)

Prepacked sterile Sephadex NAP-10 columns can also be used for this desalting step (*Protocol 7*).

Protocol 6. Desalting by conventional gel filtration

1. Load the sample of oligonucleotide on a Sephadex G25F column (30 × 1 cm) made up in sterile water (Note that the glass column should be acid washed and the Sephadex should be autoclaved).
2. Elute the column with water by gravity flow or by a peristaltic pump.
3. Collect fractions in sterile microfuge tubes.
4. Measure the UV absorbance at 260 nm. The first eluting peak at the void volume is the oligonucleotide.
5. Evaporate fractions using a SpeedVac Concentrator.

Protocol 7. Desalting using Sephadex NAP-10 columns

1. Freeze the oligoribonucleotide sample and lyophilize just long enough to reduce the volume to less than 1 ml.
2. Dilute with sterile water to exactly 1 ml.

Protocol 7. *Continued*

3. Apply to a prewashed (15 ml of sterile water) NAP-10 column.
4. Elute with sterile water (1.5 ml) by gravity flow and collect the eluate.
5. Evaporate using a SpeedVac Concentrator.

5.3 HPLC purification of oligoribonucleotides

5.3.1 Ion exchange HPLC

The advantage of ion exchange systems in purification of oligonucleotides is that resolution is by formal negative charge which increases as the length of the oligonucleotide increases. Thus, the longer the oligonucleotide the later it is eluted. This makes identification of the desired product oligonucleotide in solid phase synthesis quite easy. We have found that the best analytical ion exchange columns to date for both oligodeoxyribo- and oligoribo-nucleotides are Partisphere 5-SAX (7, 16) (*Protocol 8*) and for preparative isolation, Partisil 10-SAX (7, 17) (*Protocol 9*). A typical separation on an analytical column is shown in *Figure 5A* for the 13mer r(GCCUGU)d(C)r(AGUCCC) (see Section 7.1 for mixed ribo–deoxyribooligonucleotide synthesis). A preparative separation of the same 13mer is shown in *Figure 5B* and an analytical check of purity after storage of the 13mer for 3 months at −20 °C as a solid is shown in *Figure 5C*.

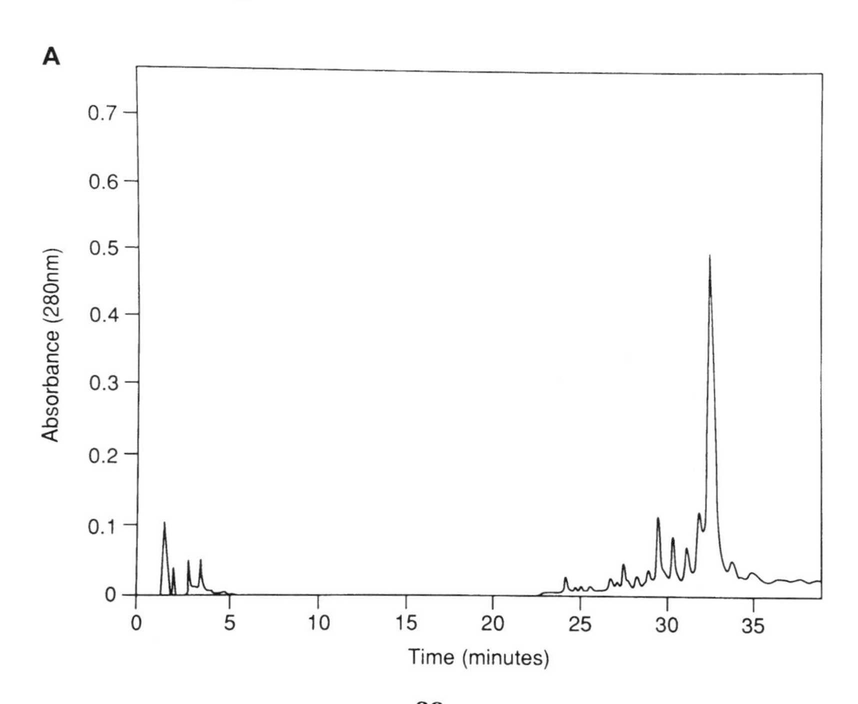

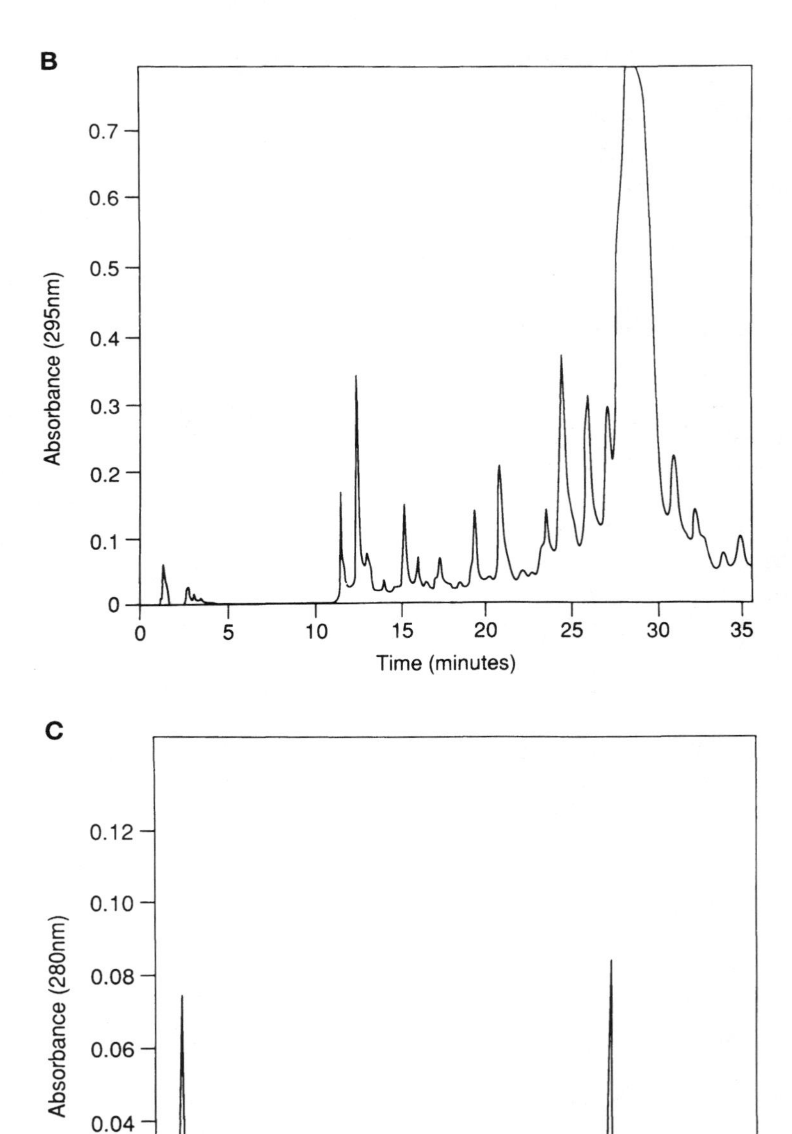

Figure 5. Ion exchange separations of the 13mer r(GCCUGU)d(C)r(AGUCCC). (**A**) Analytical separation on Partisphere 5-SAX. Conditions as in text. (**B**) Preparative separation on Partisil 10-SAX. Conditions as in text. (**C**) Analytical check on purity after storage using Partisphere 5-SAX. Conditions as in text except salt gradient flattened.

Since oligoribonucleotides are retained on ion exchange columns considerably longer than oligodeoxyribonucleotides, we have found the practical limit of resolution to be up to about 27 residues. This is in contrast to 2′-methoxytetrahydropyranyl-protected oligoribonucleotides which are retained to about the same extent as oligodeoxyribonucleotides (7).

Before HPLC the sample should be prepared as follows:

- Dissolve the oligoribonucleotide sample from a 1 μmol scale synthesis in 1 ml of sterile water just prior to purification.
- Clarify the sample by microfuge centrifugation using an Ultrafree-MC 0.22 μm filter unit (Millipore, UFC3 OGV 00), which takes 400 μl aliquots per centrifugation, or by passage through a 0.2 μm Acrodisc syringe filter (Gelman Sciences). In the latter case an extra 0.5–1 ml water must be passed through to flush the filter unit.

Protocol 8. Analytical separations of oligoribonucleotides on Partisphere 5-SAX

1. Set the flow rate to 1.0 ml/min.
2. Set the UV monitor to 280 nm on the 1.0 scale. Wavelengths lower than 270 nm cannot be used because of the UV absorption of formamide.
3. Use a starting gradient of 0–100% B over 30 min initially to gauge the success of assembly, but the gradient subsequently can be flattened if desired.
4. Inject 10 μl of sample and run the gradient. The last peak to emerge should be the desired oligonucleotide if the synthesis has proceeded correctly.

Protocol 9. Preparative separations of oligoribonucleotides on Partisil 10-SAX

1. Set the flow rate: to 2.5 ml/min.
2. Set the UV monitor to 280 nm on the 0.2 scale.
3. Use a starting gradient of 0–100% B over 30 min for the first injection.
4. Inject about 30 μl of sample and run the gradient.
5. Observe the elution position of the major (last peak) and decide if the gradient should be flattened to achieve optimal separation.
6. Adjust the UV monitor sensitivity to the 1.0 or 2.0 scale at 295 nm and inject about 200–250 μl of sample.
7. Collect the product in the desired peak as it is eluted. 4–5 injections

Protocol 9. *Continued*

should be sufficient for purification of all the material from a 1 μmol scale synthesis.

8. Remove the salt and formamide by dialysis against water (see Section 5.2) and isolate the product by lyophilization.

5.3.3 Reversed phase HPLC

This method of separation is not recommended for fully deprotected oligonucleotides as a first step procedure because the elution position of the desired oligonucleotide cannot be predicted with accuracy. However it may be useful as a second purification step after an initial separation using strong anion-exchange and especially if ultra-high purity oligoribonucleotides are required (*Protocol 10*). *Warning*: it is our experience that some oligoribonucleotides with stable secondary structures do not give single sharp peaks on reversed phase HPLC.

Protocol 10. Reversed phase separations of oligoribonucleotides

1. Set the flow rate to 1.5 ml/min.
2. Set the UV monitor to 260 nm.
3. Try initially a gradient from 0–100% B over 30 min to establish the approximate elution position.
4. Inject a sample of oligoribonucleotide (after purification by ion exchange HPLC or by PAGE) and run the gradient.
5. Observe the elution profile and decide if the product is eluting mostly as a single peak. If desired, the gradient may then be flattened considerably for optimal resolution.
6. Preparative separations should only be attempted if analytical HPLC shows predominantly a single component with no sign of secondary structural interference. In this case use a flow rate of 3 ml/min and a semi-preparative column.

5.4 Polyacrylamide gel electrophoresis (PAGE)

Methods for purification of oligoribonucleotides by PAGE (*Protocol 11*) are very similar to those used for oligodeoxyribonucleotides.

Protocol 11. Preparative separations of oligoribonucleotides by PAGE

1. Pour a 200 × 400 mm gel between glass plates having 1.5 mm spacers. We recommend 20 mm slots as being most convenient for the comb. For oligoribonucleotides up to 20 residues, a 20% gel should be used, whereas for 20–40 residues a 15% gel is preferable (7 M urea, 1 × TBE, acrylamide/bis-acrylamide (100: 2.5)).
2. Dilute the desalted oligoribonucleotides obtained from a 1 μm scale synthesis (approximately 40–50 A_{260} units in 1 ml sterile water) with 1 ml dye mix (8 ml formamide, 100 μl 0.5 M EDTA, 2 mg bromophenol blue made up to 10 ml with sterile water).
3. Load on to the gel (at least 10 × 20 mm slots).
4. Electrophorese in 1 × TBE buffer at 30–40 W for 2–3 h until the bromophenol blue dye reaches close to the bottom of the gel.
5. Remove gel from both glass plates by transferring on to a single sheet of Saranwrap.
6. Locate the bands by UV shadowing. This involves shining UV light at 254 nm on to the gel placed on an autoradiography screen (e.g. Dupont Cronex Lightning Plus-ZK screen) using a special dark box to avoid the health hazard posed by short-wave UV light. Permanent records can be obtained by Polaroid photography. *Figure 6* shows the UV shadowing of a 29mer oligoribonucleotide.
7. Excise bands and transfer to microfuge tubes.
8. Soak the gel pieces (without crushing) in sterile 0.5 M ammonium acetate, 1 mm EDTA, 0.5% sodium dodecyl sulphate (SDS) at room temperature for 6–18 h. Be sure all the gel pieces are covered by the elution buffer.
9. Carefully suck off the liquid and repeate the soaking.
10. Save and combine the gel extract solutions and centrifuge or filter to remove any small gel pieces.

The oligoribonucleotide can then be recovered by a butanol extraction procedure (20) (*Protocol 12*). This procedure removes all buffers, urea, and SDS. If desired, the oligoribonucleotide can then be precipitated with ethanol or subjected to reversed phase cartridge purification.

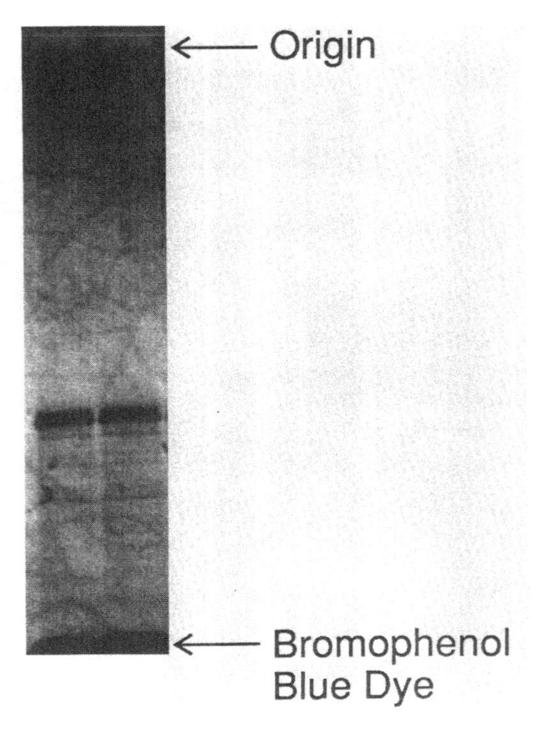

Figure 6. Polaroid photograph of UV (254 nm) shadowed preparative 15% polyacrylamide gel showing two adjacent 20 mm slots of separation of a crude 29mer oligoribonucleotide synthesis.

Protocol 12. Butanol extraction procedure for oligoribonucleotide recovery

1. Fill the microfuge tube with n-butanol.
2. Vigorously shake for 30 s and centrifuge for 1 min.
3. Remove the butanol (top) phase.
4. Repeat the extraction several times, each time filling the tube with n-butanol. Finally a pellet will be formed.
5. Redissolve the pellet in water (200 μl) and dry again to a pellet by a single n-butanol extraction.
6. Repeat step 5 three times.

---

6. Enzymatic analysis of oligoribonucleotides

This should no longer prove necessary on a routine basis since the quality of materials now available commercially is such that under normal circumstances wrong linkages (2′–5′) or modified bases should not be present in the final oligoribonucleotide. However, if there is a doubt in any particular circumstance, we recommend total enzymatic digestion with ribonucleases followed by phosphatase treatment, and then separation of the resultant nucleosides by reversed phase HLPC. The choice of ribonucleases is somewhat dependent on the sequence of the oligoribonucleotide. We prefer to use a mixture of RNase A, RNase T1 and RNase T2 which appears to be sufficient for most cases (*Protocol 13*). It is a good idea to check the quality of the ribonuclease preparations by carrying out a control digestion of a 2′–5′ linked di- or trinucleotide (e.g. rA (2′–5′)A which is available from Sigma). Such compounds should be totally resistant to digestion.

Protocol 13. Enzymatic digestion and analysis of oligoribonucleotides

1. Treat 0.2 A_{260} units of oligoribonucleotide dissolved in 0.05 M ammonium acetate, 0.002 M EDTA (pH 4.5; 40 µl) with RNase A (Sigma, 0.25 mg/ml, 5 µl), RNase T1 (Sigma, 50 units/ml, 5 µl), and RNase T2 (50 units/ml; 5 µl) at 37 °C for 16 h).
2. Evaporate to dryness (SpeedVac).
3. Dissolve the residue in 0.1 M Tris–HCl, 0.01 M $MgCl_2$ (pH 8.5; 50 µl).
4. Add calf intestinal alkaline phosphatase (Boehringer, 28 units/ml, 1 µl) and leave at 37 °C for 16 h.
5. Carry out reversed phase HPLC of the product essentially as described in section 5.3 except use isocratic elution (5% buffer B) to separate the four ribonucleosides followed by gradient elution to check for the absence of longer oligonucleotides. A typical analysis of total digestion of a 27mer oligoribonucleotide is shown in *Figure 7*.

---

7. Special applications

7.1 Mixed RNA–DNA oligonucleotides

These can be made very simply without significant changes to the protocols or reagents. In chain assembly steps whenever a deoxyribonucleotide amidite is required to be added, the normal deoxy cycle of addition (shorter coupling time) is programmed (Chapter 1). However, the synthesizer must contain sufficient ports for all the different deoxy and ribo amidites required for the

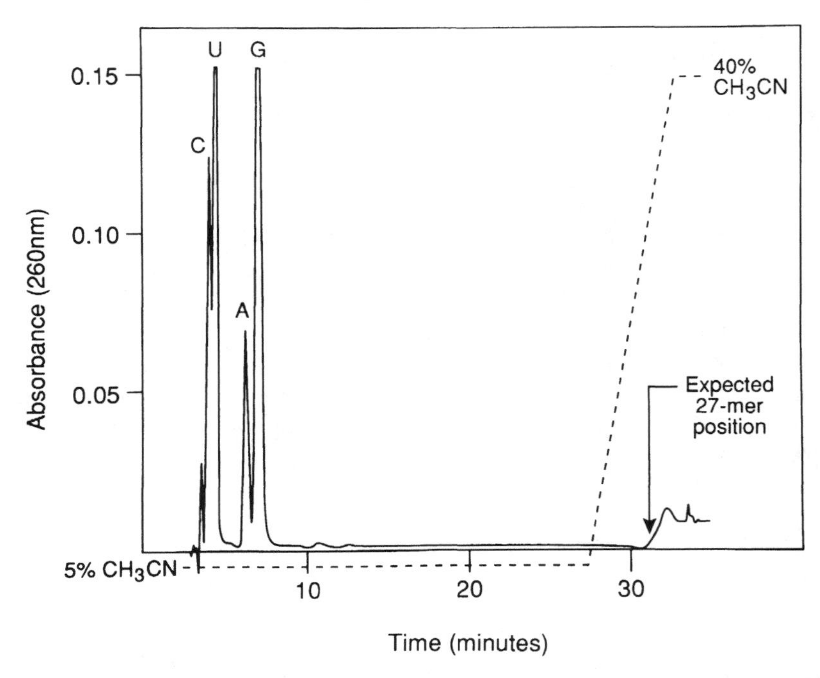

Figure 7. Reversed phase HPLC of the enzymatic digestion products of a 27mer oligoribonucleotide. Conditions of digestion and chromatography are as described in the text.

assembly. Similarly if a deoxyribonucleotide is to be introduced at the 3′-end, use a prepacked deoxynucleoside column (Chapter 1). For deprotection and work-up, follow the procedures outlined for oligoribonucleotides.

7.2 Phosphorothioate oligoribonucleotides (see Chapter 4)

An individual phosphodiester bond can be substituted for a phosphorothioate bond (racemic mixture of R_p and S_p isomers) by interrupting the synthesizer programme at the appropriate cycle just before the step of addition of the oxidation reagent and by carrying out manual sulphurization (*Protocol 14*).

Protocol 14. Manual sulphurization of oligoribonucleotides

1. Remove the column from the synthesizer.
2. Connect a syringe filled with a Luer adaptor, containing 0.4 g elemental sulphur (Aldrich gold label) in carbon disulphide/2,6-lutidine (6 ml).
3. Inject 1 ml of the sulphurization solution every hour until all the solution has passed through.

Protocol 14. *Continued*

4. Wash the column with carbon disulphide/2,6-lutidine (10 ml) and then with aceonitrile (20 ml).

5. Reconnect the column to the synthesizer and continue the assembly cycle as though starting a new cycle (i.e. acidic deprotection step).

After deprotection of the oligonucleotide in the normal way, ion exchange HPLC should show a major peak the elution position of which should be slightly retarded compared with the natural oligoribonucleotide. In favourable cases, the R_p and S_p isomers may be resolved by subsequent reversed phase HPLC (21).

We do not recommend carrying out the sulphurization on the synthesizer since elemental sulphur tends to precipitate in the delivery line thus clogging it. A 1 hour sulphurization using the new reagent tetraethylthiuram disulphide (Lancaster Synthesis, 500 mg) in acetonitrile (5 ml) has recently obviated the clogging problem (21).

8. Future developments

The currently available range of commercial ribo amidites and support-bound nucleosides is adequate for the synthesis of medium length oligoribonucleotides (up to perhaps about 30–40 residues). It is likely that for the synthesis of longer oligoribonucleotides significant improvements in chemistry will be needed. The use of base-protecting groups removable under milder ammoniacal conditions (18, 19) will be helpful and phenyoxyacetyl-protected A and G ribo amidites have very recently become available (American Bionetics and Glen Research). However, better coupling yields will be essential and this can only come through development of more active ribo amidites or more active activating agents. In order to avoid possible cleavage by ribonucleases during purification, it would also be very useful if oligoribonucleotides could be purified at the 2′-protected stage, as is possible using 2′-Mthp protected oligoribonucleotides (7). It remains to be seen whether procedures can be found for either 2′-TBDMS (described here) or 2′-Fpmp (13) containing oligoribonucleotides to be routinely purified after assembly.

One exciting prospect which should now become possible is the ability to incorporate modifications to either phosphate, sugar, or base or combinations of modifications into specific locations within oligoribonucleotides for use for example in the study of RNA–protein interactions or of self-cleaving RNA domains and ribozymes.

References

1. Griffin, B. E. and Reese, C. B. (1964). *Tetrahedron Lett.*, 2925.
2. Reese, C. B., Saffhill, R., and Sulston, J. (1967). *J. Amer. Chem. Soc.*, **89**, 3366.
3. Christodoulou, C., Agrawal, S., and Gait, M. J. (1986). *Tetrahedron Lett.*, **27**, 1521.
4. Reese, C. B. and Thompson, E. A. (1988). *J. Chem. Soc. Perkin I*, 2881.
5. Beijer, B. Sulston, I., Sproat, B. S., Rider, P., Lamond, A. I., and Neuner, P. (1990). *Nucleic Acids Res.*, **18**, 5143.
6. Iwai, S. and Ohtsuka, E. (1988). *Nucleic Acids Res.*, **16**, 9443.
7. Lehmann, C., Xu, Y.-Z., Christodoulou, C., Tan., Z.-K., and Gait, M. J. (1989). *Nucleic Acids Res.*, **17**, 2379.
8. Ogilvie, K. K., Sadana, K. L., Thompson, E. A., Quilliam, M. A., and Westmore, J. B. (1974). *Tetrahedron Lett.*, 2861.
9. Hakimelahi, G. H., Proba, Z. A., and Ogilvie, K. K. (1981). *Tetrahedron Lett.*, **22**, 5243.
10. Jones, S. J. and Reese, C. B. (1979). *J. Chem. Soc. Perkin I.*, 2762.
11. Usman, N., Ogilvie, K. K., Jiang, M. Y., and Cedergren, R. L. (1987). *J. Am. Chem. Soc.*, **109**, 7845.
12. Scarringe, S. A., Francklyn, C., and Usman, N. (1990). *Nucleic Acids Res.*, **18**, 5433.
13. Sproat, B. S. and Gait, M. J. (1984). In *Oligonucleotide Synthesis: A Practical Approach*, (ed. M. J. Gait), pp. 96–97. IRL Press, Oxford.
14. Stawinski, J., Strömberg, R., Thelin, M., and Westman E. (1988). *Nucleosides and Nucleotides*, **7**, 779.
15. Maniatis, T., Fritsch, E. F., and Sambrook, J. (1982). *Molecular Cloning. A Laboratory Manual*, p. 456. Cold Spring Harbor Laboratory Press, Cold Spring Harbor, NY.
16. Misiura, K., Durrant, I., Evans, M. R., and Gait, M. J. (1990). *Nucleic Acids Res.*, **18**, 4345.
17. Gait, M. J. and Sheppard, R. C. (1977). *Nucleic Acids Res.*, **4**, 4391.
18. Wu, T., Ogilvie, K. K., and Pon, R. T. (1989). *Nucleic Acids Res.*, **17**, 3501.
19. Chaix, C., Duplaa, A. M., Molko, D., and Teoule, R. (1989). *Nucleic Acids Res.*, **17**, 7381.
20. Cathala, G. and Brunel, C. (1990). *Nucleic Acids Res.*, **18**, 201.
21. Slim, G. and Gait, M. J. (1991). *Nucleic Acids Res.*, **19**, 1183.

3

2′-O-Methyloligoribonucleotides: synthesis and applications

B. S. SPROAT and A. I. LAMOND

1. Introduction

2′-O-Methyloligoribonucleotides (1–7) are proving to be useful reagents for a variety of biological experiments. Their usefulness stems from the following properties:

- a 2′-O-methyloligoribonucleotide–RNA duplex is thermally more stable than the corresponding oligodeoxyribonucleotide–RNA one (2) and
- the former duplex is not a substrate for RNase H (6).

This enzyme specifically cleaves RNA in RNA–DNA heteroduplexes (8). In addition, 2′-O-methyloligoribonucleotides are chemically more stable than either oligodeoxyribonucleotides or oligoribonucleotides and moreover are totally resistant to degradation by either RNA- or DNA-specific nucleases (7).

The first part of this chapter describes in detail the synthesis of appropriately protected 2′-O-methylribonucleoside-3′-O-phosphoramidites (9) and ancillary reagents required for solid phase assembly of 2′-O-methyloligoribonucleotides. This part also includes procedures for synthesis, biotinylation (10), deprotection, and purification of the polymers. The methods described here have been limited to the most commonly used solid phase synthesis method, the so-called phosphite triester method (11–13). However, with minor changes to the protocols, H-phosphonate or phosphodiester building blocks can be easily prepared.

The second part of the chapter describes some of the applications of 2′-O-methyloligoribonucleotides, in particular their usage for affinity chromatography of RNA–protein complexes. This has recently proved important for *in vitro* studies of RNA processing (14–16). (See also Chapter 10 for oligodeoxynucleotides for affinity chromatography and Chapter 11 for oligodeoxynucleotides, with reporter groups attached to the base.)

Finally, the last part lists commercial sources of specialized equipment and the more unusual reagents.

2. Synthesis of protected 2′-O-methylribonucleoside-3′-O-phosphoramidites, coupling agent, succinates, supports, and biotinylation reagent

2.1 Preparation of 2′-O-methylribonucleoside-3′-O-phosphoramidites

The syntheses described in the protocols that follow can either be performed on a large scale as described, or scaled down according to wish. Preparative liquid chromatography (Waters Prep 500) on silica gel cartridges (5 × 30 cm, 300 g) enables the fastest recovery of high purity materials prepared on large scale. However for scaled down syntheses ordinary column chromatography is perfectly effective. Note that you will need to adjust the composition of the eluant mixture when switching from preparative liquid chromatography to ordinary column chromatography. This is best done by referring to the TLC characteristics of the compound to be purified. When petrol is mentioned you should use the material with b.p. 40–60 °C. The monomers should be stored dry under nitrogen in a sealed container at −20 °C. Useful references for the chemistry in *Protocols 1–25* are contained in reference 9.

2.1.1 Synthesis of the 2′-O-methyluridine monomer

The synthesis route that will be followed is illustrated in *Figure 1*.

Protocol 1. Synthesis of 3′,5′-O-(tetraisopropyldisiloxane-1,3-diyl) uridine (compound **1**); mol. wt 486.71

1. Dry uridine (48.84 g, 200 mmol) by addition and evaporation of dry pyridine *in vacuo*, and then dissolve in dry pyridine (500 ml).
2. To the ice-cooled solution under stirring add dropwise a solution of 1,3-dichloro-1,1,3,3-tetraisopropyldisiloxane (70 g, 223 mmol) in dry 1,2-dichloroethane (50 ml). On completion of the addition allow the reaction to warm to room temperature. Silica gel TLC in ethanol/chloroform (1:9 v/v) should show complete reaction after 2–3 h with a product spot of R_f 0.58.
3. Quench with methanol (30 ml) and evaporate the solvent *in vacuo*.
4. Dissolve the residue in dichloromethane (1.5 litres) and wash the solution with 1 M aqueous sodium bicarbonate (2 × 1 litre).
5. Dry the organic phase over anhydrous sodium sulphate, filter, and evaporate *in vacuo*.

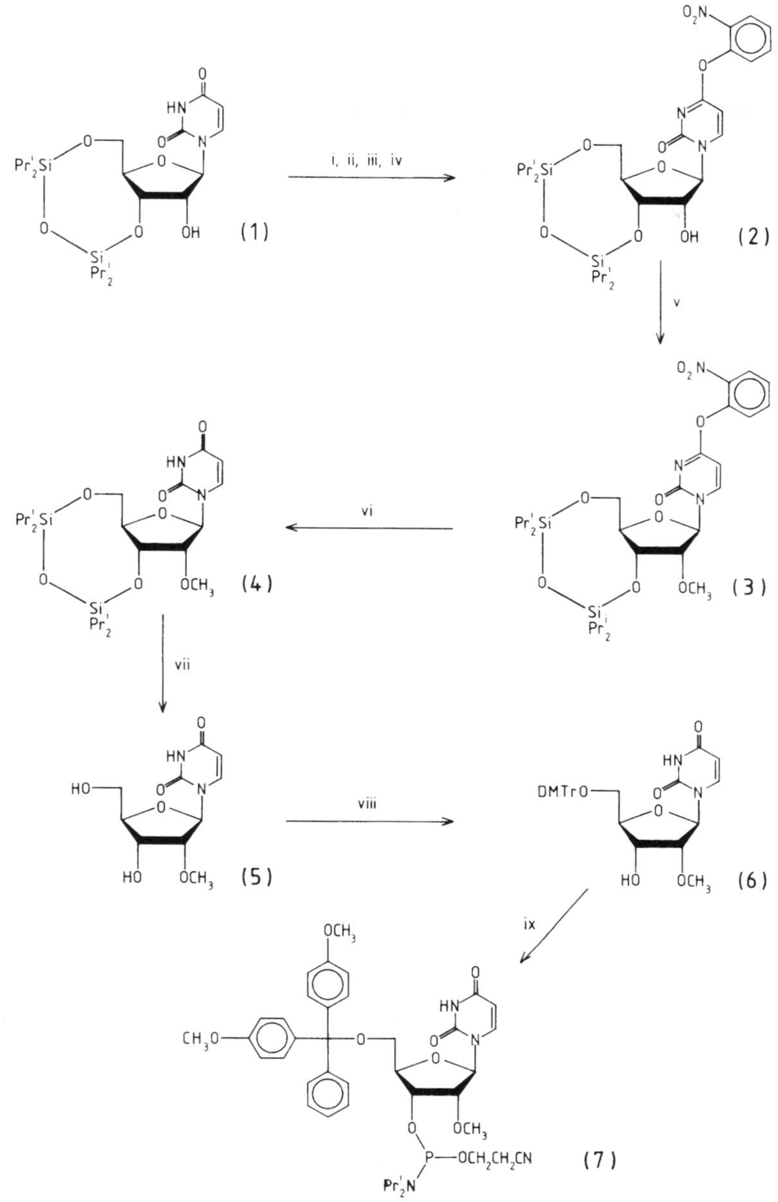

Figure 1. Reaction scheme for the preparation of the 2′-O-methyluridine building block. Reagents: (i) chlorotrimethylsilane and triethylamine in dichloromethane; (ii) 2-mesitylenesulphonylchloride, triethylamine, and 4-dimethylaminopyridine in dichloromethane; (iii) 2-nitrophenol, triethylamine and 1,4-diazabicyclo[2.2.2]octane in dichloromethane; (iv) *p*-toluene sulphonic acid in dioxan/dichloromethane; (v) methyl iodide and silver oxide in 2-butanone; (vi) 2-nitrobenzaldoxime and 1,1,3,3-tetramethylguanidine in acetonitrile; (vii) tetrabutylammonium fluoride in THF; (viii) 4,4′-dimethoxytrityl chloride and triethylamine in pyridine; (ix) 2-cyanoethoxy N,N-diisopropylaminochlorophosphine and N,N-diisopropylethylamine in 1,2-dichloroethane.

Protocol 1. *Continued*

6. Dry the residue by addition and evaporation of toluene *in vacuo* to leave the title compound **1** as a solid white foam (100 g, 100%). ^{13}C NMR spectrum ($CDCl_3$) δ: 163.24 (C-4), 150.03 (C-2), 139.98 (C-6), 101.95 (C-5), 90.99 (C-1′), 81.98 (C-4′), 75.17 (C-2′), 69.07 (C-3′), 60.38 (C-5′), 17.44-16.83 (isopropyl CH_3s), 13.41, 13.12, 12.96, and 12.56 p.p.m. (isopropyl CHs).

Protocol 2. Synthesis of 3′,5′-O-(tetraisopropyldisiloxane-1,3-diyl)-O^4-(2-nitrophenyl)uridine (compound **2**); mol. wt 607.81

1. Dissolve compound **1** (100 g, 200 mmol) in dry dichloromethane (1 litre) and add with stirring and exclusion of moisture, triethylamine (200 ml, 1.43 mol) and chlorotrimethylsilane (100 ml, 0.79 mol). A white precipitate will form. TLC in petrol/ethyl acetate (2:1 v/v) should show complete reaction after 20 min with a new spot of R_f 0.39.
2. Pour the mixture into vigorously stirred 1 M aqueous sodium bicarbonate (2 litres). Separate the organic phase, dry it over sodium sulphate, filter it, and evaporate it *in vacuo*.
3. Dry the residual foam by addition and evaporation of toluene (2 × 200 ml). Dissolve the residue in dry dichloromethane (1 litre) and add triethylamine (140 ml, 1 mol), 2-mesitylenesulphonyl chloride (65 g, 297 mmol) and 4-dimethylaminopyridine (6 g, 50 mmol) with stirring and exclusion of moisture. A deep red solution will form. TLC in petrol/ethyl acetate (2:1 v/v) should show complete reaction after 30 min with a new spot of R_f 0.63.
4. Add 1,4-diazabicyclo[2.2.2]octane (4.48 g, 40 mmol) and 2-nitrophenol (56 g, 400 mmol). TLC in petrol/ethyl acetate (2:1 v/v) should show complete reaction after 1 h with a new spot of R_f 0.45 due to the 2′-O-trimethylsilyl ether of the desired product.
5. Dilute with dichloromethane (1 litre) and wash the solution with 1 M aqueous sodium bicarbonate (2 litres). Backwash the aqueous phase with dichloromethane (2 × 500 ml). Combine the organic layers, dry over sodium sulphate, filter, and evaporate the solvent *in vacuo*.
6. Dissolve the above residue in dichloromethane (600 ml), add a solution of *p*-toluene sulphonic acid monohydrate (76 g, 400 mmol) in dioxan (600 ml) and stir for 2 min only to cleave the 2′-O-TMS ether. Quench immediately by addition of triethylamine (60 ml, 430 mmol).
7. Dilute with dichloromethane (2 litres) and pour the solution into

Protocol 2. *Continued*

vigorously stirred 1 M aqueous sodium bicarbonate (3 litres). Separate the organic layer, dry it over sodium sulphate, filter it, and evaporate solvent *in vacuo*.

8. Dissolve the crude product in ethyl acetate/petrol (1 litre 3:2 v/v) and purify in six aliquots by preparative liquid chromatography on two silica cartridges eluted at 500 ml min^{-1} with ethyl acetate/petrol (3:2 v/v).

9. Pool pure product fractions and evaporate solvent *in vacuo* to give compound **2** as a solid pale yellow foam (84 g, 69.1%) of R_f 0.24 on TLC in ethyl acetate/petrol (3:2 v/v). ^{13}C NMR spectrum ($CDCl_3$) δ: 170.75 (C-4), 154.36 (C-2), 144.86 (C-6), 144.76 (phenyl C-1), 141.38 (phenyl C-2), 134.71 (phenyl C-5), 126.53, 125.75, and 125.37 (phenyl C-3, C-4, and C-6), 94.36 (C-5), 91.98 (C-1′), 81.81 (C-4′), 74.73 (C-2′), 68.65 (C-3′), 60.13 (C-5′), 17.21-16.64 (isopropyl CH_3s), 13.17, 12.72 and 12.28 p.p.m. (isopropyl CHs).

Protocol 3. Synthesis of 3′,5′-O-(tetraisopropyldisiloxane-1,3-diyl)-2′-O-methyl-O^4-(2-nitrophenyl)uridine (compound **3**); mol. wt 621.84

1. Dry compound **2** (77.8 g, 128 mmol), from *Protocol 2*, by addition followed by evaporation of dry acetonitrile *in vacuo*.

2. Dissolve the residue in dry 2-butanone (1 litre), add fresh silver oxide (158.8 g, 0.68 mol) and methyl iodide (172 ml, 2.76 mol) in two portions over 24 h and stir the mixture under argon at room temperature for 72 h. TLC in petrol/ethyl acetate (1:1 v/v) should show a single spot of R_f 0.33.

3. Filter the mixture carefully to remove most of the silver oxide and remove solvent *in vacuo*.

4. Purify the crude product in four aliquots by preparative liquid chromatography on two silica cartridges using petrol/ethyl acetate (1:1 v/v) as eluant.

5. Remove solvent *in vacuo* from the product fractions to leave compound **3** as a solid pale cream coloured foam (60.4 g, 75.9%). ^{13}C NMR spectrum ($CDCl_3$) δ: 170.72 (C-4), 154.30 (C-2), 144.73 (phenyl C-1), 144.33 (C-6), 141.39 (phenyl C-2), 134.60 (phenyl C-5), 126.46, 125.68 and 125.32 (phenyl C-3, C-4, and C-6), 94.08 (C-5), 89.26 (C-1′), 83.08 (C-2′), 81.48 (C-4′), 67.63 (C-3′), 59.25 (C-5′), 58.87 (CH_3O), 17.25-16.59 (isopropyl CH_3s), 13.21, 12.83, 12.67, and 12.11 p.p.m. (isopropyl CHs).

Protocol 4. Synthesis of 3′,5′-O-(tetraisopropyldisiloxane-1,3-diyl)-2′-O-methyluridine (compound **4**); mol. wt 500.74

1. Dissolve 2-nitrobenzaldoxime (29.9 g, 180 mmol) and 1,1,3,3-tetramethylguanidine (20.4 ml, 162 mmol) in dry acetonitrile (400 ml) and add the deep red solution to compound **3** (28 g, 45.1 mmol), from *Protocol 3*.
2. Stir solution under anhydrous conditions. TCL should show complete reaction after 3 h.
3. Remove solvent *in vacuo*, dissolve the residual oil in ethyl acetate (1.5 litres), wash the solution with water (5 × 1 litre), dry over anhydrous sodium sulphate, filter, and evaporate the solvent *in vacuo*.
4. Purify the crude product in five aliquots by preparative liquid chromatography using petrol/ethyl acetate (1:1 v/v) as eluant.
5. Compound **4** will be obtained as a pale yellow solid (25 g, *c*. 85%), contaminated with approximately one equivalent of 2-nitrobenzonitrile. Product R_f 0.27 on silica gel TLC in ethyl acetate/petrol (1:1 v/v). ^{13}C NMR spectrum ($CDCl_3$) δ: 163.98 (C-4), 150.09 (C-2), 139.48 (C-6), 101.51 (C-5), 88.34 (C-1′), 83.83 (C-2′), 81.53 (C-4′), 68.13 (C-3′), 59.36 (C-5′), 59.07 (CH_3O), 17.43-16.72 (isopropyl CH_3s), 13.39, 13.08, 12.83, and 12.30 p.p.m. (isopropyl CHs). The presence of the 2-nitrobenzonitrile has no influence on the next reaction, after which it will be removed.

Protocol 5. Synthesis of 2′-O-methyluridine (compound **5**; mol. wt 258.23

1. Dissolve compound **4** (25 g, *c*. 38.5 mmol), from *Protocol 4*, in THF (100 ml) and add a 1 M solution of tetrabutylammonium fluoride in THF (80 ml).
2. Stir 5 min at room temperature. TLC in methanol/chloroform (1:4 v/v) will show complete reaction with a new spot of R_f 0.36.
3. Evaporate solvent *in vacuo*, dissolve the residual oil in water (600 ml) and wash the solution with dichloromethane (3 × 600 ml) followed by diethyl ether (2 × 600 ml).
4. Evaporate the aqueous phase *in vacuo* to leave an oil and dry this by addition followed by evaporation of dry methanol.
5. Purify the crude product by preparative liquid chromatography using methanol/dichloromethane (1:9 v/v) as eluant, to obtain 2′-O-methyluridine as a solid white foam (7.6 g, 76.4%). ^{13}C NMR spectrum (DMSO-d_6) δ: 163.00 (C-4), 150.45 (C-2), 140.34 (C-6), 101.79 (C-5), 85.93 (C-1′), 85.07 (C-4′), 82.61 (C-2′), 68.23 (C-3′), 60.47 (C-5′), 57.57 p.p.m. CH_3O).

Protocol 6. Synthesis of 5′-O-dimethoxytrityl-2′-O-methyluridine (compound **6**); mol. wt 560.60

1. Dry 2′-O-methyluridine (7.6 g, 29.4 mmol) by addition and evaporation of dry pyridine (2 × 100 ml) *in vacuo*.
2. To the residue add anhydrous pyridine (120 ml), triethylamine (9.7 ml, 70 mmol), and 4,4′-dimethoxytrityl chloride (19.9 g, 58.8 mmol) and stir under anhydrous conditions at room temperature.
3. TLC in ethanol/chloroform/triethylamine (10:89:1 by vol.) should show complete reaction within 90 min.
4. Quench the reaction with methanol (10 ml) and remove solvent *in vacuo* at room temperature.
5. Dissolve the syrup in chloroform (500 ml), wash the solution with 1 M aqueous sodium bicarbonate (2 × 500 ml), dry it over sodium sulphate, filter it, and evaporate solvent *in vacuo*.
6. Remove most of the residual pyridine by coevaporation with toluene (2 × 100 ml).
7. Purify the crude product in three aliquots by preparative liquid chromatography using ethyl acetate/dichloromethane (4:1 v/v) containing 1% triethylamine as eluant.
8. Evaporation of solvent from the product fractions will yield pure compound **6** as a cream coloured foam (16.3 g, 99%) of R_f 0.17 on TLC in ethyl acetate/dichloromethane (4:1 v/v) containing 0.5% triethylamine and R_f 0.53 in ethanol/chloroform/triethylamine (10:89:1 by vol.).

Protocol 7. Synthesis of 5′-O-dimethoxytrityl-2′-O-methyluridine-3′-O-(2-cyanoethyl N,N-diisopropylphosphoramidite) (compound **7**); mol. wt 760.83

1. Dissolve compound **6** (5.6 g, 10 mmol) from above in dry 1,2-dichloroethane (30 ml) containing N,N-diisopropylethylamine (3.5 ml, 20 mmol).
2. Cool the solution under a slow stream of dry argon in an ice bath.
3. Add dropwise with stirring 2-cyanoethoxy N,N-diisopropylamino-chlorophosphine (3 g, 12.7 mmol) during 2–3 min from a gas-tight syringe.
4. After 5 min on ice stir the reaction mixture for 1 h at room temperature. TLC in triethylamine/dichloromethane (1:19 v/v) should show complete reaction with a new spot of R_f 0.35. If the reaction is not complete add a

Protocol 7. *Continued*

further portion (5–10 mmol) of the chlorophosphine and more (10–20 mmol) of N,N-diisopropylethylamine.

5. When the reaction is complete, dilute the solution with dichloromethane (150 ml) and wash with 5% aqueous sodium bicarbonate (2 × 250 ml) followed by saturated brine (250 ml).

6. Add a few drops of triethylamine to the organic layer, dry over anhydrous sodium sulphate, filter, and evaporate solvent *in vacuo* at room temperature.

7. Purify the crude product by preparative liquid chromatography using petrol/dichloromethane/triethylamine (5:15:1 by vol.) as eluant. Carefully monitor the product fractions by ^{31}P NMR spectroscopy to ensure that they are free of any phosphonate impurities (δ + 10 p.p.m. relative to external trimethyl phosphate reference).

8. Remove solvent *in vacuo* from pure product fractions at room temperature to yield the phosphoramidite **7** as a fluffy white foam (5.3 g, 70%) after lyophilization from dry benzene. ^{31}P NMR spectrum (CH_2Cl_2, with external concentric D_2O lock) δ: + 147.06 p.p.m.

2.1.2 Synthesis of the 2′-O-methylcytidine monomer

The 2′-O-methylcytidine monomer is synthesized in several steps, see *Figure 2*, from the versatile intermediate, compound **3** (see *Protocol 3*) by first displacing 2-nitrophenoxide with anhydrous ammonia. Subsequent acylation, desilylation, dimethoxytritylation and phosphitylation yield the desired monomer **11**.

Protocol 8. Synthesis of 3′,5′-O-(tetraisopropyldisiloxane-1,3-diyl)-N^4-isobutyryl-2′-O-methylcytidine (compound **8**); mol. wt 569.85

1. Dissolve 3′,5′-O-(tetraisopropyldisiloxane-1,3-diyl)-2′-O-methyl-O^4-(2-nitrophenyl)uridine (32.4 g, 52.1 mmol), **3** (see *Protocol 3*) in dry THF (150 ml) and transfer the solution into the Teflon liner (capacity *c.*700 ml) of a stainless steel pressure vessel.

2. Connect a dry ammonia cylinder (lecture bottle is fine) to the gas inlet of the pressure vessel and transfer over about 50–100 g of liquid ammonia by opening the appropriate valves and by cooling the pressure vessel in a solid carbon dioxide/2-propanol bath. This will normally take about 1–2 h, and is easily monitored by shutting off the ammonia cylinder, disconnecting it, and weighing it. During the transfer the outlet valves of the pressure vessel must be kept closed.

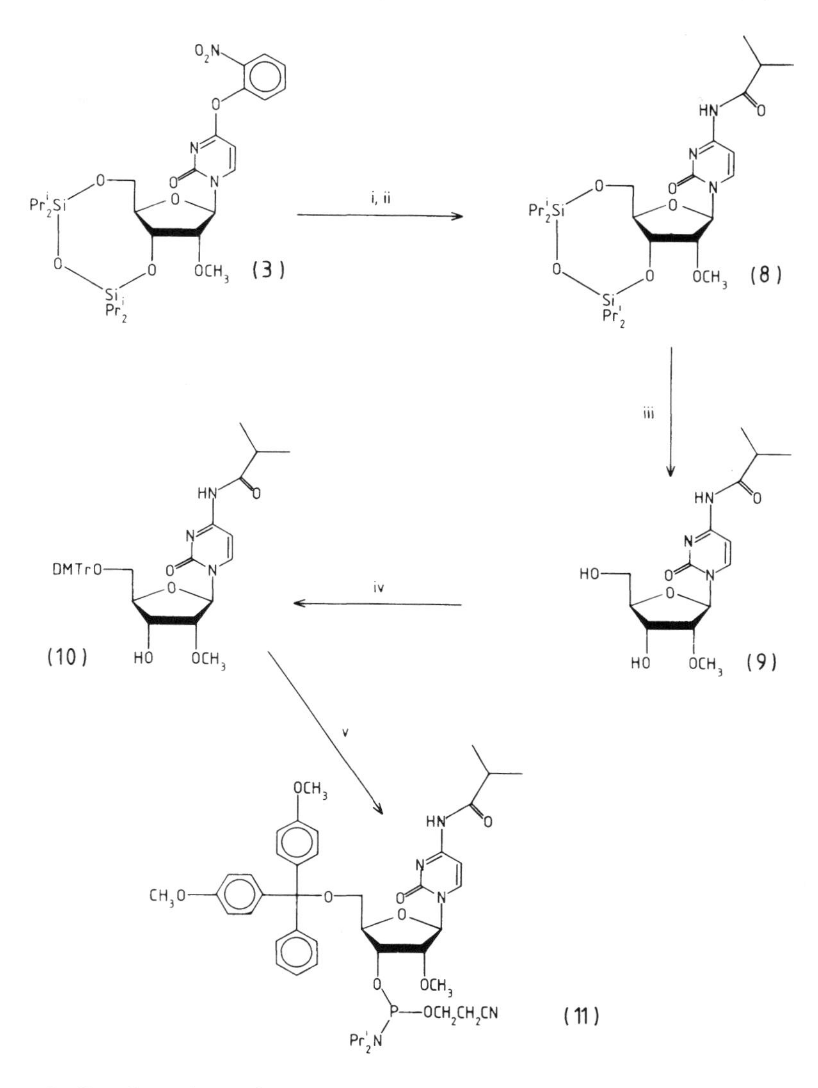

Figure 2. Reaction scheme for the preparation of the 2′-O-methylcytidine building block. Reagents: (i) liquid ammonia/THF; (ii), isobutyryl chloride in pyridine followed by dilute aqueous ammonia/pyridine; (iii) tetrabutylammonium fluoride in THF; (iv) 4,4′-dimethoxytrityl chloride and triethylamine in pyridine; (v) 2-cyanoethoxy N,N-diisopropylaminochlorophosphine and N,N-diisopropylethylamine in 1,2-dichloroethane.

Protocol 8. *Continued*

3. When the transfer is complete, close all valves, disconnect the ammonia cylinder, allow the pressure vessel to warm to room temperature, and leave it for 72 h.
4. Carefully open the pressure release valve of the pressure vessel in a very well ventilated fume cupboard and leave for several hours. TLC in ethanol/chloroform (1:9 v/v) should show a single spot of R_f 0.31.
5. Remove the deep orange solution from the pressure vessel and remove solvent *in vacuo*, to leave a foam.
6. Dissolve the foam in chloroform (1 litre), wash the solution with 1 M aqueous sodium bicarbonate (3 × 500 ml), dry it over anhydrous sodium sulphate, filter it, and remove solvent *in vacuo*.
7. Dry the foam by addition and evaporation of dry pyridine (2 × 100 ml) *in vacuo*.
8. Add anhydrous pyridine (200 ml) and isobutyryl chloride (11.4 ml, 110 mmol) and stir the mixture under anhydrous conditions for 1 h. TLC will show that no starting material remains.
9. Cool the reaction mixture in an ice-bath and quench it by addition of water (10 ml), followed after 10 min by 25% aqueous ammonia (10 min). This procedure ensures that the heterocycle is only mono-acylated.
10. After 5 min evaporate the mixture to dryness *in vacuo*, dissolve the residue in ethyl acetate (1 litre) and work up as usual (wash, dry, filter, and evaporate solvent).
11. Purify the crude product by preparative liquid chromatography in four aliquots using ethyl acetate/petrol (2:1 v/v) as eluant.
12. You will obtain pure **8** as a solid, generally light brown coloured foam (22.9 g, 77%) of R_f 0.23 on TLC in petrol/ethyl acetate (1:2 v/v) and R_f 0.65 in ethanol/chloroform (1:9 v/v). ^{13}C NMR spectrum ($CDCl_3$) δ: 177.79 (isobutyryl CO), 163.27 (C-4), 154.75 (C-2), 143.70 (C-6), 96.39 (C-5), 89.00 (C-1′), 83.25 (C-2′), 81.44 (C-4′), 67.69 (C-3′), 59.20 (C-5′), 58.82 (CH_3O), 35.72 (isobutyryl CH), 19.26 and 18.51 (isobutyryl CH_3s), 17.21–16.55 (isopropyl CH_3s), 13.15, 12.93, 12.64, and 12.12 p.p.m. (isopropyl CHs).

Protocol 9. Synthesis of N^4-isobutyryl-2′-O-methylcytidine (compound **9**); mol. wt 327.34

1. Dissolve compound **8** (22.9 g, 40.2 mmol), from *Protocol 8*, in dry THF (80 ml) and add a 1 M solution of tetrabutylammonium fluoride in THF

Protocol 9. *Continued*

(85 ml) with stirring and exclusion of moisture. Reaction is complete within 5 min.

2. Dilute the solution with pyridine/methanol/water (200 ml, 3:1:1 by vol.) and add to it Dowex 50 × 4 – 200 resin (150 g) in the pyridinium form, suspended in pyridine/methanol/water (400 ml, 3:1:1 by vol.). Stir for 20 min.
3. Filter off the resin and collect the filtrate. Wash the resin with pyridine/methanol/water (3 × 200 ml, 3:1:1 by vol.).
4. Combine filtrate and washings and remove solvent *in vacuo* to leave a glass. Dry the glass by addition and evaporation of methanol (2 × 100 ml).
5. Dissolve the crude product in methanol/chloroform (300 ml, 8:92 v/v) and purify the solution in two aliquots by preparative liquid chromatography using methanol/chloroform (8:92 v/v) as eluant.
6. Remove solvent *in vacuo* from pure product fractions to obtain N^4-isobutyryl-2′-O-methylcytidine **9** as a solid white or off-white foam (12.06 g, 91.6%) of R_f 0.50 on TLC in methanol/chloroform (1:4 v/v). ^{13}C NMR spectrum (DMSO-d_6) δ: 177.78 (isobutyryl CO), 162.64 (C-4), 154.43 (C-2), 145.02 (C-6), 95.34 (C-5), 87.90 (C-1′), 84.20 (C-4′), 83.62 (C-2′), 67.49 (C-3′), 59.46 (C-5′), 57.77 (CH_3O), 34.90 (isobutyryl CH), 19.02 and 18.86 p.p.m. (isobutyryl CH_3s).

Protocol 10. Synthesis of 5′-O-dimethoxytrityl-N^4-isobutyryl-2′-O-methylcytidine (compound **10**); mol. wt 629.71

1. Dimethoxytritylate compound **9** (12 g, 36.66 mmol) according to the first six steps of *Protocol 6*.
2. Purify the crude product by preparative liquid chromatography in two aliquots using ethyl acetate/dichloromethane (6:1 v/v) containing 2% triethylamine as eluant.
3. Remove solvent *in vacuo* at room temperature from pure product fractions to give compound **10** as a solid white foam (21.9 g, 96%) of R_f 0.12 on TLC in ethyl acetate/dichloromethane/triethylamine (78:20:2 by vol.) and R_f 0.50 in ethanol/chloroform/triethylamine (10:89:1 by vol.).

Protocol 11. Synthesis of 5′-O-dimethoxytrityl-N^4-isobutyryl-2′-O-methylcytidine-3′-O-(2-cyanoethyl N,N-diisopropylphosphoramidite) (compound **11**); mol. wt 829.93

1. Phosphitylate compound **10** (9.44 g, 15 mmol) according to the

Protocol 11. *Continued*

procedure described in the first six steps of *Protocol 7*. You should not need to use more than 18 mmol of the chlorophosphine reagent if everything is anhydrous.

2. Purify the crude product by preparative liquid chromatography on silica gel using dichloromethane/petrol (1:1 v/v) containing 10% triethylamine as eluant. The product elutes well before the phosphonate by-product so that ^{31}P NMR spectroscopy monitoring of the fractions is not required.
3. Remove solvent *in vacuo* from product fractions at room temperature. Add benzene and evaporate *in vacuo* to remove residual triethylamine and leave the desired phosphoramidite **11** as a solid white foam (9.73 g, 78%) of R_f 0.38 and 0.30 on TLC in ethyl acetate/dichloromethane (4:1 v/v) containing 5% triethylamine. ^{31}P NMR spectrum (CH_2Cl_2, external concentric D_2O lock) δ: + 146.99 and 146.91 p.p.m. The two spots on TLC and the two ^{31}P resonances are a result of the stereoisomerism at the phosphorus atom.

2.1.3 Synthesis of the 2′-O-methyladenosine monomer

The synthesis route that you will use is outlined in *Figure 3*. Compound **13** is a versatile synthetic intermediate and can be used to prepare the 2′-O-methylinosine monomer **9**.

Protocol 12. Synthesis of 3′,5′-O-(tetraisopropyldisiloxane-1,3-diyl)-6-chloropurine riboside (compound **12**); mol. wt 529.19

1. Silylate 6-chloropurine riboside (70.2 g, 245 mmol) according to the procedure given in the first five steps of *Protocol 1*.
2. Purify the crude product (deep purple coloured) in six aliquots by preparative liquid chromatography on two silica cartridges using petrol/ethyl acetate (2:1 v/v) as eluant.
3. Evaporate solvent *in vacuo* to leave an oil which will crystallize spontaneously. Compound **12** is obtained as a white crystalline solid (107.6 g, 83%) of R_f 0.21 on TLC in petrol/ethyl acetate (2:1 v/v). ^{13}C NMR spectrum ($CDCl_3$) δ: 151.79 (C-2), 151.19 (C-6), 150.68 (C-4), 144.07 (C-8), 132.41 (C-5), 89.85 (C-1′), 82.12 (C-4′), 74.86 (C-2′), 70.29 (C-3′), 61.16 (C-5′), 17.24-16.87 (isopropyl CH_3s), 13.82–12.54 p.p.m. (isopropyl CHs).

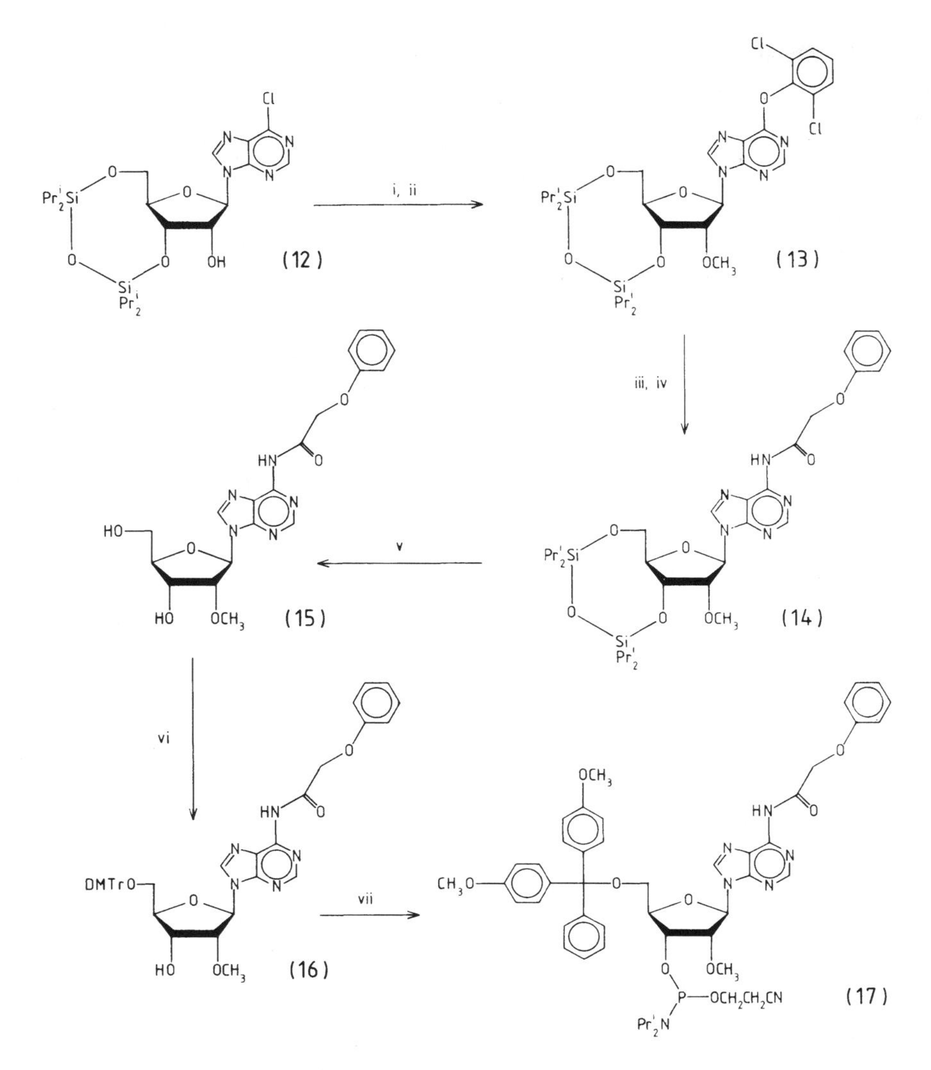

Figure 3. Reaction scheme for the preparation of the 2′-O-methyladenosine building block. Reagents (i) 2,6-dichlorophenol, triethylamine and 1,4-diazabicyclo[2.2.2]octane in dichloroethane; (ii) methyl iodide and BDDDP in acetonitrile; (iii) ammonia/THF; (iv) phenoxyacetyl chloride in pyridine/THF followed by dilute aqueous ammonia; (v) tetrabutylammonium fluoride in THF; (vi) 4,4′-dimethoxytrityl chloride and triethylamine in pyridine; (vii) 2-cyanoethoxy N,N-diisopropylaminochlorophosphine and N,N-diisopropylethylamine in 1,2-dichloroethane. Reproduced from ref. 9.

Protocol 13. Synthesis of 3′,5′-O-(tetraisopropyldisiloxane-1,3-diyl)-2′-O-methyl-6-(2,6-dichlorophenoxy) purineriboside (compound **13**); mol. wt 669.76

1. Dissolve compound **12** (107.6 g, 203.3 mmol) in 1,2-dichloroethane (800 ml) and add to it a solution of 2,6-dichlorophenol (33.1 g, 203.3 mmol), 1.4-diazabicyclo[2.2.2]octane (2.24 g, 20 mmol) and triethylamine (28.3 ml, 203.3 mmol) in 1,2-dichloroethane (200 ml).
2. Stir for 12 h at room temperature under anhydrous conditions. TLC in petrol/ethyl acetate (2:1 v/v) should show complete reaction with a new spot of R_f 0.32. A white precipitate due to triethylamine hydrochloride will be present.
3. Wash the solution with 1 M aqueous sodium bicarbonate (1 litre), dry it over anhydrous sodium sulphate, filter it, and evaporate solvent *in vacuo*. Dry the residue by evaporation of anhydrous acetonitrile (2 × 100 ml) *in vacuo*.
4. Dissolve the foam in dry acetonitrile (250 ml) under argon, and cool in an ice-bath.
5. Add 2-*tert*-butylimino-2-diethylamino-1,3-dimethylperhydro-1,3,2-diazaphosphorin (BDDDP, 58.8 ml, 203.3 mmol) and iodomethane (12.6 ml, 203.3 mmol) with stirring and exclusion of moisture. Seal the flask to prevent loss of the volatile iodomethane.
6. Keep 5 min at 0 °C then stir 5 h at room temperature. TLC in petrol/ethyl acetate (2:1 v/v) will show about 80% methylation with a new spot of R_f 0.38.
7. Add extra BDDDP (5.9 ml, 20.3 mmol) and iodomethane (1.26 ml, 20.3 mmol) and leave reaction overnight.
8. Remove solvent *in vacuo* to leave a foam.
9. Add ethyl acetate (200 ml) with vigorous stirring to the residual foam, followed by petrol (600 ml). The hydroiodide salt of the BDDDP precipitates out and should be removed by filtration.
10. Purify the filtrate in five aliquots by preparative liquid chromatography on two silica cartridges, using petrol/ethyl acetate (3:1 v/v) as eluant, to obtain pure compound **13** as a solid white foam (85.4 g, 62.7%). Some unmethylated material is recovered during the purification process and can be recycled to give an overall yield for (13) of 72–74%. ^{13}C NMR spectrum ($CDCl_3$) δ: 158.18 (C-6), 152.26 (C-4), 151.73 (C-2), 145.31 (phenyl C-1), 141.90 (C-8), 129.36 (phenyl C-2 and 6), 128.67 (phenyl C-3 and 5), 126.99 (phenyl C-4), 121.72 (C-5), 88.45 (C-1′), 83.58 (C-2′), 81.34 (C-4′), 69.58 (C-3′), 59.71 (C-5′), 59.47 (CH_3O), 17.37–16.81 (isopropyl CH_3s), 13.37, 12.88, 12.83, and 12.50 p.p.m. (isopropyl CHs).

Protocol 14. Synthesis of 3′,5′-O-(tetraisopropyldisiloxane-1,3-diyl)-N^6-phenoxyacetyl-2′-O-methyladenosine (compound **14**); mol. wt 657.92

1. Dissolve compound **13** (33.5 g, 50 mmol) in dry THF (200 ml) and place in a pressure vessel with liquid ammonia (*c.* 50–100 g), see *Protocol 8*.
2. Keep at 90 °C inside temperature for 10 days (pressure vessel equipped with stirring and heating is necessary) to achieve complete reaction. TLC should show a single spot of R_f 0.42 in ethanol/chloroform (1:9 v/v).
3. Work up as in steps 5 and 6 of *Protocol 8*, however note that the solution will be more or less colourless.
4. Dissolve the intermediate 3′,5′-O-(tetraisopropyldisiloxane-1,3-diyl)-2′-O-methyladenosine in THF/pyridine (250 ml, 3:2 v/v) and cool in an ice-bath.
5. Add phenoxyacetyl chloride (20.7 ml, 150 mmol) with stirring and exclusion of moisture.
6. Stir 1 h at room temperature and check that reaction is complete by TLC.
7. Cool mixture in ice-bath, add water (10 ml) followed after 5 min by 25% aqueous ammonia (15 ml).
8. Follow steps 10 and 11 of *Protocol 8* for work-up and product purification.
9. You should obtain pure compound **14** as a solid cream coloured foam (28.6 g, 87%) of R_f 0.34 on TLC in ethyl acetate/petrol (2:1 v/v) and R_f 0.67 in ethanol/chloroform (1:9 v/v). ^{13}C NMR spectrum ($CDCl_3$) δ: 166.88 (phenoxyacetyl CO), 156.99 (phenyl C-1), 152.10 (C-2), 150.54 (C-4), 148.17 (C-6), 141.73 (C-8), 129.37 (phenyl C-3 and 5), 122.94 (C-5), 121.85 (phenyl C-4), 114.65 (phenyl C-2 and 6), 88.27 (C-1′), 83.23 (C-2′), 81.20 (C-4′), 69.51 (C-3′), 68.07 (CH_2 of phenoxyacetyl), 59.63 (C-5′), 59.28 (CH_3O), 17.02–16.65 (isopropyl CH_3s), 13.16, 12.59, and 12.34 p.p.m. (isopropyl CHs).

Protocol 15. Synthesis of N^6-phenoxyacetyl-2′-O-methyladenosine (compound **15**); mol. wt 415.41

1. Desilylate compound **14** (28.4 g, 43.2 mmol) according to steps 1–4 of *Protocol 9*.
2. Purify the crude product following step 5 of *Protocol 9*.
3. Remove solvent *in vacuo* from pure product fractions to obtain N^6-phenoxyacetyl-2′-O-methyladenosine as an off-white foam (13.9 g, 77%). R_f 0.48 on TLC in ethanol/chloroform (1:4 v/v). ^{13}C NMR spectrum (DMSO-d_6) δ: 167.39 (phenoxyacetyl CO), 157.76 (phenyl C-1), 151.70

Protocol 15. *Continued*

(C-2), 151.52 (C-4), 148.99 (C-6), 142.83 (C-8), 129.40 (phenyl C-3 and 5), 123.34 (C-5), 121.06 (phenyl C-4), 114.56 (phenyl C-2 and 6), 86.25 (C-1′), 85.75 (C-4′), 82.75 (C-2′), 68.70 (C-3′), 67.33 (CH_2 of phenoxy-acetyl), 61.12 (C-5′) and 57.57 p.p.m. (CH_3O).

Protocol 16. Synthesis of 5′-O-dimethoxytrityl-N^6-phenoxyacetyl-2′-O-methyladenosine (compound **16**); mol. wt 717.78

1. Dimethoxytritylate compound **15** (13.85 g, 33.3 mmol) according to the first six steps of *Protocol 6*.
2. Purify the crude product in four aliquots by preparative liquid chromatography using ethyl acetate/dichloromethane/triethylamine (86:12:2 by vol.) as eluant.
3. Remove solvent *in vacuo* at room temperature to give compound **16** as an off-white foam (22.2 g, 93%). R_f 0.54 on TLC in ethanol/chloroform/triethylamine (10:89:1 by vol.).

Protocol 17. Synthesis of 5′-O-dimethoxytrityl-N^6-phenoxyacetyl-2′-O-methyladenosine-3′-O-(2-cyanoethyl N,N-diisopropyl-phosphoramidite) (compound **17**); mol. wt 918.00

1. Phosphitylate compound **16** (10.77 g, 15 mmol) according to the procedure described in the first six steps of *Protocol 7*.
2. Purify the crude product by preparative liquid chromatography on silica using petrol/dichloromethane (6:5 v/v) containing 10% triethylamine as eluant.
3. Pool product fractions and evaporate solvent *in vacuo* at room temperature. Lyophilize from benzene to give the title compound **17** as a solid white foam (10.27 g, 74.6%). R_f 0.51 and 0.44 on TLC in ethyl acetate/dichloromethane (4:1 v/v) containing 5% triethylamine. ^{31}P NMR spectrum (CH_2Cl_2, external concentric D_2O lock) δ: + 147.62 and 147.11 p.p.m.

2.1.4 Synthesis of the 2′-O-methylguanosine monomer

Refer to *Figure 4* to see the synthesis route used here.

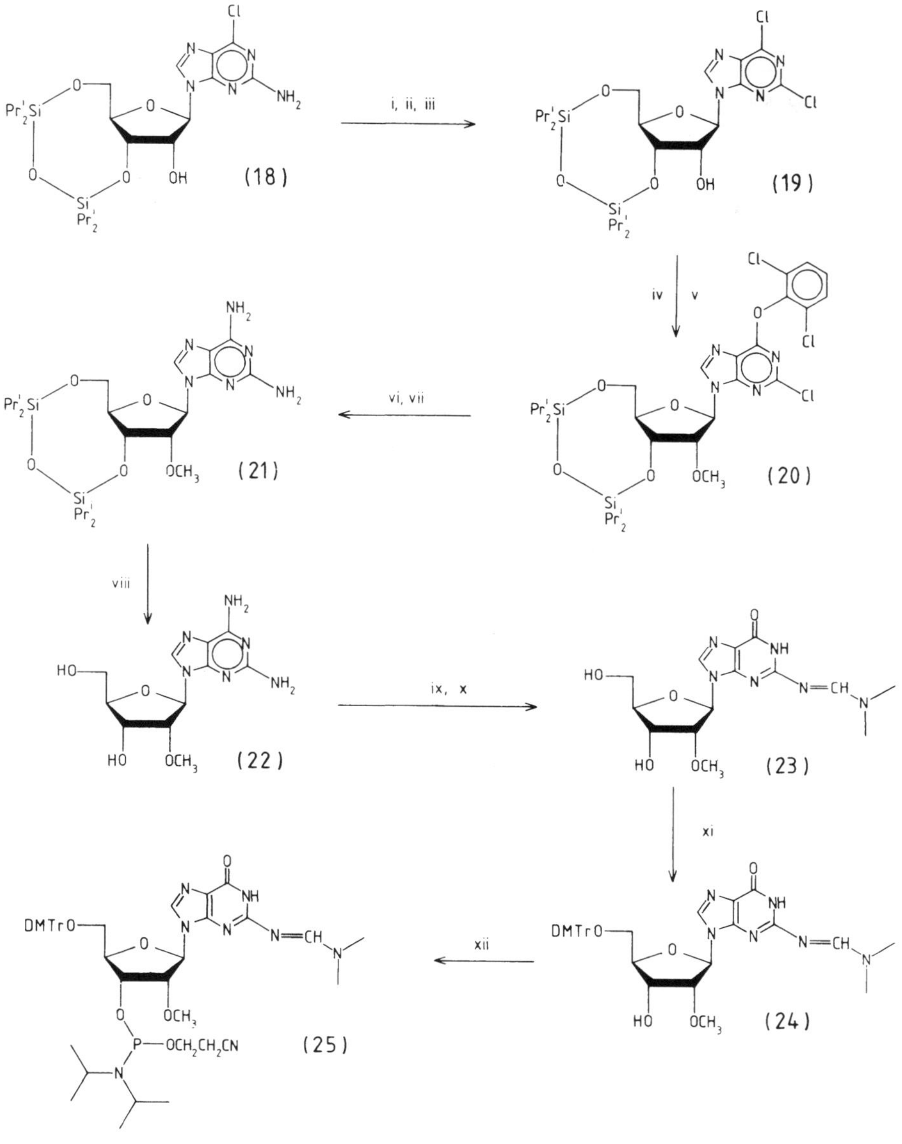

Figure 4. Reaction scheme for the synthesis of the 2′-O-methylguanosine building block. Reagents: (i) chlorotrimethylsilane and triethylamine in 1,2-dichloroethane; (ii) *tert*-butyl nitrite and antimony trichloride in 1,2-dichloroethane at −14 °C; (iii) *p*-toluene sulphonic acid in dioxan/dichloromethane; (iv) 2,6-dichlorophenol, triethylamine and 1,4-diazabicyclo[2.2.2]octane in dichloroethane; (v) methyl iodide and BDDDP in acetonitrile; (vi) sodium azide in DMF; (vii) hydrogen/Lindlar catalyst in ethyl acetate containing quinoline; (viii) tetrabutylammonium fluoride in THF; (ix) adenosine deaminase in queous phosphate buffer pH 7.4/DMSO; (x) DMF dimethyl acetal in methanol; (xi) 4,4′-dimethoxytrityl chloride and triethylamine in pyridine; (xii) 2-cyanoethoxy N,N-diisopropylaminochlorophosphine and N,N-diisopropylethylamine in 1,2-dichloroethane. Reproduced from ref. 9.

Protocol 18. Synthesis of 3′,5′-O-(tetraisopropyldisiloxane-1,3-diyl)-2-amino-6-chloropurine riboside (compound **18**); mol. wt 544.20

1. Silylate 2-amino-6-chloropurine riboside (60.33 g, 200 mmol) according to the procedure given in the first five steps of *Protocol 1*. In the work up use ethyl acetate instead of dichloromethane.
2. Purify the crude product by preparative liquid chromatography in five aliquots on two silica cartridges using petrol/ethyl acetate (6:4 v/v) as eluant.
3. Removal of solvent will give the title compound as a cream coloured foam (73.15 g, 67.2%). R_f 0.33 on TLC in petrol/ethyl acetate (1:1 v/v). ^{13}C NMR spectrum ($CDCl_3$) δ: 159.15 (C-2), 152.72 (C-4 or C-6), 151.54 (C-6 or C-4), 140.34 (C-8), 125.79 (C-5), 88.96 (C-1′), 81.84 (C-4′), 74.84 (C-2′), 70.34 (C-3′), 61.21 (C-5′), 17.35-16.86 (isopropyl CH_3s), 13.35, 12.99, 12.90 and 12.62 p.p.m. (isopropyl CHs).

Protocol 19. Synthesis of 3′,5′-O-(tetraisopropyldisiloxane-1,3-diyl)-2,6-dichloropurine riboside (compound **19**); mol. wt 563.63

1. Dissolve compound **18** (73.15 g, 134.4 mmol) in dry 1,2-dichloroethane (400 ml) and add triethylamine (90.4 ml, 650 mmol) and chlorotrimethylsilane (66 ml, 520 mmol) with stirring and exclusion of moisture. TLC in petrol/ethyl acetate (2:1 v/v) should show complete reaction after 30 min with a new spot of R_f 0.45 due to the 2′-O-TMS ether of **18**.
2. Pour the reaction mixture into vigorously stirred 1 M aqueous sodium bicarbonate (1 litre).
3. Separate the organic layer, dry it, filter it, and remove solvent *in vacuo*.
4. Dry the foam by addition and evaporation of dry toluene.
5. Dissolve the foam in pre-cooled, *c.* −15 °C, dry 1,2-dichloroethane (1.6 litre) under argon and cool the solution to *c.* −14 °C in an ice–salt bath.
6. Dissolve anhydrous antimony trichloride (43 g, 187.6 mmol) in dry 1,2-dichloroethane (200 ml) and add the solution with stirring to the above solution followed after 5 min by *tert*-butyl nitrite (82.5 ml, 680 mmol).
7. Gas will be evolved as the diazotisation proceeds so keep the reaction mixture stirred for 3 h whilst maintaining the temperature between −10 and −14 °C. TLC in petrol/ethyl acetate (2:1 v/v) should show complete reaction with a major new spot of R_f 0.60.
8. Dilute the reaction mixture with chloroform (2 litres) and pour the solution into vigorously stirred ice-cold 1 M aqueous sodium bicarbonate (2.5 litres). A copious precipitate, due to insoluble antimony salts, will form.

Protocol 19. *Continued*

9. After 5 min filter the mixture through a sintered glass funnel.
10. Separate the organic layer, wash with saturated brine (2 litres), dry it and remove the solvent *in vacuo*.
11. Dry the residue by addition and evaporation of dry toluene (2 × 100 ml).
12. To a stirred solution of the foam in dichloromethane (300 ml) add a solution of *p*-toluenesulphonic acid monohydrate (34 g, 180 mmol) in dioxan (300 ml), to cleave the 2′-O-TMS ether.
13. After precisely 2 min quench the reaction by addition of triethylamine (25 ml, 180 mmol).
14. Dilute with dichloromethane (2 litres) and pour the solution into vigorously stirred 1 M aqueous sodium bicarbonate (2 litres). Work up in the usual way.
15. Purify the crude product in three aliquots by preparative liquid chromatography on two silica cartridges using petrol/ethyl acetate (2:1 v/v) as eluant.
16. You will obtain the title compound **19** as a solid cream coloured foam (61.9 g, 81.8%). R_f 0.32 on TLC in petrol/ethyl acetate (2:1 v/v). ^{13}C NMR spectrum ($CDCl_3$) δ: 152.66 (C-4), 151.73 and 151.66 (C-6 and C-2), 144.85 (C-8), 131.38 (C-5), 89.73 (C-1′), 82.01 (C-4′), 74.70 (C-2′), 70.63 (C-3′), 61.38 (C-5′), 17.18-16.64 (isopropyl CH_3s), 13.02, 12.76, 12.60 and 12.36 p.p.m. (isopropyl CHs).

Protocol 20. Synthesis of 3′,5′-O-(tetraisopropyldisiloxane-1,3-diyl)-2′-O-methyl-2-chloro-6-(2,6-dichlorophenoxy)purine riboside (compound **20**); mol. wt 704.20

1. Convert compound **19** (61.9 g, 109.8 mmol) into the 6-(2,6-dichlorophenoxy) derivative using the procedure given in the first three steps of *Protocol 13*. In this case reaction should be complete after 30 min with a new spot of R_f 0.41 on TLC in petrol/ethyl acetate (2:1 v/v).
2. Perform the 2′-O-methylation and product purification as described in steps 4–10 of *Protocol 13*.
3. Pure product will be obtained as a solid white foam (52.11 g, 67.4%). R_f 0.57 on TLC in petrol/ethyl acetate (1:1 v/v). ^{13}C NMR spectrum ($CDCl_3$) δ: 158.12 (C-6), 153.13 and 152.54 (C-2 and C-4), 144.66 (phenyl C-1), 142.43 (C-8), 128.77 (phenyl C-2 and C-6), 128.52 (phenyl C-3 and C-5), 127.12 (phenyl C-4), 120.51 (C-5), 88.26 (C-1′), 83.10 (C-2′), 81.16 (C-4′), 69.54 (C-3′), 59.61 (C-5′), 59.29 (CH_3O), 17.19–16.63 (isopropyl CH_3s), 13.16, 12.70, and 12.29 p.p.m. (isopropyl CHs).

Protocol 21. Synthesis of 3′,5′-O-(tetraisopropyldisiloxane-1,3-diyl)-2′-O-methyl-2,6-diaminopurine riboside (compound **21**); mol. wt 538.80

1. Dissolve compound **20** (12 g, 17.04 mmol) and sodium azide (2.76 g, 42.5 mmol) in dry N,N-dimethylformamide (300 ml) and stir for 5 h at 55 °C under argon.
2. When cool, remove solvent *in vacuo* to leave a yellow syrup.
3. Work up as usual and purify the residue by silica gel (200 g) column chromatography, eluting with petrol/ethyl acetate (5:1 v/v). You will obtain the 2,6-diazidopurine riboside intermediate as a colourless oil (7.44 g, 73.5%). R_f 0.55 on TLC in petrol/ethyl acetate (3:1 v/v). ^{13}C NMR spectrum ($CDCl_3$) δ: 155.71 (C-6), 153.27 (C-2), 152.21 (C-4), 141.35 (C-8), 121.71 (C-5), 87.68 (C-1′), 83.47 (C-2′), 81.25 (C-4′), 68.87 (C-3′), 59.48 (C-5′), 59.07 (CH_3O), 17.17–16.56 (isopropyl CH_3s), 13.89, 13.18, 12.70, and 12.25 p.p.m. (isopropyl CHs).
4. Dissolve the above 2,6-diazido compound in ethyl acetate (120 ml) containing redistilled quinoline (2 ml), add Lindlar catalyst (1.4 g) and stir overnight under three atmospheres of hydrogen in a pressure vessel. Ensure that there is enough hydrogen to fully reduce both azide groups. TLC in ethanol/dichloromethane (5:95 v/v) will show a single spot of R_f 0.28 (quinoline R_f 0.54).
5. Filter off the catalyst on a bed of Celite, remove solvent *in vacuo*, and purify the product by column chromatography on silica gel (150 g) eluting with a gradient of 0–10% ethanol in dichloromethane.
6. You will obtain the title compound as a solid white foam (6.7 g, 99% based on the diazido compound). ^{13}C NMR spectrum ($CDCl_3$) δ: 159.88 (C-6), 155.77 (C-2), 150.30 (C-4), 134.87 (C-8), 113.69 (C-5), 87.27 (C-1′), 83.28 (C-2′), 80.72 (C-4′), 69.21 (C-3′), 59.55 (C-5′), 58.98 (CH_3O), 17.43–16.44 (isopropyl CH_3s), 13.09, 12.58, and 12.18 p.p.m. (isopropyl CHs).

Protocol 22. Synthesis of 2′-O-methyl-2,6-diaminopurine riboside (compound **22**); mol. wt 296.29

1. Desilylate compound **21** (5.77 g, 10.6 mmol) according to the first four steps of *Protocol 9*.
2. Remove residual pyridine by addition and evaporation of toluene *in vacuo*.
3. Add absolute ethanol (30 ml), swirl and keep at 4 °C overnight.

Protocol 22. *Continued*

4. Filter off the white crystalline product and dry it *in vacuo*.
5. 2′-O-Methyl-2,6-diaminopurine riboside should be obtained as a white solid (3.17 g, 87%) containing *c.* one mole equivalent of ethanol of crystallization. R_f 0.12 on TLC in ethanol/dichloromethane (1:9 v/v). ^{13}C NMR spectrum (DMSO-d_6) δ: 160.10 (C-6), 156.25 (C-2), 151.36 (C-4), 136.10 (C-8), 113.46 (C-5), 86.21 (C-1′), 85.09 (C-4′), 82.36 (C-2′), 68.96 (C-3′), 61.67 (C-5′), 57.47 p.p.m. (CH_3O).

Protocol 23. Synthesis of N^2-dimethylaminomethylidene-2′-O-methylguanosine (compound **23**); mol. wt 352.35

1. Dissolve compound **22** (3.17 g, *c.* 9.3 mmol) in a mixture of 0.1 M aqueous sodium phosphate (267 ml, pH 7.4) and DMSO (106 ml).
2. Add crude adenosine deaminase (71 mg) and stir the solution gently at room temperature for 48 h. TLC in ethanol/dichloromethane (1:4 v/v) should show complete reaction with a new spot of R_f 0.15 due to 2′-O-methylguanosine. If the reaction is not complete add more enzyme.
3. Evaporate the solution to dryness *in vacuo* and dry the residue by repeated addition followed by evaporation of dry methanol. The yield of the deamination is quantitative and the presence of phosphate does not interfere with the next reaction step.
4. Suspend the material in dry methanol (30 ml) and add N,N-DMF dimethyl acetal (3.9 ml, 29.6 mmol). Stir overnight with exclusion of moisture. TLC in ethanol/dichloromethane (1:4 v/v) will show complete reaction with a new spot of R_f 0.32.
5. Remove solvent *in vacuo* and partition the residue between 1 M aqueous sodium bicarbonate (200 ml) and chloroform (200 ml). Separate the chloroform layer, wash it with saturated brine, dry it, filter it, and remove solvent *in vacuo*.
6. Purify the crude product by column chromatography on silica gel (100 g) eluting with a gradient of ethanol from 5 to 20% in chloroform.
7. Remove solvent from product fractions to leave the title compound as a white solid (2.6 g, 80%). ^{13}C NMR spectrum (DMSO-d_6) δ: 157.89 (amidine CH), 157.49 (C-6), 157.33 (C-2), 149.79 (C-4), 136.58 (C-8), 119.69 (C-5), 85.93 (C-1′), 84.69 (C-4′), 82.80 (C-2′), 68.66 (C-3′), 61.26 (C-5′), 57.40 (CH_3O), 40.61 and 34.61 p.p.m. (N-CH_3s).

Protocol 24. Synthesis of 5′-O-dimethoxytrityl-N^2-dimethylamino-methylidene-2′-O-methylguanosine (compound **24**); mol. wt 654.72

1. Dimethoxytritylate compound **23** (2.6 g, 7.4 mmol) according to the first six steps of *Protocol 6*. You only need to use a 20% excess of 4,4′-dimethoxytrityl chloride.
2. Purify the crude product by column chromatography on silica gel (120 g), eluting with a gradient of ethanol from 0 to 5% in dichloromethane/triethylamine (199:1 v/v).
3. Evaporate product fractions to leave the title compound as a white foam (4.81 g, 99%). R_f 0.32 on TLC in ethanol/chloroform/triethylamine (10:89:1 by vol.).

Protocol 25. Synthesis of 5′-O-dimethoxytrityl-N^2-dimethylamino-methylidene-2′-O-methylguanosine-3′-O-(2-cyanoethyl N,N-diisopropylphosphoramidite) (compound **25**); mol. wt 854.95

1. Phosphitylate compound **24** (3.26 g, 5 mmol) according to the procedure given in the first six steps of *Protocol 7*. A 40% excess of the phosphitylating agent should suffice.
2. Purify the crude product by column chromatography on silica gel (100 g). Elute first with dichloromethane/petrol (3:1 v/v) containing 2% triethylamine to remove the phosphonate impurity.
3. Elute the product with a gradient from 0 to 3% ethanol in dichloroethane/triethylamine (49:1 v/v).
4. Evaporate solvent *in vacuo* at room temperature from the pooled product fractions. Remove residual triethylamine by addition and evaporation of benzene, to leave the 2′-O-methylguanosine monomer as a solid white foam (3.58 g, 84%). R_f 0.46 on TLC in ethanol/chloroform/triethylamine (10:89:1 by vol.). ^{31}P NMR spectrum (CH_2Cl_2, concentric external D_2O lock) δ: + 147.50 p.p.m.

2.2 Synthesis of the condensing agent 5-(4-nitrophenyl)-1H-tetrazole

Protocol 26. Synthesis of 5-(4-nitrophenyl)-1H-tetrazole; mol. wt 191.15

1. Add 4-nitrobenzonitrile (74 g, 0.5 mol), sodium azide (65 g, 1 mol) and ammonium chloride (53.5 g, 1 mol) to dry N,N-dimethylformamide (200 ml).

Protocol 26. *Continued*

2. Stir the mixture at 100 °C for 8 h under nitrogen and allow to cool overnight.
3. Remove solvent *in vacuo*, add ice-cold water (1 litre) to the residue cooled in an ice-bath.
4. Add concentrated hydrochloric acid with stirring to pH 3. Take care as hydrazoic acid is liberated.
5. Filter off the crystals and wash with ice-cold water.
6. Dry the product *in vacuo* and then recrystallize it from ethanol.
7. You should obtain 5-(4-nitrophenyl)-1H-tetrazole as pale yellow needles (89 g, 93%) of m.p. 218 °C.

2.3 Preparation of protected 2′-O-methylribonucleoside-3′-O-succinates and loading of controlled pore glass

Protocol 27. Synthesis of 3′-O-succinates

1. Dissolve a 5′-O-dimethoxytrityl base protected 2′-O-methylribonucleoside (2 mmol), *viz* compounds **6**, **10**, **16**, and **24** in dry pyridine (5 ml).
2. Add 4-dimethylaminopyridine (122 mg, 1 mmol) and succinic anhydride (250 mg, 2.5 mmol) and stir overnight under anhydrous conditions. TLC in ethanol/chloroform/triethylamine (10:89:1 by vol.) should show complete reaction.
3. Pass the reaction mixture through a 20 ml column of pyridinium form Dowex 50 × 2 – 200 resin packed in pyridine/methanol/water (8:1:1 by vol.) and elute the column with this solvent mixture.
4. Evaporate the eluate to dryness *in vacuo* at room temperature. Remove excess pyridine by co-evaporation with toluene.
5. Dissolve the residue in dichloromethane (50 ml), wash the solution with water (2 × 50 ml), dry it, and remove solvent *in vacuo*.
6. Purify the crude product by column chromatography on silica gel (40 g) and elute with a gradient from 0 to 10% ethanol in dichloromethane/pyridine (199:1 v/v).
7. Remove solvent *in vacuo* from pure product fractions to yield the 3′-O-succinate as a solid white foam (yield 65–70%).

The 3′-O-succinate is then converted to its 4-nitrophenyl ester which is then reacted with aminopropyl controlled pore glass (CPG) to give the functionalized supports for solid phase synthesis. The procedure is as follows.

Protocol 28. Derivatization of aminopropyl CPG

1. Dissolve a dry 3′-O-succinate (1 mmol) in dry dioxan (4 ml) containing pyridine (0.2 ml).
2. Add 4-nitrophenol (140 mg, 1 mmol) followed by N,N′-dicyclohexyl-carbodiimide (516 mg, 2.5 mmol) and stir 5 h under anhydrous conditions. A fine precipitate of the urea will appear.
3. Filter off the precipitate and add the filtrate containing the 2′-O-methylribonucleoside-3′-O-(4-nitrophenylsuccinate) to a suspension of 500 Å pore diameter aminopropyl CPG (5 g) in N,N-dimethylformamide/triethylamine (6 ml, 5:1 v/v/).
4. Stopper the vessel, agitate the mixture gently and then leave overnight.
5. Filter off the support, wash it with methanol (5 × 20 ml), followed by ether (5 × 20 ml), and then dry it briefly *in vacuo*.
6. Treat the glass beads with pyridine/acetic anhydride/N-methylimidazole (20 ml, 8:1:1 by vol.) for 30 min to cap off any residual amino groups.
7. Filter off the CPG, wash it sequentially with pyridine (5 × 20 ml), methanol (5 × 20 ml) and finally diethyl ether (5 × 20 ml).
8. Finally air dry the CPG and then dry it *in vacuo*.
9. Carry out a dimethoxytrityl analysis to determine the loading of the support, which should be in the range of 35–45 μmol of nucleoside g^{-1}. Weigh accurately about 10 mg of the CPG, place it in a 25 ml volumetric flask, make up to the mark with 70% aqueous perchloric acid/ethanol (3:2 v/v), and agitate for 5 min. Measure the absorbance of the orange solution at 498 nm in a 1 cm path length cuvette, and calculate the CPG loading assuming that 1 μmol of dimethoxytrityl cation corresponds to 71.7 absorbance units at 498 nm.

2.4 Synthesis of a building block for direct biotinylation during solid phase synthesis

The synthesis route to be followed is illustrated in *Figure 5*.

Protocol 29. Synthesis of 3′,5′-O-bis(thexyldimethylsilyl)-O^4-(2-nitrophenyl)-2′-deoxyuridine (compound **26**); mol. wt 633.94

1. Dry 2′-deoxyuridine (11.41 g, 50 mmol) by addition and evaporation of dry DMF *in vacuo*.
2. Dissolve the residue in dry DMF (100 ml), add imidazole (15 g, 220 mmol) and thexyldimethylchlorosilane (24.5 ml, 125 mmol) and stir under

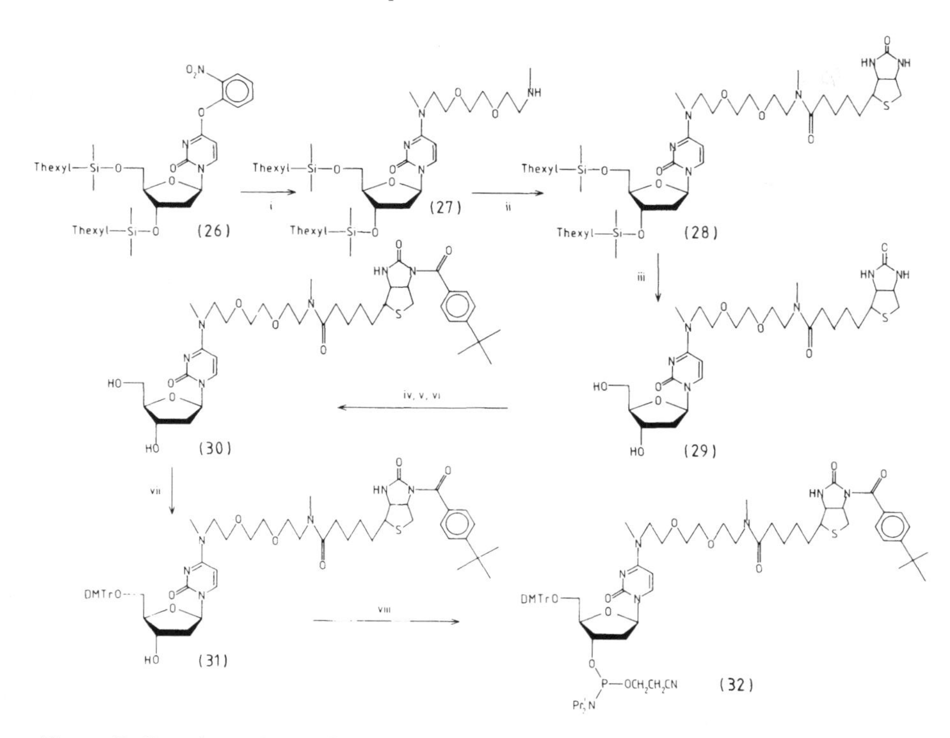

Figure 5. Reaction scheme for the synthesis of the biotinylated building block. Thexyl denotes 2,3-dimethyl-2-butyl. Reagents: (i) 1,2-bis(2-methylaminoethoxy)ethane and triethylamine in DMF; (ii) (+)-biotin 2-nitrophenyl ester in DMF; (iii) tetrabutylammonium fluoride in THF; (iv) chlorotrimethylsilane in pyridine; (v) 4-*tert*-butylbenzoyl chloride in pyridine; (vi) dilute aqueous ammonia/pyridine; (vii) 4,4′-dimethoxytrityl chloride in pyridine; (viii) bis(diisopropylamino)-2-cyanoethoxyphosphine and tetrazole in acetonitrile. Reproduced from ref. 10.

Protocol 29. *Continued*

dry nitrogen overnight. TLC in ethanol/dichloromethane (5:95 v/v) will show a spot of R_f 0.55.

3. Quench the reaction with methanol (10 ml) and remove solvent *in vacuo*.
4. Work up the residue with chloroform/1 M aqueous sodium bicarbonate.
5. The intermediate 3′,5′-O-bis(thexyldimethylsilyl)-2′-deoxyuridine will be obtained as a white semi-solid (25.4 g, 99%) after addition and evaporation of toluene *in vacuo*.
6. Convert this to the O^4-(2-nitrophenyl) derivative, R_f 0.81 on TLC in ethanol/dichloromethane (5:95 v/v), according to the procedure given in steps 3–5 of *Protocol 2*.

Protocol 29. *Continued*

7. Purify the crude product (a yellow oil) by chromatography on silica gel (300 g) eluting with a gradient of ethanol from 0 to 3% in dichloromethane.
8. Remove solvent *in vacuo* to leave the title compound as a colourless semi-solid (27.5 g, 87%). ^{13}C NMR spectrum ($CDCl_3$) δ: 170.44 (C-4), 154.63 (C-2), 144.95 (phenyl C-1), 144.80 (C-6), 141.59 (phenyl C-2), 134.69 (phenyl C-5), 126.48, 125.82, and 125.48 (phenyl C-3, C-4, and C-6), 94.14 (C-5), 87.6 (C-1′), 86.76 (C-4′), 69.83 (C-3′), 61.58 (C-5′). 42.05 (C-2′), 34.02, and 33.38 (CHs of thexyls), 25.29 and 24.30 (quaternary Cs of thexyls), 20.21 and 20.17 (SiCMe_2), 18.44 and 18.39 (CHMe_2 of thexyls), −2.5 to −3.5 p.p.m. (Si-CH_3s).

Protocol 30. Synthesis of (+)-biotin-2-nitrophenyl ester; mol. wt 365.40

1. Stir (+)-biotin (19.5 g, 79.81 mmol), 2-nitrophenol (27.95 g, 156.62 mmol), and N,N′-dicyclohexylcarbodiimide (41.4 g, 199.52 mmol) in THF/DMF/pyridine (450 ml, 6:2:1 by vol.) for five days under anhydrous conditions.
2. Filter the mixture and wash the precipitate with methanol (2 × 200 ml).
3. Combine filtrate and washings and evaporate to dryness *in vacuo*. Evaporate toluene (2 × 200 ml) and ethanol (100 ml) from the residue.
4. Wash the yellow flaky residue with ether (2 × 100 ml) and crystallize it from 2-propanol.
5. When cool, filter off the crystalline product and wash it with ether.
6. Evaporate the mother liquor to dryness and crystallize the residue.
7. Repeat steps 5 and 6.
8. You should obtain the desired product as a white solid (24.5 g, 83%). ^{13}C NMR spectrum (DMSO-d_6) δ: 170.80 (C-10), 162.61 (C-2), 142.99, 141.56, 135.36, 127.10, 125.44, and 125.16 (phenyl), 60.97 (C-3), 59.16 (C-4), 55.21 (C-2), (C-5) under DMSO signals, 33.00 (C-9), 27.91 (C-7), 27.76 (C-6), and 24.02 p.p.m. (C-8).

Protocol 31. Synthesis of 1,2-bis(2-methylaminoethoxy)ethane

1. Place 1,2-bis(2-chloroethoxy)ethane (16.83 g, 90 mmol) and a 40% solution of methylamine in water (500 ml, i.e. 5.9 mol) in a pressure vessel. Heat for 5–7 days at 160 °C.

Protocol 31. *Continued*

2. Cool the pressure vessel and make the solution basic (pH 14) with 5 M sodium hydroxide solution.
3. Extract the product into chloroform (1 litre) and dry the solution over anhydrous sodium sulphate.
4. Filter and evaporate solvent *in vacuo* to leave the desired product as a yellow oil (15.3 g, 87%), which is pure enough for use without distillation. ^{13}C NMR spectrum ($CDCl_3$) δ: 69.97 and 69.82 (O-CH_2s), 50.90 (N-CH_2), 35.92 p.p.m. (CH_3N).

Protocol 32. Synthesis of compound **27**; mol. wt 669.07

1. Dissolve compound **26** (14.8 g, 23.4 mmol, from *Protocol 29*), 1,2-bis(2-methylaminoethoxy)ethane (19.36 g, 110 mmol) and dry triethylamine (10 ml, 70 mmol) in dry DMF (100 ml) and stir 24 h under anhydrous conditions.
2. Evaporate the yellow reaction mixture to dryness *in vacuo* and remove residual DMF by addition and evaporation of toluene and ethanol (100 ml).
3. Dissolve the residual oil in dichloromethane (200 ml) and work up as usual.
4. Purify the crude product by column chromatography on silica gel (250 g) eluting with a gradient from 5–10% ethanol in dichloromethane/triethylamine (99:1 v/v).
5. Evaporate product fractions to obtain the title compound as a pale yellow oil (12.5 g, 80%), R_f 0.41 on TLC in ethanol/dichloromethane/triethylamine (20:175:5 by vol.). ^{13}C NMR spectrum (DMSO-d_6) δ: 162.88 (C-4), 155.02 (C-2), 140.19 (C-6), 90.82 (C-5), 86.74 (C-1′), 85.27 (C-4′), 70.67-69.25 (C-3′ and O-CH_2s of spacer), 61.54 (C-5′), 50.69 (N-CH_2), 48.76 (N-CH_2), 41.63 (C-2′), 37.38 (N-CH_3), 35.67 (N-CH_3), 33.69 and 33.55 (CHs of thexyls), 24.90 and 24.39 (quaternary Cs of thexyls), 19.84 (Si-CMe_2), 18.14 and 18.10 (CHMe_2 of thexyls), −2.84, −3.31, −3.37 and −3.92 p.p.m. (Si-CH_3s).

Protocol 33. Synthesis of compound **28**; mol. wt 895.36

1. Dissolve compound **27** (12.5 g, 18.68 mmol) and (+)-biotin-2-nitrophenyl ester (7.32 g, 19.95 mmol, from *Protocol 30*) in dry DMF (100 ml). Add triethylamine (5 ml) and stir overnight under anhydrous conditions.

Protocol 33. *Continued*

2. Evaporate the yellow solution to dryness *in vacuo*. Remove residual DMF by coevaporation with toluene and ethanol (100 ml).
3. Dissolve the yellow residue in dichloromethane (30 ml) and apply the solution directly to a column of silica gel (300 g). Elute first with dichloromethane and then with a gradient from 5 to 10% ethanol in dichloromethane.
4. Evaporate solvent *in vacuo* to yield the desired product as a white foam (10.5 g, 63%). R_f 0.51 on TLC in ethanol/dichloromethane (1:9 v/v). ^{13}C NMR spectrum ($CDCl_3$) δ: 173.21 and 172.85 (biotin C-10), 163.64 (C-4), 163.09 (biotin C-2′), 155.28 (C-2), 140.53 (C-6), 91.06 (C-5), 87.07 (C-1′), 85.60 (C-4′), 70.67–68.80 (C-3′ and O-CH_2s of spacer), 61.86 (C-5′), 61.75 (C-3 biotin), 60.09 (biotin C-4), 55.37 (biotin C-2), 49.56 (N-CH_2 spacer), 48.78 (N-CH_2), 41.93 (C-2′), 40.39 (N-CH_3 spacer), 37.71 (N-CH_3), 36.97 (biotin C-5), 34.04 and 33.86 (CHs of thexyls), 32.91 (biotin C-9), 28.32 (biotin C-7), 28.22 (biotin C-6), 25.10 and 24.87 (quaternary Cs of thexyls), 24.72 (biotin C-8), 20.14 (SiC*Me*$_2$), 18.44 (CH*Me*$_2$ of thexyls), −2.52, −3.00, −3.41, and −3.60 p.p.m. (Si-CH_3s).

Protocol 34. Synthesis of compound **29**; mol. wt 612.74

1. Desilylate compound **28** (30 g, 33.5 mmol) according to the first four steps of *Protocol 9*, however a reaction time of 2 h is required to remove fully the thexyldimethylsilyl groups.
2. Dissolve crude product in 30 ml of 4% ethanol in dichloromethane and load on a column of silica gel (500 g). Elute with a gradient from 4 to 30% of ethanol in dichloromethane.
3. Remove solvent from pure fractions to yield the desired product as a white foam (15.7 g, 76.5%). R_f 0.29 on TLC in ethanol/dichloromethane (1:4 v/v). ^{13}C NMR spectrum ($CDCl_3$) δ: 173.07 and 172.68 (biotin C-10), 163.38 (C-4), 163.23 (biotin C-2′), 155.43 (C-2), 141.40 (C-6), 91.10 (C-5), 87.25 (C-1′), 84.78 (C-4′), 70.56–67.93 (C-3′ and O-CH_2s of spacer), 61.37 (C-5′), 60.43 (biotin C-3), 59.99 (biotin C-4), 55.36 (biotin C-2), 49.44 (N-CH_2 spacer), 47.37 (N-CH_2), 40.56 (C-2′), 40.31 (N-CH_3 spacer), 37.70 (N-CH_3), 36.70 (biotin C-5), 32.80 (biotin C-9), 28.22 (biotin C-6 and C-7), 24.62 p.p.m. (biotin C-8).

Protocol 35. Synthesis of compound **30** mol. wt 772.96

1. Dissolve compound **29** (16.87 g, 27.5 mmol) in dry pyridine (150 ml) and add chlorotrimethylsilane (14.1 g, 130 mmol).

Protocol 35. *Continued*

2. Stir 2 h under anhydrous conditions, add 4-*tert*-butylbenzoyl chloride (6.13 g, 31.2 mmol) and stir for a further 2 h.
3. Cool the reaction mixture in ice and quench by adding water (50 ml) followed after 5 min by 25% aqueous ammonia (50 ml).
4. Keep 20 min at room temperature then evaporate to dryness *in vacuo*. Dry the residue by evaporation of toluene.
5. Work up as usual with dichloromethane/aqueous sodium bicarbonate.
6. Purify the crude product by chromatography on silica gel (300 g), eluting with a gradient from 0 to 10% ethanol in dichloromethane.
7. You will obtain pure product as a solid white foam (14.5 g, 68.2%). R_f 0.7 on TLC in ethanol/dichloromethane (1:4 v/v). ^{13}C NMR spectrum ($CDCl_3$) δ: 173.55 and 173.10 (biotin C-10) 169.14 (benzoyl CO), 163.22 (C-4), 156.25 (biotin C-2′), 155.72 (C-2), 154.77 (benzoyl C-4), 141.85 (C-6), 131.73 (benzoyl C-1), 128.82 and 124.38 (benzoyl C-2, 3, 5 and 6), 92.15 (C-5), 87.28 (C-1′ and C-4′), 70.83-68.51 (C-3′ and O-CH_2s of spacer), 62.44 (biotin C-3), 61.61 (C-5′), 58.10 (biotin C-4), 55.24 (biotin C-2), 49.69 (N-CH_2 spacer), 47.71 (N-CH_2), 40.60 (C-2′ and N-CH_3 spacer), 38.10 (N-CH_3), 36.90 (biotin C-5), 34.88 (quaternary C of *tert*-Bu), 32.66 (biotin C-9), 31.09 (CH_3s of *tert*-Bu), 28.35 (biotin C-6), 28.11 (biotin C-7), 24.57 p.p.m. (biotin C-8).

Protocol 36. Synthesis of compound **31**; mol. wt 1075.33

1. Dimethoxytritylate compound **30** (14.5 g, 18.76 mmol) according to the first six steps of *Protocol 6*, using only a 20% excess of tritylating agent.
2. Purify the crude product on a column of silica gel (350 g), eluting with a gradient from 0 to 5% ethanol in dichloromethane/triethylamine (99:1 v/v).
3. Evaporate solvent from pure product fractions *in vacuo* at room temperature to give the title compound as a solid white foam (16 g, 79.3%). R_f 0.75 on TLC in ethanol/dichloromethane/triethylamine (5:94:1 by vol.).

Protocol 37. Synthesis of the biotin containing phosphoramidite, (compound **32**); mol. wt 1275.55

1. Dissolve compound **31** (16 g, 14.88 mmol) in dry acetonitrile (100 ml). Add tetrazole (0.63 g, 8.6 mmol) and 2-cyanoethoxy bis(diisopropyl-

Protocol 37. *Continued*

amino)phosphine (5.2 g, 17.3 mmol) and stir under argon for 2 h. Reaction should be complete, check by TLC.

2. Quench the reaction by addition of ethanol (50 ml) and evaporate the mixture to dryness.
3. Work up the residue in the usual way with dichloromethane/aqueous sodium bicarbonate.
4. Purify the crude product by chromatography on silica gel (250 g) packed in ethyl acetate/dichloromethane/triethylamine (97:97:6 by vol.) and elute with this solvent until all phosphonate has been washed from the column (check by ^{31}P NMR spectroscopy).
5. Elute the product with ethyl acetate/dichloromethane/triethylamine/ ethanol (45:45:5:5 by vol.).
6. Evaporate solvent *in vacuo* at room temperature to give the desired phosphoramidite as a white foam (15.3 g, 80.6%). R_f 0.41 on TLC in ethyl acetate/dichloromethane/triethylamine (45:45:10 by vol.). ^{31}P NMR spectrum ($CDCl_3$) δ: + 145.50 and 145.30 p.p.m.

3. Synthesis of 2′-O-methyloligoribonucleotides

Nucleotide monomers, viz. compounds **7**, **11**, **17**, **25**, and **32** as well as the condensing agent, 5-(4-nitrophenyl)-1H-tetrazole must be dried thoroughly before use, preferably overnight, under high vacuum (<0.1 mm Hg) over phosphorus pentoxide and potassium hydroxide pellets. This procedure is necessary to ensure removal of traces of moisture and triethylamine, both of which will impair coupling efficiency. The monomers and condensing agent are to be used as 0.1 M solutions in anhydrous acetonitrile (<30 p.p.m. water). The 5-(4-nitrophenyl)-1H-tetrazole may require gentle warming to achieve solution.

2′-O-Methyloligoribonucleotides can be assembled on controlled pore glass either manually or on any commercial DNA synthesizer (see Chapter 1) using a DNA synthesis cycle with one simple change: the condensation (coupling step) must be increased to 6 min using 5-(4-nitrophenyl)-1H-tetrazole (0.1 M in acetonitrile) in place of tetrazole (normally 0.5 M).

The special biotin-containing reagent, compound **32** should be incorporated singly or multiply (more than four residues are not an advantage) either at the 3′- or 5′-end or at both ends of the oligonucleotide. The modified 2′-deoxycytidines should be regarded as non-hybridizing.

Upon completion of a synthesis it is generally recommended to leave the 5′-O-dimethoxytrityl group on to aid reversed phase HPLC purification. All other protecting groups and the linkage to the carrier are cleaved by treatment with 25% aqueous ammonia solution for 3–4 h at 60 °C. 5′-O-Dimethoxytrityl-protected 2′-O-methyloligoribonucleotides should be purified by reversed phase HPLC. They are slightly retarded relative to the corresponding oligodeoxyribonucleotides, and biotinylation causes a further slight increase in retardation. After purification the dimethoxytrityl group is cleaved in the usual way with 80% acetic acid.

4. Antisense affinity selection and affinity depletion of RNA–protein complexes

4.1 General points

For the separate tasks of either isolating specific RNAs or RNA–protein complexes in pure form or efficiently removing individual RNAs or RNA–protein complexes from cellular extracts, it is best to use antisense oligoribonucleotides with different base compositions in each case. We have found that the incorporation of 2′-O-methylinosine in place of 2′-O-methylguanosine is advisable for isolating targeted complexes with minimal background. Note that the use of 2′-O-methylinosine does not increase the level of non-specific affinity selection but actually decreases it. In contrast, when the goal is to work with an extract which has been quantitatively depleted of a specific complex, it is very important to use antisense probes containing 2′-O-methylguanosine (17).

The exact amount of antisense 2′-O-methyloligoribonucleotide required to select or deplete a given complex varies according to the abundance of the complex and also according to a less well defined property which can be termed ‘accessibility’. The ‘accessibility’ of a complex is affected by the length and base composition of the stretch of RNA complementary to the antisense probe. It also depends on the secondary and tertiary structure of the particle and on the presence of protein components which may sterically hinder the interaction between biotin and avidin/streptavidin during chromatography. Thus the efficiency of antisense affinity selection can be influenced by the site of biotinylation on the probe (15).

In practical terms the different parameters found to affect affinity selection; i.e. sequence of probe, site of biotinylation, final concentration and time of incubation in a cellular extract, must be determined empirically. As a general guide however we have found that approximately 0.2–0.5 pmol of antisense probe per microgram of cell extract is sufficient to efficiently select abundant RNA–protein complexes such as U1 snRNP. The time required for probe binding is usually in the range of 30–60 min at 30 °C.

4.2 Affinity depletion

Affinity chromatography of RNAs or RNA–protein complexes bound to biotinylated antisense 2′-O-methyloligoribonucleotides depends upon achieving highly specific, biotin-dependent binding to either streptavidin, avidin, or anti-biotin antibodies mounted on a solid support. We have reproducibly obtained good results using streptavidin coupled to agarose beads as supplied by Sigma. To suppress non-specific binding to the beads the following pre-blocking procedure is recommended shortly before use.

Protocol 38. Preparation of pre-blocked streptavidin–agarose beads

1. Prepare pre-block buffer (final concentrations): 20 mM Tris–HCl pH 7.6, 0.01% NP40, 50 mM KCl, 50 μg/ml glycogen, 0.5 mg/ml BSA, and 50 μg/ml yeast tRNA, and also wash buffer: 20 mM Tris–HCl pH 7.6, 0.01% NP40, and 50 mM KCl.
2. Streptavidin–agarose beads are supplied as a slurry. Spin down the beads (Eppendorf microfuge) at 2000 r.p.m. for 2 min. Care should be taken to avoid spinning the beads too hard as this can damage them. Carefully remove the supernatant (which can be discarded) and note the packed bead volume.
3. Gently resuspend the beads in an equal volume of pre-block buffer.
4. Rotate the beads for 20 min at 4 °C.
5. Pellet the beads in an Eppendorf microfuge at 2000 r.p.m. for 2 min.
6. Resuspend the beads in an equal volume of wash buffer and then pellet again as in step 5. Repeat this three times.
7. Keep the pre-blocked beads on ice until ready for use. They can be kept on ice for at least several hours without loss of activity.

Protocol 39. Preparation of HeLa cell nuclear extracts (17, 18)

1. Prepare the following buffers (all concentrations are final concentrations): Buffer A contains 10 mM Hepes pH 7.9, 1.5 mM $MgCl_2$, 10 mM KCl, and 0.5 mM DTT. Buffer S contains 20 mM Hepes pH 7.9, 10% glycerol, 0.42 M KCl, 1.5 mM $MgCl_2$, 0.2 mM EDTA, 0.5 mM DTT, and 0.5 mM PMSF. Buffer D contains 20 mM Hepes pH 7.9, 20% glycerol, 0.1 M KCl, 0.2 mM EDTA, 0.5 mM DTT, and 0.5 mM PMSF. The DTT and PMSF should be added fresh immediately prior to use.
2. Perform all manipulations at 4 °C unless otherwise stated. Adjust buffers to pH 7.9 at 4 °C. It is important to cool buffers and other reagents to 4 °C prior to use.
3. HeLa cells are either harvested in early to mid log phase or else can be purchased as a frozen whole cell pellet from the Computer Cell Culture Centre (Mons, Belgium). Quickly thaw frozen cell pellets at 30 °C immediately before use.

Protocol 39. *Continued*

4. Measure the packed cell volume (p.c.v.) in a graduated tube and wash the cells by adding 5 × p.c.v. of buffer A. Collect cells by centrifugation at 0 °C (2500 r.p.m. for 10 min using a Sorvall GSA rotor). Repeat the washing step once.
5. Resuspend the cell pellet in 2 × p.c.v. (i.e. original p.c.v.) of buffer A, transfer to a 40 ml Dounce homogenizer (all glass type) and lyse the cells by 8–10 strokes using an A type pestle.
6. Transfer lysed cell suspension to 50 ml Oakridge type centrifuge tubes (kept on ice) and pellet the nuclei by centrifugation (at 0 °C) for 5 min at 2500 r.p.m. using a Sorvall SS34 rotor.
7. Remove the supernatant, taking care not to disturb the pelleted nuclei. Centrifuge the pellets in the same rotor for an additional 20 min at 14 500 r.p.m.
8. Discard the supernatant and resuspend the nuclei in buffer S, using approximately 4.5 ml of buffer S per 10^9 nuclei.
9. Transfer to a 40 ml Dounce homogenizer (all glass type) and lyse the nuclei by 8–10 strokes with a B type pestle. Transfer nuclear lysate to 15 ml plastic tubes and slowly rotate for 30 min at 4 °C. Do not use magnetic stirrers for this step.
10. Centrifuge the lysate (0 °C) for 30 min at 14 500 r.p.m. in a Sorvall SS34 rotor. Transfer the supernatant to dialysis tubing (Spectrapor membrane tubing, mol. wt cut-off 12 000–14 000).
11. Dialyse for 5 h at 4 °C against 120 volumes of buffer D. Renew the dialysis buffer after 1 h and again after 3 h.
12. Remove the heavy precipitate, mostly comprised of proteins which are insoluble at 0.1 M KCl, by centrifugation for 10 min at 14 000 r.p.m. (Sorvall SS34 rotor).
13. Discard the precipitate and divide the extract into aliquots (normally 50–200 μl) and snap-freeze in liquid nitrogen. Store at −80 °C or else in liquid nitrogen.

Protocol 40. Efficient antisense affinity depletion from nuclear extracts

1. In addition to buffers listed under *Protocol 39* prepare the following additional buffers: Buffer MD 0.1 contains 20 mM Hepes pH 7.9, 10% glycerol, 0.1 M KCl, 0.2 mM EDTA, 0.5 mM DTT, and 0.5 mM PMSF. Buffer MD 0.6 contains 20 mM Hepes pH 7.9, 10% glycerol, 0.6 M KCl,

Protocol 40. *Continued*

0.2 mM EDTA, 0.5 mM DTT, and 0.5 mM PMSF. Add the DTT and PMSF fresh immediately prior to use.

2. Carry out steps 1–10 from *Protocol 39*.
3. Dialyse the nuclear extract initially against 120 volumes of buffer MD 0.1 for 3.5 h at 4 °C.
4. Remove the resulting precipitate by centrifugation in a Heraeus Labofuge for 10 min at 4000 r.p.m.
5. Dialyse the supernatant for 1.5 h against 120 volumes of buffer MD 0.6 at 4 °C.
6. Incubate the nuclear extract with the antisense probe at 30 °C for 30–60 min. It is convenient to perform this step using individual volumes of 1–2 ml. Add the antisense probe in a cocktail which also contains ATP, creatine phosphate, and NP40. The total volume of this cocktail should not exceed 10% of the total volume of nuclear extract and should bring each component to a final concentration of 1.5 mM ATP, 5 mM creatine phosphate and 0.05% NP40. The concentration of antisense oligonucleotide must be determined by experiment but will probably be in the range of 1–4 nmol per ml of nuclear extract.
7. Add the extract to an equal volume of pre-blocked streptavidin–agarose beads (prepared according to *Protocol 38*). Before adding the beads it is important to spin them for 2 min at 4000 r.p.m. (Eppendorf microfuge) to remove excess buffer and hence avoid additional dilution of the nuclear extract. Do not spin the beads too hard however or they will be damaged.
8. Slowly rotate the mixture at 0 °C for 45 min. It is advisable to maintain a low temperature by placing the tube containing the nuclear extract/streptavidin–agarose beads mixture within a larger tube packed with ice.
9. Remove the beads by centrifugation at 4000 r.p.m. for 2 min (Eppendorf microfuge). Repeat the selection procedure adding another volume of pre-blocked streptavidin–agarose beads to the nuclear extract and incubating for 45 min at 0 °C. This second round of selection is advisable for high efficiency depletion.
10. Remove the second batch of streptavidin–agarose beads by centrifuging twice at 4000 r.p.m. for 2 min (Eppendorf microfuge). Return the extract to dialysis tubing and dialyse against 120 volumes of buffer D for 75 min at 4 °C.
11. Snap-freeze the depleted extract in liquid nitrogen in aliquots of 50–100 μl and store at −80 °C or else in liquid nitrogen.

An example of a HeLa cell nuclear extract specifically depleted of U4 snRNP is shown in *Figure 6*, panel A. It is possible to obtain depletion efficiencies of over 99% using this method.

4.3 Affinity selection

For efficient affinity selection it is not necessary to use long probes to form stable hybrids. We have obtained excellent results using probes ranging from 10 to 16 nucleotides in length. Low non-specific backgrounds and efficient recovery of complexes from streptavidin–agarose beads is also aided by the replacement of 2′-O-methylguanosine by 2′-O-methylinosine.

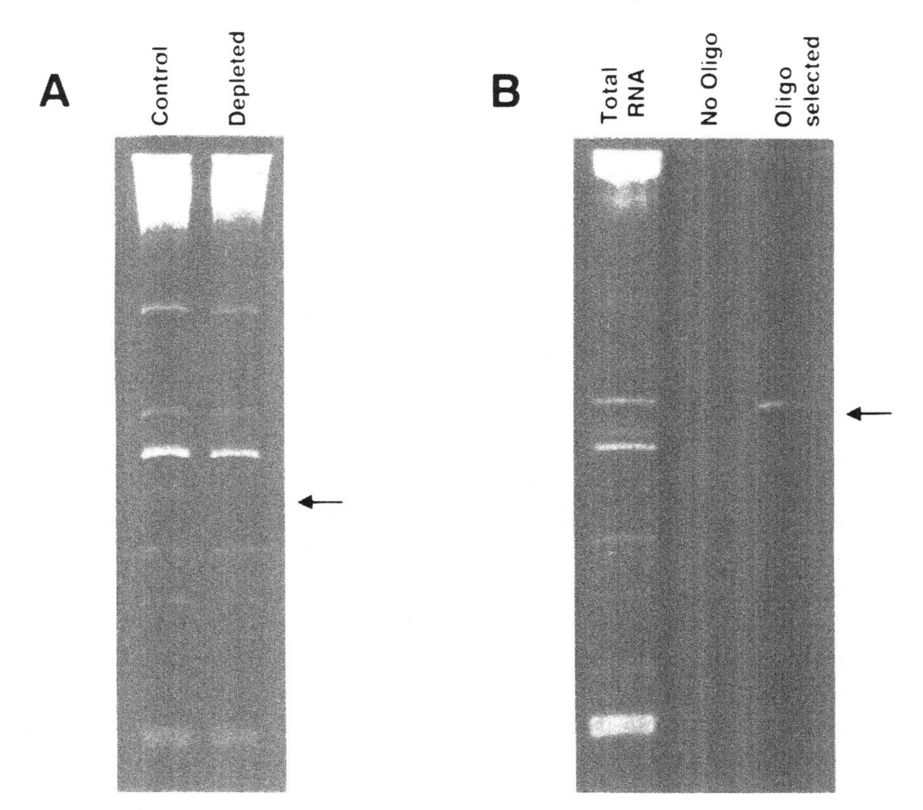

Figure 6. Examples of antisense affinity depletion and affinity selection of targeted RNA–protein complexes. Panel **A** compares the total RNA present in a control HeLa cell nuclear extract with the total RNA remaining after targeted depletion of U4 snRNP by an anti-U4 2′-O-methyloligoribonucleotide. Panel **B** shows the specific affinity selection of U2 snRNP from a crude HeLa cell nuclear extract using an anti-U2 2′-O-methyloligoribonucleotide. In both panels the RNAs are shown separated on a 10% polyacrylamide/8 M urea gel, and are detected by ethidium bromide staining. Arrows mark the depleted or selected RNA.

Protocol 41. Affinity selection of targeted RNA–protein complexes

1. Prepare nuclear extracts as described in *Protocol 38*. Affinity selection is routinely done in a 0.1–0.2 ml reaction containing 35% (v/v) nuclear extract and 2.5 mM $MgCl_2$, 1.5 mM ATP, and 5 mM creatine phosphate (all final concentrations). The inclusion of ATP and creatine phosphate in these reactions suppresses protein precipitation (by an unclear mechanism) and generally improves the efficiency of affinity selection.
2. Antisense oligonucleotides targeted to specific RNAs or RNA–protein complexes are typically used at a final concentration of 0.2–2 nmol/ml. For optimal efficiency of affinity selection this concentration must be titrated empirically.
3. Incubate the antisense probe with the extract at 30 °C for 30–60 min. We have observed a marked decrease in affinity selection efficiency when the incubation temperature is lower than 20 °C or higher than 37–40 °C. The optimum time of incubation varies for different target complexes.
4. Centrifuge the extract for 3 min at 12 000 r.p.m. (Eppendorf microfuge) to remove any insoluble material which may have precipitated during the incubation. This helps to reduce non-specific binding to the streptavidin–agarose beads.
5. Adjust supernatant to a final concentration of 0.3 M KCl and keep on ice.
6. Pre-block streptavidin–agarose beads as described in *Protocol 38*, with the exception that the final wash step is done using wash buffer containing 300 mM KCl (WB300: 20 mM Tris–HCl pH 7.6, 0.01% NP40, and 300 mM KCl).
7. Add an equal packed bead volume of pre-blocked streptavidin–agarose beads to the nuclear extract and rotate at 4 °C for 45–60 min.
8. Pellet beads by centrifugation at 4000 r.p.m. for 2 min (Eppendorf microfuge) and remove the supernatant.
9. Gently resuspend the beads in an equal volume of WB300 and pellet as before. Repeat the washing step twice.

4.3.1 Recovery of affinity selected complexes

As described below elution of affinity selected complexes from streptavidin–agarose beads (*Protocol 41*) can be done in a number of ways, according to the type of analysis which will be carried out on the selected complex. An example of the affinity selection efficiency possible using this method is shown in *Figure 6*, panel B.

Protocol 42. Recovery of RNA for analysis

1. After affinity selection (*Protocol 41*) resuspend beads in PK buffer (10 mM Tris-HCl pH 7.6, 50 mM NaCl, 1 mM EDTA, and 0.5% SDS), and digest with proteinase K (1 mg/ml final concentration) at 65 °C for 45 min.
2. Heat for 10 min at 85 °C.
3. Pellet the beads by centrifugation at 6000 r.p.m. for 5 min (Eppendorf microfuge).
4. Add glycogen to the resulting supernatant (20 µg/ml, final concentration) and recover the RNA by precipitation with 2.5 volumes of absolute ethanol.

For analysis of protein components in selected complexes, treatment with proteinase K must obviously be avoided. The use of oligonucleotides coupled to biotin through a spacer arm containing a disulphide link allows affinity selected complexes to be isolated by resuspending in a reducing buffer containing 10 mM Tris–HCl pH 7.6, 50 mM NaCl, 1 mM EDTA, and 150 mM DTT.

An alternative elution strategy is to affinity-select targeted complexes using very short probes that form hybrids which are only stable in a high salt buffer (such as WB300) and at low temperature. Elution is then performed by resuspending the beads in low salt buffer (10 mM Tris–HCl pH 7.6, 10 mM NaCl, 1 mM EDTA). This is best done by first washing the beads from *Protocol 41* in low salt buffer then resuspending the bead pellet in an equal volume of low salt buffer and incubating for 20–30 min at 30 °C. The beads are subsequently removed by centrifugation at 6000 r.p.m. for 5 min in an Eppendorf microfuge.

References

1. Inoue, H., Hayase, Y., Asaka, M., Imura, A., Iwai, S., Miura, K., and Ohtsuka, E. (1985). *Nucleic Acids Res. Symposium Series*, **16**, 165.
2. Inoue, H., Hayase, Y., Imura, A., Iwai, S., Miura, K., and Ohtsuka, E. (1987). *Nucleic Acids Res.*, **15**, 6131.
3. Mukai, S., Shibahara, S., and Morisawa, H. (1988). *Nucleic Acids Res. Symposium Series*, **19**, 117.
4. Inoue, H., Hayase, Y., Iwai, S., and Ohtsuka, E. (1988). *Nucleic Acids Res. Symposium Series*, **19**, 135.
5. Shibahara, S., Mukai, S., Nishihara, T., Inoue, H., Ohtsuka, E., and Morisawa, H. (1987). *Nucleic Acids Res.*, **15**, 4403.
6. Inoue, H., Hayase, Y., Iwai, S., and Ohtsuka, E. (1987). *FEBS Lett.*, **215**, 327.
7. Sproat, B. S., Lamond, A. I., Beijer, B., Neuner, P., and Ryder, U. (1989). *Nucleic Acids Res.*, **17**, 3373.

8. Berkower, I., Leis, J., and Hurwitz, J. (1973). *J. Biol. Chem*, **248**, 5914.
9. Sproat, B. S., Beijer, B., and Iribarren, A. (1990). *Nucleic Acids Res.*, **18**, 41.
10. Pieles, U., Sproat, B. S., and Lamm, G. M. (1990). *Nucleic Acids Res.*, **18**, 4355.
11. Matteucci, M. D. and Caruthers, M. H. (1981). *J. Am. Chem. Soc.*, **103**, 3185.
12. McBride, L. J. and Caruthers, M. H. (1983). *Tetrahedron Letters*, **24**, 245.
13. Sinha, N. D., Biernat, J., McManus, J., and Köster, H. (1984). *Nucleic Acids Res.*, **12**, 4539.
14. Lamond, A. I, Sproat, B. S., Ryder, U., and Hamm, J. (1989). *Cell*, **58**, 383.
15. Barabino, S., Sproat, B. S., Ryder, U., Blencowe, B. J., and Lamond, A. I. (1989). *EMBO J.*, **8**, 4171.
16. Blencowe, B. J., Sproat, B. S., Ryder, U., Barabino, S., and Lamond, A. I. (1989). *Cell*, **59**, 531.
17. Barabino, S. M. L., Blencowe, B. J., Ryder, U., Sproat, B. S., and Lamond, A. I. (1990). *Cell*, **63**, 293.
18. Dignam, J. D., Lebowitz, R. M., and Roeder, R. G. (1983). *Nucleic Acids Res.*, **11**, 1475.

4

Phosphorothioate oligonucleotides

GERALD ZON and WOJCIECH J. STEC

1. Introduction

Phosphorothioate analogues of DNA and RNA have sulphur in place of oxygen as one of the non-bridging ligands bonded to phosphorus. Differential electronegativity, bond length, and other factors involving sulphur vs. oxygen lead to unique physicochemical features for this class of backbone-modified analogues, relative to native DNA and RNA. Important distinctions include: (a) resistance of the internucleoside phosphorothioate linkage to enzymatic cleavage; (b) chirality at phosphorus, which is designated as R_p and S_p; (c) greater nucleophilicity for sulphur; and (d) a convenient radio-isotope (^{35}S).

```
        O                        S⁻
        ||                       |
5′O---P---O3′           5′O---P---O3′
        |                        ||
        S⁻                       O

        Rp                       Sp
```

These molecular features have been utilized for diverse applications in the fields of molecular biology, enzymology, biochemistry, biophysical chemistry, and biomedical science. Eckstein, who pioneered such work, has written several reviews (1–3) on the subject.

Much attention has been given recently to the potential use of various types of DNA or RNA analogues, including phosphorothioates, as sequence-specific agents to control transcription, translation, or other regulated processes. Prominent among these strategies is the design of so-called 'antisense oligonucleotides,' which are directed at complementary sequences of DNA or RNA targets. Because of widespread, growing interest in the antisense area (for reviews see 4–10), and increasing experimentation with antisense phosphorothioate oligonucleotide analogues, the synthesis of these compounds has recently received renewed attention. Efficency, convenience, automatability, scalability, product purification, and peripheral analytical methods are important factors to consider from a practical viewpoint. Stereo-random versus stereo-controlled coupling chemistry must also be addressed, given that there are 2^n possible diastereomers for an oligonucleotide which has n phosphorothioate linkages.

The present article tabulates reported uses of phosphorothioate analogues of DNA or RNA that are derived biochemically or chemically. Among these various uses, antisense applications are discussed in some detail. Methods for synthesis, purification, and analysis are also tabulated. Due to space limitations, protocols are given only for chemical syntheses of phosphorothioate analogues that have been automated, used extensively, and lend themselves to successful implementation by persons who are not necessarily skilled in organic synthesis (see also Chapter 9). Purification protocols focus on the use of reversed phase HPLC, which has been widely employed with success and can be scaled-up relatively easily. The ability to scale-up synthesis and purification methods readily by at least several orders of magnitude is a practical necessity in antisense applications of phosphorothioate oligonucleotide analogues *in vivo*.

2. Applications of phosphorothioate analogues of DNA and RNA

Applications of phosphorothioate analogues of DNA and RNA, excluding those which are demonstrably antisense, are grouped in *Table 1* according to the following general categories that are 'biological' in the broad sense of the word: (a) enzyme biochemistry; (b) autolytic processing of RNA (ribozymes); (c) interactions with proteins; (d) oligonucleotide-directed mutagenesis; (e) antiviral agents; and (f) miscellaneous. Specific uses and literature references are listed within each category. It is evident from *Table 1*, and not unexpected, that phosphorothioate analogues have been extensively employed to study enzymes which cleave the phosphate group in DNA or RNA. Recent attention being given to both autolytic processing of RNA and the design of useful ribozymes (41) will undoubtedly fuel further interest in phosphorothioate-modified RNA, especially in view of the reported (42) feasibility of developing therapeutic anti-HIV ribozymes. These and other ribozyme drugs could be made to have increased stability *in vivo* through strategic placement of phosphorothioate linkages. Other studies worth noting are the use of double-stranded phosphorothioates to compete for binding of proteins, such as transcriptional regulation factors (30, 31), and to induce antibody formation (91). Further applications are presented in Chapter 9 of this book.

The antisense applications of phosphorothioate analogues listed in *Table 2* are grouped according to the intended RNA target. All of these applications refer to β-anomeric oligodeoxyribonucleotides, with the exception of a

Table 1. Biological applications of phosphorothioate analogues of DNA and RNA, excluding demonstrable antisense

(a) **Enzyme biochemistry**

*Eco*RI (11–15)
*Ava*I (15)
*Bam*HI (15)
*Hin*dIII (15)
*Sal*I (15)
Exonuclease III (16)
Spleen phosphodiesterase (17, 22)
DNase I (67)
DNase II (13, 22)
Snake venom phosphodiesterase (18)
Nuclease P1 (18)
Ribonuclease A (19)
HIV and MuLV reverse transriptase (20)
DNA polymerase I (20)
α–, β–, γ-polymerses (20, 21)
HSV-1, HSV-2, EBV DNA polymerase (21)
Staphyloccal nuclease (22)
RNase L (23–25)

(b) **Autolytic processing of RNA**

Tobacco ringspot virus (26, 27)

(c) **Interactions with proteins**

*Eco*RI (13)
DNase II (13)
DNA autoantibody (28)
Bacteriophage R17 (29)
SP1 transriptional factor (30)
κB consensus sequence (31)

(d) **Oligonucleotide-directed mutagensis**

M13 RF IV DNA (32, 78, 79)

(e) **Antiviral agents**

VSV/interferon induction (33, 34)
HIV/reverse transcriptase (20)
HSV-2 (21, 45)
HIV (25, 35–39)
MuLV (400

(f) **Miscellaneous**

Recombination (87)
Splicing (88)
Immunogenicity (91)
Stability in biological milieu (61, 86)

Table 2. Antisense applications of phosphorothioate oligonucleotide analogues

Target RNA	**Target RNA**
HIV-1 (46–52; for related studies, see 25, 35–39, 68)	IL 1ß (58) globin (59[a], 60)
HSV-1 (53)	Vg1 (61[a])
VSV (54)	U1 small nuclear RNA (62)
Influenza A and C virus (139)	Histone H4 (63[a])
Chloramphenicol acetyl transferase (43)	An2, cyclin (64[a])
TGF-ß (55)	BCL2 (65)
c-*myc* (56) *ras* p21 (57)	

[a] Includes cellular microinjection experiments

reported (39) study with α-anomeric deoxyribophosphorothioates and another with oligo(2′-O-methyl)ribonucleotide phosphorothioates (52).

The first two published studies of antisense phosphorothioate oligonucleotides dealt with inhibition of expression of chloramphenicol acetyl transferase encoded by a plasmid (43), and inhibition of the cytopathic effect of HIV (36). These shortly followed the report by Zamecnik *et al.* (44) that replication and expression of HIV in cell culture was blocked even by unmodified oligonucleotides. Studies with HIV are discussed in Section 2.2.2.

RNase H 'recognizes' DNA–RNA heteroduplexes and selectively cuts only the ribonucleotide strand (92). This provides an enzyme-mediated mechanism for irreversible cleavage of multiple strands of the RNA target by a single molecule of antisense phosphorothioate oligonucleotide. Such enzymatic 'turnover' could be more effective than physical blockage of translation, which is reversible and occurs on a 1:1 molecular basis. That phosphorothioate-modified DNA duplexed with RNA can in fact serve as a substrate for cleavage of RNA by RNase H has been experimentally demonstrated and explored by a number of research groups (59–64). By contrast, methylphosphonate and phosphoramidate linkages can prevent this cutting process from occurring.

Analytically related studies using phosphorothioate-containing DNA and RNA are listed in *Table 3*. These include a recently published (75) ^{35}S labelling scheme which provides material for measurement of antisense phosphorothioate oligonucleotide cellular uptake, its distribution *in vitro* or *in vivo*, and degradation by nucleases. Because of growing interest in such measurements, the protocol for this labelling method is given in Section 4.

Table 3. Analytically-related studies using phosphorothioate analogues of DNA and RNA

B–Z transition (66)	^{35}S-labelling (75)
Duplex formation (57, 66, 67, 69, 70, 90)	Heavy metal staining (76)
Triplex formation (67)	Flow cytrometry (56, 77)
HPLC elution (71)	^{31}P NMR (11, 12, 69, 89)
DNA and RNA sequencing (72, 76)	Stereochemical correlation (80–85, 146)
Fluorescent labelling (73, 74, 149)	Stability in biological media (61, 86)

2.1 Issues in antisense applications of phosphorothioate oligonucleotides

2.2.1 Oligonucleotide design strategy, synthesis, and stereochemistry

At this very early stage of its evolution, the field of antisense oligonucleotides is largely exploratory. Currently there are no established 'selection rules' with regard to the target sequence, mechanism of action, type of analogue, length of oligonucleotide, etc. For example, there are no reliable guidelines for choosing between targeting cytoplasmic mRNA at, say, the start of translation, and nuclear RNA at, say, a splice site. This also applies to choosing between translation arrest and RNase H-mediated cleavage, or between oligonucleotide analogues that are entirely methylphosphonate, phosphorothioate, etc. and 'end-capped' versions, or between an 18mer and 25mer oligonucleotide. Antisense applications therefore necessarily involve a substantial amount of basic or empirical research and systematic development, all of which can be greatly facilitated by timely access to large numbers and amounts of oligonucleotide test compounds, relative to what is needed for typical molecular biological studies. These needs, especially during initial phases of research, can be met by use of currently available, multiple column, automated oligonucleotide synthesizers, which operate at the ≤ 1 μmol scale and can be adapted to accommodate unconventional chemistries needed for production of analogues. In principle, and in practice (93, 134, 140), such solid phase syntheses can be scaled up to begin to address requirements for much larger amounts of oligonucleotide analogues that will be needed for preclinical/clinical studies and, presumably, future manufacturing of these compounds as drugs. However, as discussed in detail elsewhere (94), manufacturing and economic considerations clearly indicate that substantial, highly innovative, improvements in automated solid phase synthesis are required for producing these drugs at costs which are reasonable for their successful commercialization. Detailed comparative analyses of solution phase syntheses are apparently unavailable at this time for either classic approaches (141) or novel ones (142).

Antisense phosphorothioate oligonucleotide analogues can be constructed with sulphur either at selected positions or throughout the phosphate backbone. Ideally, this construction should be carried out in a stereospecifically defined manner to allow the synthesis of any one of the 2^n possible diastereomers. Enzymatic syntheses of phosphorothioate DNA or RNA analogues using appropriate templates, polymerases, and 5′-O-(1-thiotriphosphate) nucleotides have afforded, to date, only the R_p configuration (2, 3). While a hypothetical polymerase with opposite diastereoselectivity, or a hypothetical post-enzymatic inversion scheme, could provide access to the S_p configuration, a general, stereochemically controlled method will most likely be an entirely chemical process. Among the reported stereochemical approaches given in *Table 4*, those reported by Cosstick and Eckstein (66) and Lesnikowski and co-workers (95, 96) represent complementary strategies within the current state-of-the-art. Cosstick and Eckstein prepared oligonucleotides that contained alternating phosphodiester and phosphorothioate linkages, the latter of which were incorporated as presynthesized, diastereomerically pure, dimer building blocks. Lesnikowski and co-workers investigated stereospecific formation of a P-chiral internucleotide linkage using S-protection.

Table 4. Stereo-specific (-selective) syntheses of phosphorothioate DNA or RNA

Enzymatic, RNA (17, 33, 34)
Enzymatic, DNA (67, 89, 90)
Reaction of phosphoranilidate (81)
Acid hydrolysis of phosphormorpholidothioate (147)
Coupling of a presynthesized dimer block[a] (66)
Formation of S-methylphosphorothioate triester (95)
Formation of S-(2-nitrobenzyl) phosphorothioate triester (96)
Formation of S-(2-cyanoethyl) phosphorothioate triester (101)
Reaction of aroyl-(97) or acylphosphonate (102)
Sulphurization of H-phosphonate (100)

[a] For the fully-protected 16 possible di-2′-deoxynucleoside thiophosphotriesters, see (98). For 2′-5′ A analogues, see (99) and references, cited therein.

Stereochemical composition of a phosphorothioate-containing oligonucleotide will theoretically, and as shown experimentally (66, 69), have an influence on duplex structure, conformation, and/or stability (melting temperature, T_m). In a doubly-modified DNA duplex, [d(GGsAATTCC)]$_2$, it appears that an ‘inward’ oriented P–S moiety in the R_p configuration decreased the T_m by *c.* 1 °C, relative to the native duplex, whereas the ‘outward’ oriented P–S group in the S_p configuration had essentially no effect on T_m. Interestingly, different lengths of stereo-random oligophosphorothioates (11, 14, and 27mers), when targeted to DNA, led to only *c.* 10 °C depressions of T_m relative to corresponding unmodified oligonucleotides, which suggests a ‘levelling-off’ relationship between the number of stereo-random P–S moieties and the T_m lowering. In any event, the lower binding affinity of an oligophosphorothioate can be offset simply by extending the sequence of the phosphorothioate oligonucleotide by several bases.

Due to the present unavailability of a general, stereo-specific synthetic method, there are no data concerning differential biological effects of individual diastereomers of antisense phosphorothioate analogues. There is also no information on the relative nuclease-resistance of these diastereomers. The importance or not of such effects in antisense applications will remain, unfortunately, a topic of much speculation and some debate until practical, stereo-controlled synthetic methods are devised.

Another issue of concern related to methods for synthesis is the 'chemical integrity' of the phosphorothioate linkages. In particular, the amount and distribution of inadvertently produced, unmodified phosphodiester moieties within the final product will necessarily influence both its rate of degradation by nucleases and the accompanying size-dispersion of such fragments. Consider as an example a hypothetical synthetic procedure which provides a 27mer oligonucleotide analogue with 99% of the theoretical amount of phosphorothioate linkages, based on ^{31}P NMR analysis, and the remaining 1% of linkages being randomly distributed phosphodiester 'defects'. These defects may be subject to relatively facile cleavage by nucleases. A statistical calculation can be made using the binominal expression $(x + y)^{26} = x^{26} + 26x^{25}y + 325x^{24}y^2 + \ldots$, where x = fraction of P–S and y = fraction of P–O (143). This calculation predicts that, for $x = 0.99$ and $y = 0.01$, 77.0% of the molecules are all P–S, 20.2% of the molecules will have one phosphodiester P–O, 2.5% of the material will have two of these defects per molecule, etc. This analysis underscores the importance of synthetic methods which lead to complete incorporation of sulphur, or very nearly so. Of the two methods for which this has been investigated, the H-phosphonate route (see Section 3.1.2) can apparently afford complete incorporation of sulphur, i.e., no detectable (<0.1%) phosphodiester linkages by ^{31}P NMR (143) under conditions of high signal-to-noise, whereas the phosphoramidite route (see Section 3.1.1) leads to *c*. 95–99.5% sulphurization, depending on the particular sulphurizing reagent, reaction conditions, and other details of the synthesis cycle. There is apparently no published evidence for 'wash out' of sulphur during ammoniolytic cleavage from the support, deprotection, and detritylation with acetic acid.

2.2.2 Results and issues for antisense experiments with phosphorothioate oligonucleotides

As a follow-up to initial encouraging observations by Marcus-Sekura *et al.* (43) in a transfection assay with a 15mer phosphorothioate targeted against chloramphenicol acetyl transferase mRNA, antisense experiments were conducted by Chang *et al.* (57) using a relatively simple cell-free translation system to compare a series of 11mers having different types of linkages and various degrees of complementarily with the start codon and downstream 8 bases of *ras* p21 mRNA. The phosphorothioate analogues were found to be the most potent inhibitors; however, at a 10-fold higher concentration, an apparently sequence non-specific effect was found. Cazenave *et al.* (59) obtained similar results for a 17mer phosphorothioate targeted to the coding region of *β*-globin mRNA, in both a cell-free and cellular microinjection assay. Unpublished observations were said (59) to have shown that phosphorothioate oligonucleotides bind tightly to ribosomes, which could conceivably account for the onset of non-specific inhibition of protein synthesis at higher concentrations of the analogues. In view of the more recent observation (103) that cellular microinjection of oligonucleotides complementary to ribosomal RNA can sequence-specifically abolish protein synthesis, apparently non-specific inhibitory effects might actually be derived, at least in part, by accidental hybridization, which is not easily excluded as a possibility, to ribosomal RNA or some other non-target RNA/DNA. Such inadvertent hybridization could involve partially mismatched sequences, given the inherently non-stringent, non-equilibrium conditions which pertain to all antisense applications. To the extent that these unintended hybridizations might be minimized through empirical screening of different antisense sequences, varying sequence to maximize the desired effect while minimizing side-effects would be analogous to structure–activity refinements in conventional medicinal chemistry.

Investigations of oligonucleotide phosphorothioates and other types of analogues as potential antisense inhibitors of HIV were logical developments in view of both the apparent causal relationship between HIV and AIDS, and the urgent need for new and improved anti-AIDS drugs. On the other hand, the complexities of HIV molecular biology and the replication cycle for HIV (104) necessarily complicate all antisense studies of this retroviruse, both experimentally and interpretatively. Matsukura *et al.* (36) initially employed a *de novo* infection assay that utilized a line of immortalized $T4^+$ T cells (ATH8 cells) because of the profound sensitivity of these cells to the cytopathic effect of HIV. Various antisense 14mers, their complements ('sense' oligonucleotides), and homo-oligomers were tested for inhibition of the viral cytopathic effect 7 days after infection with 500 virus particles per cell and then continuous exposure of the cells to the oligomer. Under these conditions, unmodified and methylphosphonate oligomers failed to show statistically significant levels of inhibitory activity, as measured by cell viability, whereas all of the phosphorothioates were active. Southern blot analysis (36) established that a phosphorothioate 28mer of dC prevented detectable levels of viral DNA synthesis, which suggested intervention of the oligonucleotide at one or more of the pre-integration steps: i.e. binding, internalization, uncoating, reverse transcription, migration, and integration. Experimental evidence (20) for inhibition of reverse transcription by a phosphorothioate analogue has subsequently been obtained *in vitro*, acellularly; however, the relevance of these results to infected cell cultures is unclear at this time.

Zamecnik and co-workers (35, 44) used different cell lines, different viral RNA targets, and different control sequences from those used by Matsukura *et al.* (36) to investigate the anti-HIV effects of various types of antisense oligonucleotides. These included phosphodiesters, phosphorothioates, methylphosphonates, and phosphoramidates. The results of their *de novo* infection assays indicated that *all* of the test compounds had some measure of antiviral activity. These findings and those of others (37, 38) using phosphorothioate analogues underscore the limited utility of *de novo* infection assays, which are efficient for initial screening of compounds for antiviral activity, but inappropriate for defining mechanisms of action, especially in the case of antisense oligonucleotides.

The aforementioned mechanistic ambiguity prompted Matsukura *et al.* (46) to develop the use of chronically HIV (strain III_B)-infected H9 cells as the basis for an assay aimed at establishing whether a phosphorothioate oligomer could exhibit sequence-specific suppression of HIV expression. In these investigations, which used an extended (28mer) version of the previously described (36) anti-*rev* sequence, antiviral activity (measured by ^{3}H-labelled nucleotide uptake and p24 levels) was found with the anti-*rev* phosphorothioate but not with the unmodified (phosphodiester) control, most likely due to degradation of the latter (46). More importantly, neither the sense, randomized sequence, dC_{28}, nor anti-*rev* with four N_3-dT residues showed statistically significant levels of activity. The latter control compound shows no measurable duplex formation with complementary DNA as a model for the RNA target. The altered HIV mRNA profile (46) afforded by treatment of the infected cells with the anti-*rev* phosphorothioate oligomer was consistent with interference of *rev* protein function (104). Analogous experiments by Zamecnik and co-workers (47) have been reported to confirm the sequence-specific anti-HIV inhibitory effects of phosphorothioates. There are some recent preliminary indications (144), however, of possible host cell and HIV strain-dependence such that not all HIV strains, when used to infect H9 or other types of cells, are subject to measurable inhibition of expression by the anti-*rev* phosphorothioate (46). It should also be noted with regard to mechanisms of action that endoribonucleolytic cleavage of the *gag* region of HIV RNA duplexed with an oligonucleotide has been recently shown to be mediated *in vitro* by the RNase H-like activity associated with HIV reverse transcriptase (145). Thus, the HIV's own enzyme activity can possibly lead to a suicide-like cleavage reaction which is triggered by an antisense phosphorothioate.

3. Synthetic methods

3.1 Overview

All of the major methods for chemically synthesizing DNA or RNA have been investigated with regard to preparing phosphorothioate analogues. Procedures employing a condensation reagent, such as benzenesulphonyl chloride (106–108), or using a thionophosphotriester strategy (109, 100) did not appear to win popular support, given the superior features of Caruthers' phosphoramidite method (see preceding chapters) for chain assembly, and the ability, with this method, to sulphurize any or all of the intermediate phosphite linkages (111, 112). Two research groups independently investigated automated versions of this stepwise approach to sulphurization. Eckstein and co-workers (11, 113) used a 0.2 M solution of elemental sulphur (S_8) in CS_2-pyridine (1:1) at room temperature for 30 min, while Stec and Zon and co-workers (12) used 0.4 M S_8 in 2,6-lutidine at 60 °C for 15 min. (A reported variation (114) that included addition of 0.15 M Et_3N to the CS_2-pyridine was accompanied by accelerated discoloration of the reagent). CS_2-pyridine as the solvent required a somewhat longer cycle time but avoided the need for a specially made, heated synthesis column; however, prolonged usage of CS_2-pyridine was found to cause deterioration of key parts in a synthesizer (115). These technical difficulties, plus the inflammability and toxicity of CS_2, are factors that eventually prompted consideration of an alternative sulphurizing procedure. An important additional aspect concerned the degree of sulphur incorporation, which was found by the present authors to be only *c.* 95% based on ^{31}P NMR analysis. It was unclear at the time whether the contaminating phosphodiester linkages arose from side-reaction(s) of the intermediate phosphite or loss of sulphur at the thionophosphotriester or phosphorothioate stages. The advent of practical, automated H-phosphonate chemistry (116) for DNA synthesis, which was first extended by Froehler (117) to phosphorothioate analogues, provided for the use of an attractive one-step sulphurization reaction that could be easily and safely performed manually. Andrus *et al.* (118) developed an improved method for H-phosphonate chain assembly by use of triethylammonium isopropylphosphonate (IPP) as the capping reagent, and adamantane carbonyl chloride (AdCOCl) as the activator for both coupling and capping. The protocol for this widely used method is given below. A diverse array of activators have been investigated (116, 119, 120); however, trimethylacetyl (pivaloyl) chloride (116) and AdCOCl are employed most widely, with AdCOCl offering the advantage of being inherently more stable in solution.

Key features of the H-phosphonate route to phosphorothioates include it being one-step, cost-effective, and very efficient with regard to the degree of sulphurization, i.e., the product is essentially free of ^{31}P NMR-detectable phosphodiester linkages. These factors and the availability of $^{35}S_8$ provided for an efficient ^{35}S-labelling protocol (75), as detailed in Section 4. Alternative reagents for sulphurizing H-phosphonate linkages have been investigated (12, 122), as well as H-phosphonothioate monomers (123–125), use of which precludes the need for sulphurization after coupling.

The H-phosphonate method has significantly lower coupling efficiency (*c*. 95%) than phosphoramidites (*c*. 99%), partly due to side-reactions during H-phosphonate diester formation (119). 'Hydrolysable phosphite' approaches to constructing internucleoside H-phosphonate linkages may prove to be more efficient in this regard. This strategy has been investigated using either phosphormorpholidite (126; see also 123 and 148), phosphordiethyl amidite (127), and perfluoropropylphosphite (128) internucleoside linkages. The morpholidite protocol (126) given below, as well as its relatives (127, 128), requires use of commercially unavailable monomers rather than the standard β-cyanoethyl phosphoramidites. This situation and the aforementioned shortcoming of the H-phosphonate method, plus the widespread interest (43, 62–64, 70, 73, 130) in site-specific sulphurization, led to several independent investigations of a substitute for S_8 in stepwise sulphurization of internucleoside β-cyanoethyl phosphite linkages. These include the 'Beaucage reagent,' [3H]1,2-benzodithiol-3-one, 1,1-dioxide (129, 130), van Boom's (121) application of phenacetyl or benzoyl disulphide, via the Schönberg reaction (131), and tetraethylthiuram ('TETD'), $[Et_2NC(S)S\text{-}]_2$ (132). The protocol provided here has been thoroughly tested and employs TETD, a reagent that combines good performance characteristics (*c*. 99% average coupling yield and *c*. 99% sulphurization, at 1 μmol) with the availability of a commodity chemical, which is also soluble and stable in acetonitrile. The 'Beaucage reagent' appears to be faster reacting than TETD, although cycle time is generally not a key issue in the synthesis of phosphorothioates.

3.1.1 Stepwise sulphurization

The following protocol, which is based upon a reported study (132) and extensive unpublished investigations (133), pertains to Applied Biosystems DNA Synthesizer Models 380, 381, 392, and 394 using a modified β-cyanoethyl phosphoramidite cycle that features capping after sulphurization (130) with TETD. (The standard cycle with I_2-H_2O-pyridine involves capping before oxidation.) This sulphurization cycle can be used in conjunction with the standard cycle employing I_2-H_2O-pyridine (see Chapter 1) to synthesize analogues with phosphorothioate linkages wherever desired. Extensions to another manufacturer's equipment, or the use of manual procedures, requires suitable cycle modification(s). Use of TETD with methoxy phosphoramidites is not recommended due to the possibility of demethylation by $Et_2NC(S)S^-$, an anion which is presumably present during the sulphurization reaction. For detailed procedures using 'Beaucage's reagent,' see reference 130.

Protocol 1. Automated stepwise sulphurization with TETD and ß-cyanoethyl phosphoramidite chemistry

Materials

- dA, dG, dC, and dT controlled pore glass (CPG) columns (1 μmol) and β-cyanoethyl phosphoramidite reagents in dry CH_3CN (0.1 M)
- [1H]tetrazole (sublimed, 0.5 M) in dry CH_3CN
- 1-methylimidazole (NMI, redistilled, 2.4 M) in dry tetrahydrofuran (THF)
- Acetic anhydride (1 M) and 2,6-lutidine (1M) in dry THF
- TETD (0.5 M) in CH_3CN
- Trichloroacetic acid (TCA) in CH_2Cl_2, 3%w/w
- Concentrated NH_4OH
- CH_3CN for washing

Method

1. Detritylation: TCA, 60 s, wash with CH_3CN.
2. Coupling: phosphoramidite + tetrazole (1:1 v/v), 15 s, wash with CH_3CN.
3. Sulphurization: TETD, 900 s, wash with CH_3CN.
4. Capping- NMI + Ac_2O (1:1 v/v), 30 s, wash with CH_3CN.
5. Repeat steps 1–4.
6. Cleavage: concentrated NH_4OH, 6 × 60 s (3 ml total).
7. Deprotection: concentrated NH_4OH, 55 °C, 12–18 h.
8. Concentration: add 50 μl Et_3N, vortex and then remove NH_3 or concentrate to dryness in a vacuum centrifuge.
9. Purify the resultant 5′-dimethoxytrityl (DMT) phosphorothioate DNA according to *Protocol 4*.

3.1.2 One-step sulphurization

Protocol 2 refers to automated H-phosphonate chemistry using an Applied Biosystems Model 380B DNA Synthesizer and triethylammonium H-phosphonate monomers, with AdCOCl activation and IPP capping (118). Gaffney and Jones (134) have reported detailed procedures using the DBU salt form of the monomers with pivaloyl chloride on an alternative synthesizer. *Protocol 3* refers to the use of bismorpholidites (126). Note that eight times more monomer is used, relative to the H-phosphonate method. Also note that sulphurization of an H-phosphonate oligomer bound to CPG leads to an oligonucleotide with phosphorothioate linkages throughout the molecule.

Protocol 2. One-step sulphurization following automated H-phosphonate chemistry.

Materials

- dA, dG, dC, and dT CPG columns (1 μmol) and triethylammonium H-phosphonate monomers (0.025 M) in dry pyridine (pyr)-CH_3CN (1:1)
- AdCOCl (0.1 M) in pyr-CH_3CN (1:1)
- IPP (0.05 M) in pyr-CH_3CN (1:1)
- Dichloroacetic acid (DCA) in CH_2Cl_2, 2% w/w
- pyr-CH_3CN (1:1)
- S_8 (0.25 g) in CS_s (2.4 ml)/pyr (2.4 ml)/Et_3N (0.2 ml), freshly prepared
- Concentrated NH_4OH

Method

1. Detritylation: DCA, 100 s, wash with CH_3CN then pyr-CH_3CN.
2. Coupling: H-phosphonate + AdCOCl (1:1 v/v), 40 s
3. Capping: IPP + AdCOCl (1:1 v/v) 30 s then pyr-CH_3CN wash.
4. Repeat steps 1–3.
5. Sulphurization: add solution of S_8, wait 1 h (or up to 24 h), wash thoroughly with CS_2 and then CH_3CN before drying CPG with air or N_2.
6. Cleavage, deprotection, and concentration are the same as in *Protocol 1*.
7. Purify the resultant 5′-DMT phosphorothioate DNA according to *Protocol 4*.

Protocol 3. One-step sulphurization following automated bismorpholidite chemistry

Materials

- dA, dG, dC, and dT CPG columns (1 μmol) and bismorpholidite monomers (126) (0.2 M) in dry CH_3CN.
- [1H]tetrazole (sublimed, 0.5 M) in dry CH_3CN
- 20% v/v H_2O in 0.5 M tetrazole-CH_3CN
- DCA in CH_2Cl_2, 2% w/w
- CH_3CN for washing
- S_8 (saturated) in diisopropylethylamine
- Concentrated NH_4OH

Protocol 3. *Continued*

Method

1. Detritylation: DCA, 60 s, wash with CH_3CN.
2. Coupling: bismorpholidite + tetrazole (1:1 v/v), 180 s wash with CH_3CN.
3. Hydrolysis: H_2O–tetrazole–CH_3CN, 300 s, wash with CH_3CN.
4. Repeat steps 1–3.
5. Sulphurization: add solution of S_8, wait 8 h (or up to 24 h), wash thoroughly with CS_2 or pyridine and then CH_3CN before drying CPG with air or N_2.
6. Cleavage, deprotection, and concentration are the same as in *Protocol 1*.
7. Purify the resultant 5′-DMT phosphorothioate DNA according to *Protocol 4*.

4. Product isolation

Isolation of small amounts (*c.* 5–25 OD_{260} units) of a phosphorothioate analogue via its 5′-DMT derivative using an Applied Biosystems OPC affinity purification cartridge (135) is achieved according to the manufacturer's protocol for DNA, with only one change: increase the proportion of CH_3CN from 20 to 35% CH_3CN in the final elution step following detritylation *in situ* with dilute aqueous trifluoroacetic acid. *Protocol 4* employs reversed-phase HPLC and then precipitation to obtain 5–25 mg quantities of product in its Na^+ form. The purity of this product with regard to truncated sequences reflects, to a large extent, the inherent homogeneity of the 5′-DMT material (135). A detailed account of HPLC of 5′-DMT and 5′-hydroxyl oligonucleotides with one or more phosphorothioate linkages is given elsehwere (71). Ion exchange HPLC has not been reported for oligonucleotides with a high ratio of phosphorothioate to phosphodiester linkages. Reviews of the isolation and purification of oligodeoxynucleotides (136) and analogues (136, 137) may be consulted for further information.

Protocol 4. HPLC of the 5′-DMT derivative and isolation of the Na^+ form of the phosphorothioate oligonucleotide

Materials

- Polystyrene-based HPLC column (e.g. ABI Polypore PRP or Hamilton PRP-1)
- Triethylammonium acetate (TEAA, pH 7, 0.1 M) in H_2O
- CH_3CN

Protocol 4. *Continued*

- Acetic acid in H_2O (80% v/v)
- Ethyl acetate (EtOAc)
- NaCl in H_2O (1 M)
- EtOH

Method

1. Trial analysis: depending on the HPLC column dimensions and UV detector (260 nm) sensitivity, inject an appropriately small, analytical portion of a solution of the crude 5′-DMT material (see *Protocols 1*, *2*, or *3*) using an appropriate flow-rate (*c*. 3 ml/min for 10 × 250 mm column or *c*. 11 ml/mm for 25 × 250 mm column) with the following conditions (linear gradient).

Time (min)	% CH_3CN	% TEAA
0	20	80
10	20	80
40	50	50
50	50	50

The product elutes at a CH_3CN: TEAA ratio of ~40:60 and is often seen as a *c*. 1:1 'doublet' due to the R_p and S_p configurations of the phosphorothioate linkage immediately adjacent to the 5′-DMT group (71).

2. Preparative collection: depending on the capacity of the column, inject an appropriate portion of the sample (*c*. 1 μmol or more for 10 × 250 mm column and *c*. 10 μmol or more for 25 × 250 mm) using an appropriate detector attenuation and/or a less sensitive, longer wavelength (e.g. 280 nm) for detection. Collect the central region of the peak(s), discarding its leading and trailing edges.
3. Concentration: concentrate the collected fraction *in vacuo* using a rotary evaporator and gentle heating (*c*. 45 °C).
4. Detritylation: to the oily residue add enough 80% acetic acid to cause dissolution, which may be somewhat slow (a pale orange colour may be seen due to formation of the DMT cation). After *c*. 10–30 min, rapidly remove the acetic acid *in vacuo* using a rotary evaporator either without heating or with minimal heat.
5. Extraction: dissolve the residue in H_2O, extract DMT-OH twice with an equal volume of EtOAc, and reconcentrate *in vacuo*, after removal of a small analytical portion which should be taken to dryness then and quantified by UV (OD_{260}).

Protocol 4. *Continued*

6. Precipitation: dissolve the residue in 1 M NaCl using *c.* 1 ml/500 OD_{260} units as a rough rule-of-thumb. Cool in an ice bath and add 3 volumes of chilled EtOH with stirring. Collect the precipitate by centrifugation or filtration, and repeat the precipitation twice more.
7. Lyophilization: the final precipitate is diluted to an appropriate volume with H_2O, filtered, and then lysophiliated in the usual manner.
8. Analysis: Evaluate the homogeneity of the final product by use of conventional procedures (e.g. 138) for polyacrylamide gel electrophoresis of oligonucleotides with UV-shadowing or Stains-All (Aldrich) visualization. Conventional procedures (e.g. 138) for oligonucleotide ^{32}P-labelling and nuclease-mediated base-composition analysis may not be efficient when applied to analogues with a high ratio of phosphorothioate: phosphodiester linkages.

5. ^{35}S-Labelling

The following protocol for uniform labelling of a phosphorothioate oligonucleotide throughout its backbone was adapted from a recent publication (75) that should be consulted for additional details. In particular, this article discusses how to conduct serial ^{35}S-labelling reactions of multiple oligomers (with either the same or different sequence) to maximize utilization of the $^{35}S_8$ reagent.

Protocol 5. ^{35}S-Labelling via H-phosphonate chemistry

Materials

- 5′-DMT H-phosphonate oligomer (1 μmol) attached to CPG, as derived by executing *Protocol 2* steps 1–4.
- $^{35}S_8$ (*c.* 5 mCi, *c.* 10–15 mg, Amersham)
- Pyridine
- CS_2
- OPC cartridges (Applied Biosystems)
- TEAA (0.1 M)
- CH_3CN
- Concentrated NH_4OH
- Trifluoroacetic acid (TFA)

Method

1. Transfer air- or N_2-dried CPG from the synthesis column to a glass

Protocol 5. *Continued*

scintillation vial and set aside. Dissolve $^{35}S_8$ in CS_2-pyr (1 ml, 1:1), add Et_3N (50 μl), and transfer to the vial containing to CPG. After 45 min, carefully remove the sulphurization solution (without CPG), which can be reacted with another H-phosphonate sample (75) or discarded.

2. Wash the CPG first with CS_2 (1 ml) and then with CH_3CN (1 ml) before addition of concentrated NH_4OH (1 ml). After 1 h, transfer the solution to a 4 ml screw-cap via and combine with 2 ml of additional concentrated NH_4OH, which is used to remove residual product from the CPG.
3. After heating the concentrated NH_4OH solution overnight at 55 °C, follow the manufacturer-supplied OPC procedure for unmodified oligonucleotides, which involves preparing the cartridge with TEAA, directly loading the sample in NH_4OH solution, elution of capped shorter sequences, detritylation of 5′-DMT material in the cartridge with dilute TFA, and then elution of the product. Note that the last step should employ 35% CH_3CN in H_2O. The product fraction is taken to dryness in a vacuum centrifuge.
4. Dissolve the sample in H_2O and remove small portions for radioactivity quantification by scintillation counting and DNA quantification by UV (OD_{260}).

References

1. Eckstein, F. (1983). *Angew. Chem., Intl. Ed. Engl.*, **22**, 423.
2. Eckstein, F. (1985). *Annu. Rev. Biochem.*, **54**, 367.
3. Eckstein, F., and Gish, G. (1989). *Trends Biochem. Sci.*, **14**, 97.
4. Ts'o, P. O. P., Miller, P., Aurelian, L., Murakami, A., Agris, C., Blake, K. R., Lin, S., Lee, B. L., and Smith, C. C. (1987). *Ann. N.Y. Acad. Sci.*, **507**, 220.
5. Miller, P. S., and Ts'o, P. O. P. (1988). *Annu. Rep. Med. Chem.*, **23**, 295.
6. Hélene, C. and Toulmé, J.-J. (1988). *Gene*, **72**, 51.
7. van der Krol, A. R., Mol, J. N. M., and Stuitje, A. R. (1988). *BioTechniques*, **6**, 958.
8. Stein, C. A. and Cohen, J. S. (1988). *Cancer Res.*, **48**, 2659.
9. Zon, G. (1988). *Pharmacol. Res.*, **5**, 539.
10. Uhlmann, E. and Peyman, A. (1990). *Chem. Rev.*, **90**, 544.
11. Connolly, B., Potter, B. V. L., Eckstein, F., Pingoud, A., and Grotjahn, L. (1984). *Biochemistry*, **23**, 3443.
12. Stec., W. J., Zon, G., Egan, W., and Stec, B. (1984). *J. Am. Chem. Soc.*, **106**, 6077.
13. Koziolkiewicz, M., Niewiarowski, W., Uznanski, B., and Stec, W. J. (1986). *Phosphorus and Sulfur*, **27**, 81.
14. Koziolkiewicz, M., Uznanski, B., and Stec, W. J. (1989). *Nucleosides & Nucleotides*, **8**, 185.

15. Potter, B. V. L. and Eckstein, F. (1984). *J. Biol. Chem.*, **259**, 14243.
16. Putney, S. D., Benkovic, S. J., and Schimmel, P. R. (1981). *Proc. Natl. Acad. Sci. USA*, **78**, 7350.
17. Matzura, H. and Eckstein, F. (1968). *Eur. J. Biochem.*, **3**, 448.
18. Stec, W. J. and Zon, G. (1984). *Tetrahedron Lett.*, **25**, 5275.
19. Usher, D. A., Erenrich, E. S., and Eckstein, F. (1972). *Proc. Natl. Acad. Sci. USA*, **69**, 115.
20. Majumdar, C., Stein, C. A., Cohen, J. S., Broder, S., and Wilson, S. H. (1989). *Biochemistry*, **28**, 1340.
21. Gao, W., Stein, C. A., Cohen, J. S., and Dutschman, G. E. (1989). *J. Biol. Chem.*, **264**, 11521.
22. Spitzer, S. and Eckstein, F. (1988). *Nucleic Acids Res.*, **16**, 11691.
23. Nelson, P. S., Bach, C. T., and Verheyden, J. P. H. (1984). *J. Org. Chem.*, **49**, 2314.
24. Charachon, G., Sobol, R. W., Bisbal, C., Salehzada, T., Silhol, M., Charubala, R., Pfleiderer, W., LeBlue, B., and Suhadolnik, R. J. (1990. *Biochemistry*, **29**, 2550.
25. Montefiori, D., Sobol, Jr, R. W., Li, S. W., Reichenback, N. L., Suhadolnik, R. J., Charubala, R., Pfleiderer, W., Modliszewski, A., Robinson, Jr, W. E., and Mitchell, W. M. (1989). *Proc. Natl. Acad. Sci. USA*, **86**, 7191.
26. Buzayan, J., Feldstein, P. A., Segrelles, C., and Bruening, G. (1988). *Nucleic Acids Res.*, **16**, 4009.
27. van Tol, H., Buzayan, J. M., Feldstein, P. A., Eckstein, F., and Bruening, G. (1990). *Nucleic Acids Res.*, **18**, 1971.
28. Stoller, D., Zon, G., and Pastor, R. W. (1986). *Proc. Natl. Acad. Sci. USA*, **83**, 4469.
29. Milligan, J. F. and Uhlenbeck, O. C. (1989). *Biochemistry*, **28**, 2849.
30. Wu, H., Holcenberg, J. S., Tomich, J., Chen, J., Jones, P. A., Huang, S.-H., and Calame, K. L. (1990). *Gene*, **89**, 2030.
31. Bielinska, A., Shivdasani, R. A., Zhang, L. Q., and Nabel, G. J. (1990). *Science*, **250**, 997.
32. Taylor, J. W., Ott, J., and Eckstein, F. (1985). *Nucleic Acids Res.*, **13**, 8765.
33. de Clercq, E., Eckstein, F., Sternbach, H., and Merigan, T. C. (1970). *Virology*, **42**, 421.
34. Black, D. R., Eckstein, F., and DeClerq, E. (1973). *Antimicrob. Agents Chemother.*, **3**, 198.
35. Agrawal, S., Goodchild, J., Civeira, M. P., Thornton, A. H., Sarin, P. S., and Zamecnik, P. C. (1988). *Proc. Natl. Acad. Sci. USA*, **85**, 7079.
36. Matsukura, M., Shinozuka, K., Zon, G., Mitsuya, H., Reitz, M., Cohen, J. S., and Broder, S. (1987). *Proc. Natl. Acad. Sci. USA*, **84**, 7706.
37. Stein, C. A., Matsukura, M., Subasinghe, C., Broder, S., and Cohen, J. S. (1989). *AIDS Research and Human Retroviruses*, **5**, 1639.
38. Letsinger, R. L., Zhang, G., Sun, D. K., Ikeuchi, T., and Sarin, P. S. (1989). *Proc. Natl. Acad. Sci. USA*, **86**, 6553.
39. Rayner, B., Matsukura, M., Morvan, F., Cohen, J. S., and Imbach, J-L. (1990). *C. R. Acad. Sci. Paris*, **310**, Ser III, 61.
40. Covey, J., personal communication.

41. Maddox, J. (1989). *Nature*, **342**, 609.
42. Sarver, N., Cantin, E. M., Chang, P. S., Zaia, J. A., Ladne, P. A., Stephens, D. A., and Rossi, J. J. (1990). *Science*, **247**, 1222.
43. Marcus-Sekura, C. J., Woerner, A. M., Shinozuka, K., Zon, G., and Quinnan, Jr, G. V. (1987). *Nucleic Acids Res.*, **15**, 5749.
44. Zamecnik, P. C., Goodchild, J., Taguchi, Y., and Sarin, P. S. (1986). *Proc. Natl. Acad. Sci. USA*, **83**, 4143.
45. Gao, W.-Y., Hanes, R. N., Vazquez-Padua, M. A., Stein, C. A., Cohen, J. S., and Cheng, Y.-C. (1990). *Antimicrob. Agents Chemother.*, **34**, 808.
46. Matsukura, M., Zon, G., Shinozuka, K., Robert-Guroff, M., Shimada, T., Stein, C. A., Mitsuya, H., Wong-Staal, F., Cohen, J. S., and Broder, S. (1989). *Proc. Natl. Acad. Sci. USA*, **86**, 4244.
47. Agrawal, S., Ikeuchi, T., Sun, D., Sarin, P. S., Konopka, A., Maizel, J., and Zamecnik, P. (1989). *Proc. Natl. Acad. Sci. USA*, **86**, 7790.
48. Kinchington, D., Galpin, S., Jaroszewski, J., Subasinghe, C., and Cohen, J. S., (1990). *Antiviral Res.* (Suppl. 1), Abstract No. 81.
49. Kemal, O., Brown, T., Burgess, S., Bishop, J. D., and Leigh-Brown, A. J. (1990). *9th International Round Table Nucleosides, Nucleotides & Their Biological Applications*, Uppsala, Sweden, Abstract No. 114.
50. Mag, M., Muth, J., Lucking, S., Biesert, L., and Engels, J. (1990). *Nucleosides & Nucleotides*, manuscript submitted for publication.
51. Mag, M., Muth, J., Lucking, S., Biesert, L., and Engels, J. (1990). *Biol. Chem. Hoppe Seyler*, **371**, 801.
52. Shibahara, S., Mukai, S., Morisawa, H., Nakashima, H., Kobayashi, S., and Yamamoto, N. (1989). *Nucleic Acids Res.*, **17**, 239.
53. Ceruzzi, M. and Draper, K. (1989). *Nucleotides & Nucleotides*, **8**, 815.
54. Shea, R. G., Marsters, J. C., and Bischofberger, N. (1990). *Nucleic Acids Res.*, **18**, 3777.
55. Gehron-Robey, P. (1988) *ASCB/ASBMB* Meeting, August 1988. Abstract No. A27.
56. Loke, S. L., Stein, C., Zhang, X., Avigan, M., Cohen, J., and Neckers, L. M. (1988). *Curr. Top. Microbiol. Immunol.*, **141**, 282.
57. Chang, E. H., Yu, Z., Shinozuka, K., Zon, G., Wilson, W. D., and Stekowska, A. (1989). *Anti-Cancer Drug Design*, **4**, 221.
58. Manson, J., Brown, T., and Duff, G. (1990). *Lymphokine Res*, **9**, 35.
59. Cazenave, C., Stein, C. A., Loreau, N., Thuong, N. T., Neckers, L. M., Subasinghe, C., Hélene, C., Cohen, J. S., and Toulmé, J.-J. (1989). *Nucleic Acids Res.*, **17**, 4255.
60. Furdon, P. J., Dominiski, Z., and Kole, R. (1989). *Nucleic Acids Res.*, **17**, 9193.
61. Woolf, T. M., Jennings, C. G. B., Rebagliati, M., and Melton, D. A. (1990). *Nucleic Acids Res.*, **18**, 1763.
62. Agrawal, S., Mayrand, S. H., Zamecnik, P. C., and Pederson, T. (1990). *Proc. Natl. Acad. Sci. USA*, **87**, 1401.
63. Baker, C., Holland, D., Edge, M., and Colman, A. (1990). *Nucleic Acids Res.*, **18**, 3537.
64. Dagle, J. M., Walder, J. A., and Weeks, D. L. (1990). *Nucleic Acids Res.*, **18**, 4751.

65. Reed, J. C., Stein, C., Subasinghe, C., Haldar, S., Croce, C. M., Yum, S., and Cohen, J. (1990). *Cancer Res.*, **50**, 6565.
66. Cosstick, R. and Eckstein, F. (1985). *Biochemistry*, **24**, 3630.
67. Latimer, L. J. P., Hampel, K., and Lee, J. S. (1989). *Nucleic Acids Res.*, **17**, 1549.
68. Iyer, R. P., Uznanski, B., Boal, J., Storm, C., Egan, W., Matsukura, M., Broder, S., Zon, G., Wilk, A., Koziolkiewicz, M., and Stec, W. J. (1990). *Nucleic Acids Res.*, **18**, 2855.
69. LaPlance, L. A., James, T. L., Powell, C., Wilson, W. D., Uznanski, B., Stec, W. J., Summers, M. F., and Zon, G. (1986). *Nucleic Acids Res.*, **14**, 9081.
70. Stein, C. A., Subasinghe, C., Shinozuka, K., and Cohen, J. S. (1988). *Nucleic Acids Res.*, **16**, 3209.
71. Stec, W. J., Zon, G., and Uznanski, B. (1985). *J. Chromatogr.*, **326**, 263.
72. Gish, G. and Eckstein, G. (1988). *Science*, **240**, 1520.
73. Fidanza, J. A. and McLaughlin, L. W. (1989). *J. Am. Chem. Soc.*, **111**, 9117.
74. Hodges, R. R. Conway, N. E., and McLaughlin, L. W. (1989). *Biochemistry*, **28**, 261.
75. Stein, C. A., Iversen, P. L., Subasinghe, C., Cohen, J. S., Stec, W. J., and Zon, G. (1990). *Anal. Biochem.*, **188**, 11.
76. Strothkamp, K. G., and Lippard, S. J. (1976). *Proc. Natl. Acad. Sci. USA*, **73**, 2536.
77. Marti, G., Egan, W., Noguchi, P., Zon, G., Iversen, P., Meyer, A., Matsukura, M., and Broder, S., manuscript in preparation.
78. Sayers, J. R., Schmidt, W., and Eckstein, F. (1988). *Nucleic Acids Res.*, **16**, 791.
79. Fritz, H.-J., Hohlmaier, J., Kramer, W., Ohmayer, A., and Wippler, J. (1988). *Nucleic Acids Res.*, **16**, 6987.
80. Burgers, P. M. J., and Eckstein, F. (1979). *Biochemistry*, **18**, 592.
81. Uznanski, B., Niewiarowski, W., and Stec, W. J. (1982). *Tetrahedron Lett.*, **23**, 4289.
82. Connolly, B. A., Eckstein, F., and Füldner, H. H. (1982). *J. Biol. Chem.*, **257**, 3382.
83. Potter, B. V. L., Connolly, B. A., and Eckstein, F. (1983). *Biochemistry*, **22**, 1369.
84. Gallo, K. A., Shao, K.-L., Phillips, L. R., Regan, J. B., Koziolkiewicz, M., Uznanski, B., Stec., W. J., Zon, G. (1986). *Nucleic Acids Res.*, **14**, 7405.
85. Hamblin, M. R., Cummins, J. H., and Potter, B. V. L. (1987). *Biochem. J.*, **241**, 827.
86. Campbell, J. M., Bacon, T. A., and Wickstrom. E. (1990). *J. Biochem. Biophys. Methods*, **2**, 259.
87. Kitts, P. A. and Nash, H. A. (1988). *Nucleic Acids Res.*, **16**, 6839.
88. Griffiths, A. D., Potter, B. V. L., and Eperon, I. C. (1988). *J. Biol. Chem.*, **263**, 12295.
89. Eckstein, F. and Jovin, T. M. (1983). *Biochemistry*, **22**, 4546.
90. Suggs, J. W. and Taylor, D. A. (1985). *Nucleic Acids Res.*, **13**, 5707.
91. Braun, R. P. and Lee, J. S. (1988). *J. Immunol*, **141**, 2084.
92. Yang, W., Hendrickson, W. A., Crouch, R. J., and Satow, Y. (1990). *Science*, **249**, 1398.

93. Andrus, A., Geiser, T., and Zon, G. (1989). *Nucleosides & Nucleotides*, **8**, 967.
94. Geiser, T. (1990). *Ann. N.Y. Acad. Sci.*, **616**, 173.
95. Lesnikowski, Z. J. and Sibinska, A. (1986). *Tetrahedon*, **42**, 5025.
96. Lesnikowski, Z. J. and Jaworska, M. M. (1989). *Tetrahedon Lett.*, **30**, 3821.
97. Fujii, M., Ozaki, K., Kume, A., Sekine, M., and Hata, T. (1986). *Tetrahedron Lett.*, **26**, 935.
98. Beiter, A. H., and Pfleiderer, W. (1989). *Synthesis*, 497.
99. de Vroom, E., Fidder, A., Saris, C. P., van der Marel, G. A., and van Boom, J. H. (1987). *Nucleic Acids Res.*, **15**, 9933.
100. Seela, F. and Kretschmer, U. (1990). *9th International Round Table Nucleosides, Nucleotides & Their Biological Applications*, Uppsala, Sweden. Abstract No. 159.
101. Cosstick, R. and Williams, D. M. (1987). *Nucleic Acids Res.*, **15**, 9921.
102. Fujii, M., Ozaki, K., Sekine, M., and Hata, T. (1987). *Tetrahedron*, **43**, 3395.
103. Saxena, S. K. and Ackerman, E. J. (1990). *J. Biol. Chem.*, **265**, 3263.
104. Haseltine, W. A. and Wong-Staal, F. (1988). *Sci. Am.*, **259**, 52.
105. Chu, B. C., and Orgel, L. E. (1990). *Nucleic Acids Res.*, **18**, 5163.
106. Eckstein, F. (1967). *Tetrahedron Lett.*, 1157.
107. Eckstein, F. (1967). *Tetrahedron Lett.*, 3495.
108. Malkievicz, A. and Smrt, J. (1973). *Tetrahedron Lett.*, 491.
109. Kemal, O., Reese, C. B., and Serafinowska, H. T. (1983). *J. Chem. Soc. Chem. Commun.*, **591**.
110. Marugg, J. E., van den Bergh, C., Tromp, M., van der Marel, G. A., van Zoest, W. J., and van Boom, J. H. (1984). *Nucleic Acids Res.*, **12**, 9095.
111. Burgers, P. M. J., and Eckstein, F. (1978). *Tetrahedron Lett.*, 3835.
112. Marlier, J. F., and Benkovic, S. J. (1980). *Tetrahedron Lett.*, **21**, 1121.
113. Ott, J. and Eckstein, F. (1987). *Biochemistry*, **26**, 8237.
114. Matsukura, M., Zon, G., Shinozuka, K., Stein, C. S., Mitsuya, H., Cohen, J. S., and Broder, S. (1988). *Gene*, **72**, 343.
115. Stein, C. A., personal communication.
116. Froehler, B. C., and Matteucci, M. D. (1986). *Tetrahedron Lett.*, **27**, 469.
117. Froehler, B. C. (1986). *Tetrahedron Lett.*, **27**, 5575.
118. Andrus, A., Efcavitch, J. W., McBride, L., and Giusti, B. (1988). *Tetrahedron Lett.*, **29**, 861.
119. Regberg, T., Stawinski, J., and Strömberg, R. (1988). *Nucleosides & Nucleotides*, **7**, 23.
120. Sakatsume, O., Yamane, H., Takaku, H., and Yamamoto, N. (1990). *Nucleic Acids Res.*, **18**, 3327.
121. Kamer, P. C. J., Roelen, H. C. P. F., van den Elst, H., van der Marel, G. A., and van Boom, J. H. (1989). *Tetrahedron Lett.*, **30**, 6757.
122. Stawinski, J., Thelin, M., and von Stedingk, E. (1990). *9th International Round Table Nucleosides, Nucleotides & Their Biological Applications*. Uppsala, Sweden, Abstract No. 168.
123. Nielsen, J., Brill, W. K.-D., and Caruthers, M. H. (1988). *Tetrahedron Lett.*, **29**, 2911.
124. Porritt, G. M., and Reese, C. B. (1989). *Tetrahedron Lett.*, **30**, 4713.
125. Stawinski, J., Thelin, M., and Zain, R. (1989). *Tetrahedron Lett.*, **30**, 2157.

126. Uznanski, B., Wilk, A., and Stec, W. J. (1987). *Tetrahedron Lett.*, **28**, 3401.
127. Yamana, K., Nishijima, Y., OKa, A., and Shimidzu, T. (1989). *Tetrahedron Lett.*, **45**, 4135.
128. Watanabe, T., Sato, H., and Takaku, H. (1989). *J. Am. Chem. Soc.*, **111**, 3437.
129. Iyer, R. P., Egan, W., Regan, J. B., and Beaucage, S. L. (1990). *J. Am. Chem. Soc.*, **112**, 1253.
130. Iyer, R. P., Phillips, L. R., Egan, W., Regan, J. B., and Beaucage, S. L. (1990). *J. Org. Chem*, **55**, 4693.
131. Schönberg, A. (1935). *Chem. Ber.*, **68**, 163.
132. Andrus, A., Vu, H., and Hirschbeim, B. (1991). *Tetrahedron Lett.*, manuscript submitted.
133. Hirschbein, B., Fearon, K, and Bergot, B. J., patent applied for.
134. Gaffney, B. L. and Jones, R. A. (1988). *Tetrahedron Lett.*, **29**, 2619.
135. McBride, L. J., McCollum, C., Davidson, S., Efcavitch, J. W., Andrus, A., and Lombardi, S. F. (1988). *BioTechniques*, **6**, 362.
136. Zon, G. and Thompson, J. A. (1986). *BioChromatography*, **1**, 22.
137. Zon, G. (1990). In *HPLC in Biotechnology* (ed. W. S. Hancock), pp. 301–97. John Wiley & Sons, New York.
138. Applied Biosystems DNA Synthesizer Model 380/381 User Bulletin Issue No. 13. Revised, April, 1987.
139. Leiter, J. M. E., Agrawal, S., Palese, P., and Zamecnik, P. C. (1990). *Proc. Natl. Acad. Sci. USA*, **87**, 3430.
140. Gebhart, F. (1988). *Gen. Engr. News*, July/August, 6.
141. Ohtsuka, E., Ikehara, M., and Soll, D. (1982). *Nucleic Acids Res.*, **10**, 6553.
142. Kamaike, K., Hasegawa, Y., Masuda, I., Ishido, Y., Watanabe, K., Hirao, I., and Miura, K. (1990). *Tetrahedron*, **46**, 163.
143. Egan, W., personal communication.
144. Mitsuya, H. and Broder, S., personal communication.
145. Dudding, L. R., Harington, A., and Mizrahi, V. (1990). *Biochem. Biophys. Res. Commun.*, **167**, 244.
146. Lesnikowski, Z., Smrt, J., Stec, W. J., and Zielinski, W. (1978). *Bull. Acad. Polon. Sci.*, **26**, 661.
147. Wilk, A., Uznanski, B., and Stec, W. J. (1990). *9th International Round Table Nucleosides, Nucleotides & Their Biological Application*. Uppsala, Sweden. Abstract No. 200.
148. Ozaki, H., Yamoto, S., Maikuma, S., Honda, K., and Shimidzu, T. (1989). *Bull. Chem. Soc. Jpn.*, **62**, 3869.
149. Agrawal, S. and Zamecnik, P. C. (1990). *Nucleic Acids Res.*, **18**, 5419.

5

Synthesis of oligonucleotide phosphorodithioates

GRAHAM BEATON, DOUGLAS DELLINGER, WILLIAM S. MARSHALL, and MARVIN H. CARUTHERS

1. Introduction

Over the past several years, the development of high yielding, rapid methods for synthesizing DNA (1–3) have led to a large number of applications for oligonucleotides (4). These include its use for cloning and synthesizing genes (5), as primers for sequencing DNA and various PCR applications (6), mutagenesis of genes in a site-specific manner (7), examination of how nucleic acids interact with proteins (8), and for studies on nucleic acid structure (9). Mainly because of these results, many have discovered that synthetic DNA needs to be further modified so that it resists hydrolysis by nucleases and contains reporter groups, haptens, cross-linking reagents, sugars, steroids, and peptides. Invariably such interest has stimulated the development of new analogues and new chemistries for synthesizing these derivatives. In this chapter, the chemical synthesis of a potentially very useful analogue for many of these applications, which we call dithioate DNA, will be discussed.

An inspection of DNA or RNA suggests several potential sites for analogue introduction. These include the sugar, base, and phosphate portions of these molecules and, additionally, the ends. The choice of analogues and methods for introducing these changes are the subject of this volume. Except for certain obvious exclusions, we favour introducing analogues at the phosphate internucleotide linkage (10–12). A particularly attractive reason is that modifications at phosphorus are positioned on the extremities of DNA or RNA duplexes and thus are accessible to solvent, reagents, and other macromolecules such as nucleic acid binding proteins, cell receptors, enzymes, or other proteins. Reporter groups and the like at other sites such as the sugars and bases can interfere with polynucleotide conformations and lead to masking of a reporter group's function by intercalation. An added feature as well is that derivatization at phosphorus is sequence non-specific.

In order to have versatility through base modification, different chemistries for each of the bases will be required.

An additional consideration is resistance toward degradation by nucleases which is a requirement for any analogue that is to be used as a potential therapeutic. Among the derivatives so far examined are the methylphosphonate, phosphoramidate, and phosphorothioate internucleotide linkages (13) (see Chapters 4 and 9). These derivatives are, however, all phosphorus chiral which inevitably leads to a large number of non-resolvable diastereomeric oligomers having variable biophysical, biochemical, and biological properties. Recent research has therefore focused on the development of stereo-selective approaches for synthesizing these derivatives (14); or, alternatively, the development of achiral analogues (10). A particularly attractive analogue in the latter category is the deoxyoligonucleotide phosphorodithioate (dithioate DNA) which has deoxynucleoside-OPS_2O-deoxynucleoside internucleotide linkages. It is isostructural and isopolar with the natural phosphate diester linkage, stable toward enzymatic and chemical hydrolysis, and can be further modified to contain reporter groups at phosphorus. Although the initial synthetic pathway for preparing this analogue has only recently been developed (15), its attractiveness for a large number of applications has stimulated considerable additional effort (16–27). In this laboratory several methods have been investigated for its synthesis. These include the use of deoxynucleoside diamidites, thioamidites, H-phosphonodithioates, and dithiophosphate triesters as synthons. Our research to date suggests that deoxynucleoside phosphorothioamidites are the most attractive and versatile. Here we will describe the use of these synthons for preparing dithioate DNA.

2. Outline of the chemistry

Although the chemistry for synthesizing phosphorodithioate DNA from deoxynucleoside 3′-phosphorothioamidites continues to evolve, our current methodology is summarized in *Figure 1* and *Table 1*. Briefly, deoxynucleoside 3′-phosphorothioamidite synthons (**4**) are prepared via a one-pot, two-step procedure from a suitably protected deoxynucleoside (**1**), *tris*(pyrrolidino)-phosphine (**2**), and tetrazole. The resulting deoxynucleoside diamidite (**3**) is converted without isolation to the deoxynucleoside phosphorothioamidite by addition of 2,4-dichlorobenzyl-mercaptan. Following an aqueous work-up and drying over sodium sulphate, these synthons are isolated by precipitation and stored as dry powders under an inert gas atmosphere without decomposition. Several alternative phosphines have been tried, with varying degrees of success, for synthesizing **4**. In some cases, such as with *bis*(pyrrolidino)-S-(4-chlorobenzyl)phosphine or *bis*(N,N-dimethylamino)-S-(4-chlorobenzyl) phosphine, the phosphine could not be purified by fractional distillation and thus complex mixtures of reaction products were obtained. Other bifunctional phosphitylating agents such as the dichloro(pyrrolidino)phosphine or dichloro-N,N-dimethylaminophosphine proved unsuccessful because symmetrical deoxydinucleoside thiophosphites were the predominant product under all reaction conditions tested (low temperatures, order of additions and mole ratios of reagents). A more successful route [Brill, W. K.-D., Nielsen, J. and Caruthers, M. H., (1991). *J. Am. Chem. Soc.*, **113**, 3972.] was based upon the use of *bis*(pyrrolidino)chlorophosphine and *bis*-(N,N-dimethylamino)chlorophosphine. The synthesis of these phosphines in pure form was possible and their use for preparing **4** involved a one pot, two step procedure analogous to the pathway shown in *Figure 1*. In this case, the amine hydrochloride from the first step catalysed the formation of **4** in the presence of a mercaptan. However, certain problems with this approach have led us to the *tris*-aminophosphines as phosphitylating reagents. For example, if amine hydrochloride is not removed entirely from the reaction product, decomposition occurs and **4** hydrolyses or disproportionates to **3** and the *bis*-mercaptylphosphite upon dissolution in solvent. Several side-products, including the phosphorothioate can then be detected during oligonucleotide synthesis.

Synthesis of dithioate DNA begins by treating a 5′-O-dimethoxytrityldeoxynucleoside linked to a silica support **5** with 3% trichloroacetic acid to yield **6**, a compound having a free 5′-hydroxyl group accessible for polynucleotide synthesis (see *Table 1* for a more complete outline of these

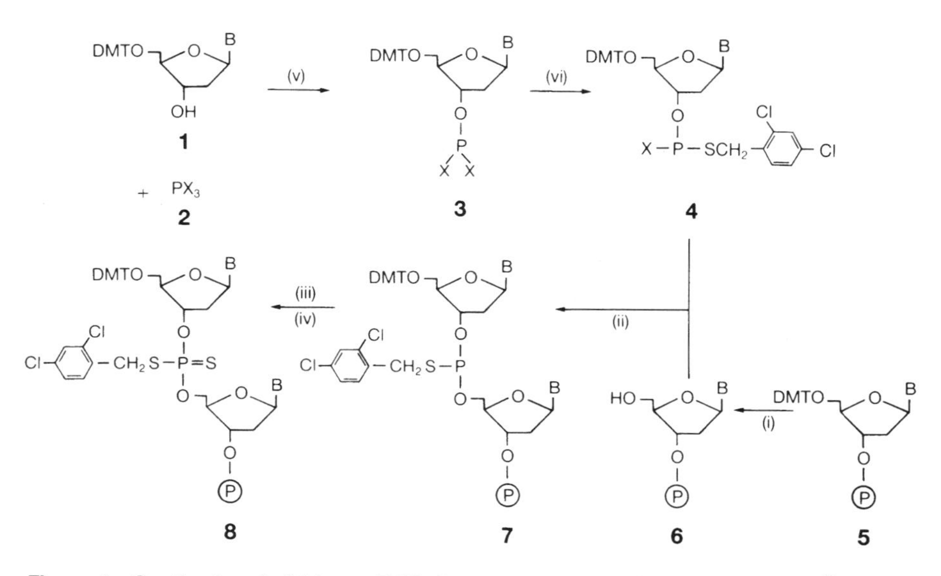

Figure 1. Synthesis of dithioate DNA from thioamidites. X = pyrrolidino; Ⓟ, silica support; DMT, dimethoxytrityl. Reagents: (i) trichloroacetic acid; (ii) tetrazole, (iii) sulphur, (iv) acetic anhydride, (v) tetrazole, (vi) 2,4-dichlorobenzylmercaptan.

Table 1. Chemical steps for synthesis of dithioate DNA on a solid support[a]

Step		Reagent or Solvent[b]	Purpose	Time (min)
(i)	a.	Trichloroacetic acid in CH_2Cl_2 (3%, w/v)	Detritylation	0.50
	b.	CH_2Cl_2	Wash	0.50
	c.	Acetonitrile	Wash	0.50
	d.	Dry acetonitrile	Wash	0.50
(ii)	a.	Activated nucleotide in acetonitrile[a]	Add nucleotide	0.75
	b.	Repeat step a	Complete nucleotide addition	1.50
(iii)	a.	Sulphur in CS_2:pyridine: TEA (95:95:10; v/v/v)[d]	Oxidation	1.00
	b.	CS_2	Wash	0.50
	c.	CH_3OH	Wash	0.50
	d.	CH_2Cl_2	Wash	0.50
(iv)	a.	NMI:THF (30:70; v/v)[e] acetic anhydride: lutidine: THF (2:2:15; v/v/v)	Capping reaction	0.50
	b.	CH_2Cl_2	Wash	0.50

[a] See *Figure 1* for an explanation of the various steps i–iv. and *Tables 2* and *3* for a programmed cycle useful for machine synthesis.

[b] Multiple washes with the same solvent are possible.

[c] For each micromole of deoxynucleoside attached to silica, 0.48 M tetrazole (0.125 ml) and 0.15 M deoxynucleoside phosphorothioamidite (0.125 ml) are premixed in acetonitrile. During this coupling step, activated nucleotide and tetrazole are flushed from the lines leading into the reaction chamber. This procedure reduces contaminating phosphorothioate internucleotide linkages.

[d] 5% sulphur by weight in CS_2:pyridine: TEA (95:95:10; v/v/v); TEA, triethylamine.

[e] NMI, N-methylimidazole; THF, tetrahydrofuran.

steps). Deoxynucleoside 3′-phosphorothioamidites are then activated with tetrazole and condensed with **6** to yield a thiophosphite triester **7**. This step is followed by sulphurization using elemental sulphur to yield the protected phosphorodithioate derivative **8**, capping or acylating unreactive silica-linked deoxynucleoside with acetic anhydride, and detritylation with trichloroacetic acid. Further repetitions of this cycle using either deoxynucleoside phosphorothioamidites or deoxynucleoside phosphoramidites and tetrazole as an activator yield oligodeoxynucleotides having normal phosphate diester and phosphorodithioate diester linkages in any combination. Condensation yields of 96–99% per cycle (based upon dimethoxytrityl cation released during detritylation) and 97–98% dithioate per linkage (the remaining 2–3% is phosphorothioate) have been obtained. The major challenge with this approach is to generate high yields of the thiophosphite and then the phosphorodithioate linkage under conditions where side-products do not form. These criteria translate into finding the right balance from among the protecting group on sulphur, the amino group that is part of the thioamidite, and the activating reagent. Thus although the 2,4-dichlorobenzyl group suffers from a requirement for thiophenol during deprotection, it is currently preferred as others such as β-cyanoethyl lead to a higher percentage of phosphorothioate internucleotide linkages. Similarly when X is N,N-diisopropylamino, activation with tetrazole occurs very slowly and stronger acids, which lead to side products, must be used.

3. Synthesis of protected deoxynucleoside-3′-O-pyrrolidino-S-(2,4-dichlorobenzyl)-phosphorothioamidites

3.1 Preparation of *tris*(pyrrolidino)phosphine

This *tris*-aminophosphine is conveniently synthesized from the reaction of excess pyrrolidine with phosphorus trichloride. The reaction is moisture-sensitive, highly exothermic, and generates hydrogen chloride. Exercise extreme caution, use a fume hood and protective eyewear and clothing at all times. The following equipment is required:

- a 3 neck 2 litre round-bottomed flask (neck size 24)
- a 1 litre round-bottomed flask (neck size 24)
- 500 ml Erlenmeyer flask
- a 250 ml quickfit addition funnel
- 2 Teflon stirrer bars
- a 100 ml round-bottomed flask (neck size 14)
- a short path vacuum jacketed micro-distilling head†
- a 3 way cow receiver†
- 3 or 4 appropriate round-buttomed receiver flasks (neck size 14)†
- an appropriate thermometer†
- glass syringes and steel needles
- an oil bath
- a magnetic stirrer hotplate
- a large dish (for an ice bath)
- a laboratory jack
- one 150 ml sintered glass funnel
- a vented argon manifold
- a 2-stage vacuum pump and efficient vacuum line

† This specialized apparatus may be obtained from Ace Glassware (Catalogue 1000, microdistilling head: page 343, catalogue no. 9317–42 or 9317–52; cow receiver: page 89, catalogue no. 9316–04). Use a single neck flask without a gas inlet tube for the crude phosphine to achieve a better vacuum. This is acceptable provided the crude phosphine is efficiently stirred.

Protocol 1. Preparation of *tris*(pyrrolidino)phosphine

1. Dry glassware in an oven overnight. Cool in a desiccator.
2. Distil pyrrolidine (b.p. 87–88 °C) from calcium hydride prior to use.
3. Set up the 2 litre flask on the stirplate. Insert stirrer bar, attach addition funnel and cover with serum caps. Flush with argon. Ensure that the reaction vessel is vented at all times using argon and one of the flask necks as an outlet.
4. Add 500 g anhydrous ether.
5. While flushing with argon, slowly add phosphorus trichloride (25 g, 16 ml, 0.18 mol) using a syringe.
6. Cool flask in an ice bath.
7. Dilute the pyrrolidine (116.5 g, 137 ml, 1.64 mol) with ether (125 ml) in the Erlenmeyer flask. Transfer to the addition funnel.
8. Add the pyrrolidine *dropwise* to the stirred phosphorus trichloride solution over a 30 min period.
9. Remove the ice bath and stir for one hour.
10. Filter solids from the mixture into a 1 litre round-bottomed flask under an argon atmosphere. Concentrate under reduced pressure to a yellow oil.
11. Transfer the oil to a 100 ml round-bottomed flask containing a dry stirrer bar. Attach the distillation head, cow adaptor and receiver flasks and set up on the stirrer hotplate with the oil bath.
12. Connect the distillation apparatus to a 2-stage vacuum pump with inline dry ice to liquid nitrogen traps. A vacuum of 0.1 mm Hg is necessary for efficient distillation of this phosphine. Attempting to distil at higher pressures (higher temperatures) will promote polymerization of the phosphine and greatly reduce yield.
13. Distil the phosphine by heating the oil bath. Retain fractions which distil at 90–100 °C.
14. Characterize the phosphine by proton and phosphorus NMR (^{1}H NMR ($CDCl_3$): multiplets at 1.8 and 3.2 p.p.m; ^{31}P NMR ($CDCl_3$): singlet at 103 p.p.m.).

3.2 Preparation of deoxyribonucleoside-3′-O-pyrrolidino-S-(2,4-dichlorobenzyl)phosphorothioamidites

This one-pot method for the synthesis of phosphorothioamidite synthons is moisture sensitive. Consequently all solvents, reagents and apparatus for the thioamidite formation must be anhydrous. Solvents and reagents used in these procedures are hazardous such that all reactions and preparation of reagents relating to this analogue chemistry must be performed in a fume hood. *Wear appropriate eye protection and protective clothing at all times.*

Protocol 2. Preparation of deoxyribonucleoside-3′-O-pyrrolidino-S-(2,4-dichlorobenzyl)phosphorothioamidites

1. Distil dichloromethane (b.p. 40 °C) from calcium hydride prior to use.
2. Distil acetonitrile (b.p. 82 °C) first from phosphorus pentoxide, then from calcium hydride.
3. 1H Tetrazole can be further purified by sublimation.
4. Dry glass syringes and flasks in an oven, then cool in a desiccator. Disposable syringes are acceptable for use in the initial phosphitylation step and the accompanying formation of the phosphorothioamidite.
5. Allow the phosphitylating reagent (stored at −18 °C) to warm to room temperature before use. Protect reagents from moisture and oxygen using argon. This can be achieved via argon-filled rubber balloons or preferably using a gas manifold (see *Figure 2*: Set up for phosphitylation step).
6. Weigh sufficient 5′-O-dimethoxytrityldeoxyribonucleoside (4 mmol) into a dry 100 ml round-bottomed flask. Cover the flask with a rubber serum cap (Aldrich Z10, 145–1 for size 24 flask necks). Pierce cap with an 18-gauge needle and dry overnight in a desiccator attached to a vaccum pump.
7. The next day, slowly introduce argon into the desiccator and flask.
8. Weigh sufficient tetrazole for a 0.5 M solution into an amber glass borosilicate vial (typically when preparing all four monomers, weigh out 210 mg, 3 mmol). Cover with a rubber cap (Aldrich Z10,079–9) and flush with argon in the usual way. After 10 min, add dry acetonitrile (6 ml) and dissolve using a heat gun or sonicator.
9. To a second 100 ml round-bottomed flask, add a magnetic stirring bar and cover with a serum cap. Place on a magnetic stirring plate and flush with argon.
10. Dissolve the protected deoxyribonucleoside in dry dichloromethane (50 ml, 70 ml for N^6-benzoyl-5′-O-dimethoxytrityl-2′-deoxyadenosine (28) due to solubility). Connect this flask to an argon line and the second flask using a cannula (*Figure 2*). Vent the second flask with an 18-gauge needle.
11. Into the second flask inject *tris*(pyrrolidino)phosphine (1.02 ml, 4.7 mmol) and tetrazole solution (1 ml, 0.5 mmol). Stir the mixture and push the end of the cannula into the deoxyribonucleoside solution such

Protocol 2. *Continued*

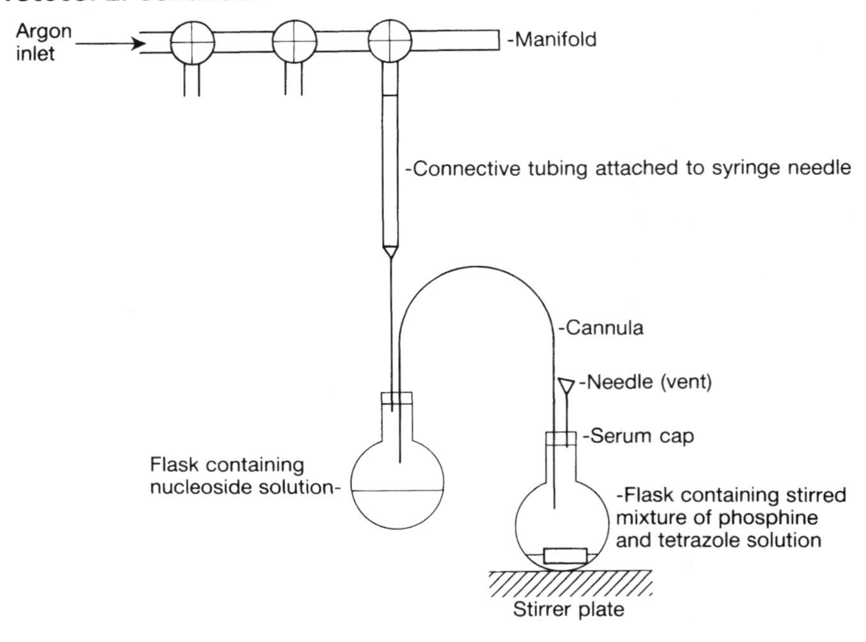

Figure 2. Set up for phosphitylation.

that it transfers to the second flask. When the transfer is complete, add dry dichloromethane (5 ml) to the first flask, rinse and transfer as before to ensure total addition of the nucleoside to the reaction. Stir for 15 min.

12. The course of the reaction can be monitored by thin layer chromatography (TLC) using ethyl acetate:dichloromethane:triethylamine (45:45:10) as solvent system. Pre-elute an aluminium-backed silica gel TLC plate with the solvent system described and allow to dry. Apply separately a sample of protected deoxyribonucleoside and the reaction mixture to the plate and redevelop. Look at the plate under short-wave UV light (*use eye protection*) to determine if all the deoxyribonucleoside has reacted (see *Figure 3*: Typical TLC of the phosphitylation of 5′-O-dimethoxytritylthymidine). The presence of the dimethoxytrityl group as part of a compound can be confirmed by spraying the plate with 20% sulphuric acid (orange coloration of the spots is observed). The least polar spot is the deoxyribonucleoside-3′-O-*bis*(pyrrolidino)phosphoramidite where the position of protected deoxyribonucleoside is determined by the marker applied. Other components in the mixture are the 3′-O-H-phosphonate and general decomposition (unknown) baseline material). It is paramount that the phosphitylation reaction is complete,

Protocol 2. *Continued*

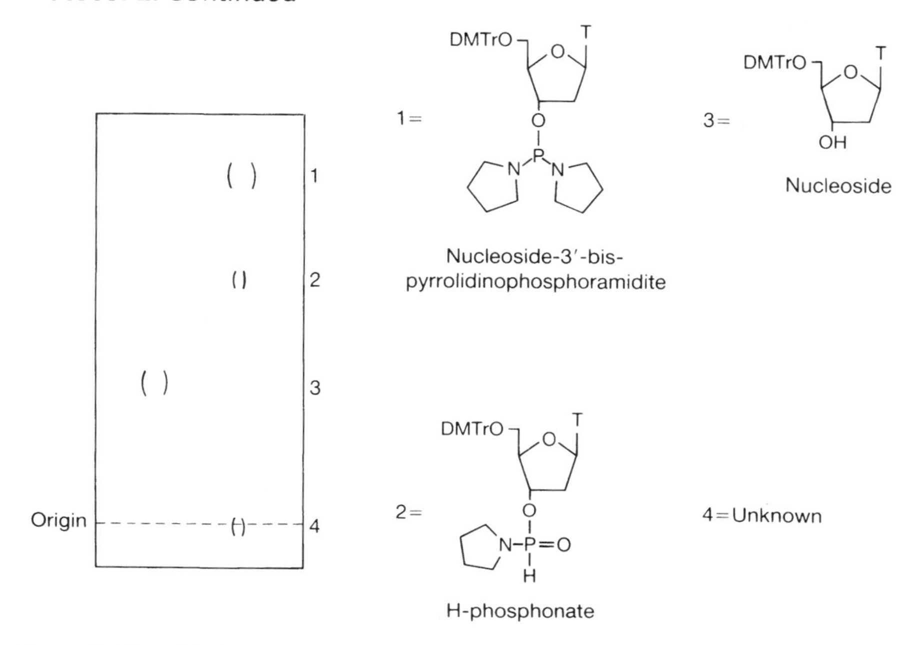

Figure 3. Phosphitylation of 5′-O-dimethoxytritylthymidine (TLC analysis).

since unreacted deoxyribonucleoside will interfere with phosphorothioamidite coupling during DNA synthesis. If the reaction is incomplete, this may be a function of the phosphine purity, in which case add another portion (0.1 ml). Alternatively, if large amounts of polar material and H-phosphonates are observed, the problem is most likely due to moisture in one of the reagents.

13. When the reaction is complete, inject 2,4-dichlorobenzylmercaptan (1.6 ml, 9.9 mmol). Destroy all mercaptan residues in bleach/ethanol (1:1). This stage of the procedure is best monitored by phosphorus NMR (^{31}P NMR, see *Figure 4*) as it is difficult to distinguish the deoxyribonucleoside-3′-O-phosphorothioamidite from the precursor *bis*(pyrrolidino) phosphoramidite by TLC. The advantage of studying this reaction by ^{31}P NMR is illustrated in *Figure 4* where the two species are easily distinguished through their differences in chemical shift. The third peak observed is due to reaction of the mercaptan with excess of the phosphine and can be removed in the reaction work up. The stirred reaction should reach $> 95\%$ completion after 30 min. If not, this reaction can be allowed to proceed for longer periods (up to 1 h) without decomposition of the monomer.

Protocol 2. *Continued*

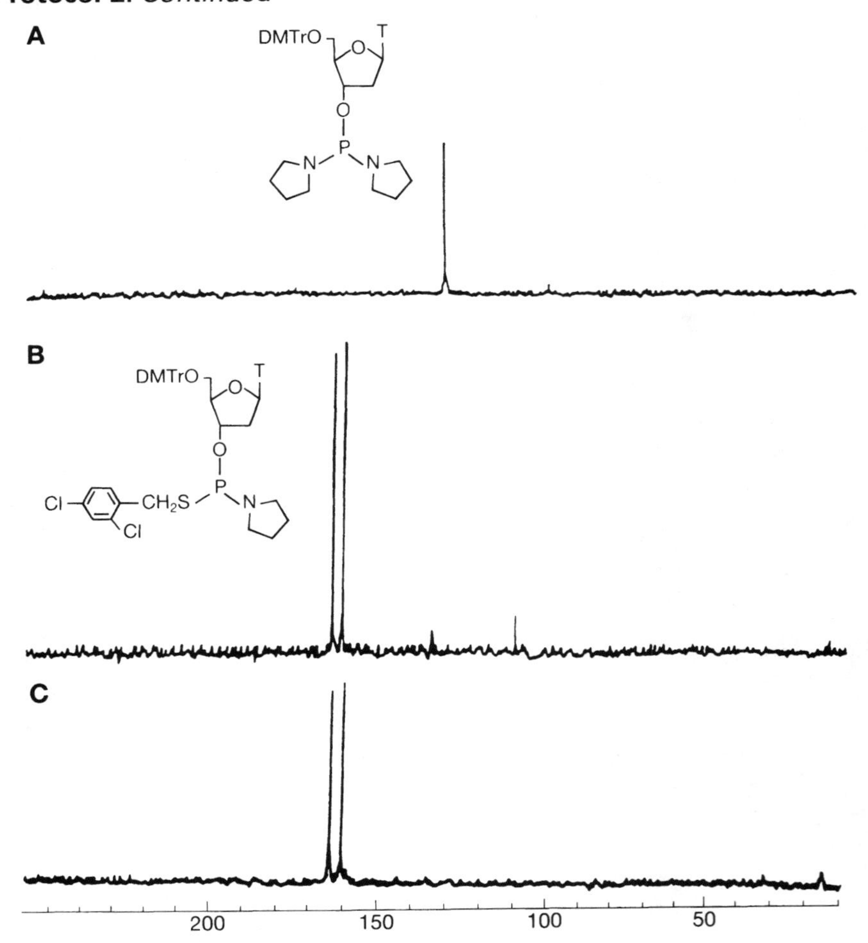

Figure 4. NMR analysis of phosphorothioamidite formation. **A**, phosphitylation; **B**, reaction after addition of mercaptan; **C**, isolated product. Chemical shift of phosphorothioamidite diastereomers are 160–165 p.p.m. ($CDCl_3$) depending on heterocyclic base (19).

14. Pour the reaction mixture into dichloromethane (200 ml, 1% triethylamine) and transfer this solution into a 1 litre separating funnel. Wash the solution with saturated sodium bicarbonate solution (200 ml), 10% sodium carbonate solution (2 × 200 ml) and saturated aqueous sodium chloride (200 ml).

15. Dry the organic phase over anhydrous sodium sulphate for 20 min. Add triethylamine (3 ml) to ensure stability of the phosphorothioamidite. Filter and wash the solids with toluene (20 ml).

Protocol 2. *Continued*

16. Because of the properties of these thioamidites, it is extremely difficult to purify them using column chromatography. These compounds are therefore isolated via precipitation. Pour heptane (500 ml, 1% in triethylamine) into a 1 litre conical flask containing a stirrer bar and degas by bubbling argon through the stirred solvent.
17. Concentrate the solution from step 15 to approximately 20 ml under reduced pressure and add the concentrate dropwise to the rapidly stirred heptane. When finished, remove the stirrer bar and isolate the precipitated product by filtration (water pump).
18. The protected deoxyribonucleoside-3′-O-pyrrolidinophosphorothioamidite is then dried in a desiccator overnight using a vacuum pump.
19. Purity of the dried phosphorothioamidite can be assessed by ^{31}P NMR. (Care must be taken during this assessment as these phosphorothioamidites are particularly susceptible to oxidation (21). Solutions of amidites for NMR or DNA synthesis should be made up under an argon atmosphere.) Routinely, these products are usually 90–100% pure with yields typically of 85–90%. Where impurities exist, they largely take the form of the H-phosphonate derived from hydrolysis of the thioamidite. Additionally, there may be traces of another impurity (seen at 30 p.p.m. in the phosphorus NMR) attributable to oxidation of the thioamidite. However, as with conventional phosphoramidites, neither species is appreciably activated with tetrazole which means that the phosphorothioamidites can be used for DNA synthesis even when minor amounts of these impurities are present.

4. Synthesis of DNA containing phosphorodithioate linkages by solid phase methods

Conventional DNA synthesizers can be readily adapted to the synthesis of compounds containing all dithioate linkages or a mixture of diester and dithioate linkages. This is possible because most machines can accommodate more than the number of reagent bottles necessary for conventional DNA synthesis. Use of additional reagents compatible with other chemistries thus increases the versatility of these machines for analogue synthesis, particularly where modification of the DNA backbone is concerned. To date, all solid phase synthesis of dithioate-modified DNA in this laboratory has been performed using the Applied Biosystems 380A automated DNA synthesizer. This machine has the capacity to accommodate reagents in 18 bottles or vials. This is sufficient for the synthesis of DNA containing combinations of normal and dithioate linkages or all dithioate linkages. The ports for these reagents can be dedicated as in *Table 2*.

Using this set-up on the 380A synthesizer, phosphodiesters can be synthesized using one cycle and dithioate linkages on another. By alternating cycles, compounds containing both linkages can be constructed.

4.1 Common synthesis reagents

Tetrazole and reagents for the capping and detritylation steps are common to the synthesis of both diester and dithioate linkages. Most of these reagents are commercially available (see Appendix 1). Anhydrous solvents can be obtained by distillation. The other reagents required besides phosphoramidite synthons are an iodine solution for oxidation to the diester and a sulphur solution for the sulphurization step in dithioate synthesis. Carbon disulphide is also required. Because of the hazardous nature of all these reagents, solutions must be prepared in fume hoods and protective eyewear and clothing worn at all times. Components required for DNA synthesis are:

- working DNA synthesizer,
- reagent bottles and vials which fit machine,
- an argon line.

Table 2. Assignment of synthesizer ports in the synthesis of dithioate DNA

Port no.	Reagent	Purpose
1–7	Conventional phosphoramidite or phosphorothioamidites	DNA building blocks
8	Sulphur solution	Sulphurization in dithioate synthesis
9	Tetrazole	Catalyst for coupling reaction
10	Ammonia—automated synthesis (not used)	Deprotection
11	Acetic anhydride/lutidine solution (cap A)	Capping
12	1-Methylimidazole solution (cap B)	Capping
13	Iodine solution	Oxidation in phosphodiester synthesis
14	Trichloroacetic acid solution	Detritylation
15	Carbon disulphide	Wash (sulphur removal)
16	Dichloromethane	Wash
17	Methanol	Wash
18	Acetonitrile	Wash

Protocol 3. Preparation of common stock solutions

1. Distil tetrahydrofuran (THF, b.p. 67 °C) from sodium and benzophenone.
2. Prepare the following cap solutions:
   (a) Cap A solution. Mix acetic anhydride (40 ml), 2,6-lutidine (40 ml) and THF (300 ml).

Protocol 3. *Continued*

(b) Cap B solution. Mix 1-methylimidazole (30 ml) and THF (70 ml).

3. Trichloroacetic acid solution. Dissolve trichloroacetic acid (9 g) in dry, distilled dichloromethane (300 ml).
4. Prepare the iodine solution by dissolving iodine (5.08 g, 20 mmol) in THF/water/2,6-lutidine (49:1:1; 200 ml).
5. Dissolve 1H tetrazole (1.05 g, 15 mmol) in dry, distilled acetonitrile (30 ml) to yield a 0.5 M solution.

4.2 The sulphur solution

The use of triethylamine in sulphur solutions greatly increases the rate of sulphurization but also promotes the decomposition of the sulphur solution with time. Sulphur also tends to crystallize from solution on standing for long periods. A stock sulphur solution should therefore be prepared relatively often and triethylamine added prior to use. To be sure of efficient synthesis after the triethylamine addition, do not use the solution for more than 24 h.

Protocol 4. Preparation of sulphur solution

1. Distil pyridine (b.p. 115 °C) from calcium hydride prior to use.
2. Distil triethylamine (b.p. 89 °C) first from *p*-toluene sulphonyl chloride and then from calcium hydride.
3. Dissolve sulphur (10 g) in pyridine/carbon disulphide (1:1; 200 ml). Store this solution *in the dark* in a 500 ml flat-bottomed flask covered with a serum cap (Aldrich Z10, 145–1).
4. Prior to machine synthesis, mix the stock solution from step 3 (37.5 ml) and triethylamine (2.5 ml) in a 45 ml amber glass borosilicate vial covered with a serum cap (Aldrich Z10,079–9). Use dry glass syringes and argon to protect these reagents from oxygen. Fit to port no. 8 on DNA synthesizer. (If not used regularly, wash the port and lines after use with carbon disulphide to prevent potential blockage.)

4.3 Preparation of deoxyribonucleoside-3′-O-phosphoramidite and deoxyribonucleoside-3′-O-phosphorothioamidite solutions

Conventional deoxyribonucleoside-3′-O-phosphoramidites can be prepared as 0.1 M solutions by standard methods. Because of the particular susceptibility of the deoxyribonucleoside-3′-O-phosphorothioamidites to oxidation and other side reactions, these synthons are less stable than conventional amidites in solution. However, the predominant side-products do not interfere with coupling of the thioamidite except by reducing its effective concentration in solution. For this reason, it is important to estimate the purity of the monomer by ^{31}P NMR and prepare solutions at a concentration of 0.15 M. Thus, should the deoxyribonucleoside-3′-O-phosphorothioamidite begin to decompose in solution, there will always be sufficient excess of the active monomer for effective coupling on solid support. Additionally, because of the oxidation phenomenon, *it is crucial* that the deoxyribonucleoside-3′-O-phosphorothioamidite be protected from oxygen with an argon atmosphere prior to solution preparation. If possible, it is advantageous to degas acetonitrile and other solvents used for these solutions.

Protocol 5. Preparation of deoxyribonucleoside-3′-O-phosphorothioamidite solutions

1. If possible degas dry, distilled acetonitrile and dichloromethane by bubbling argon through the respective solvent (approximately 2 h per 50 ml solvent).
2. Weigh sufficient deoxyribonucleoside-3′-O-phosphorothioamidite, based on purity, for the required number of couplings on a solid support (example: when synthesizing a dithioate linkage 3′ to the base cytosine (C), the molecular weight of protected C phosphorothioamidite = 924. For C thioamidite 90% pure, for three couplings this translates to (924 × 100/90 × 0.15) = 154 mg of monomer in 1.2 ml solution). Weigh into an amber glass borosilicate vial covered with a serum cap (Aldrich Z10,079–9) and flush with argon.
3. Dissolve the monomer in the appropriate volume of solvent under an argon atmosphere. Monomers derived from pyrimidines are soluble in acetonitrile, those from purines in 10% dichloromethane in acetonitrile.
4. Fit the vial containing solution to the appropriate phosphoramidite port using the synthesizer bottle change routine.

4.4 The synthesis cycle

For dithioate synthesis, oligonucleotides can be synthesized using the 380A synthesizer and commercially available polymeric supports. The desired sequence is programmed into the machine. The cycle for synthesis with the essential chemical steps is shown in *Table 3* for 1 μmol scale. This cycle requires more time than conventional diester cycles due to longer coupling and sulphurization steps. Other differences in the cycle involve the inversion of oxidation (with sulphur) and capping steps and step 7 where the residual

Table 3. Cycle for synthesis of dithioate DNA (1 μM scale)

Step no.	Function Name[a]	(no.)	Time (s)	Remarks
1	No. 18 to column	9	30	
2	Reverse flush	2	10	
3	Block flush	1	4	
4	Phos1 prep	28	5	
5	Bottle function	61	2	
6	Base + tetrazole to column	19	7	Coupling
7	Block flush	1	4	
8	Wait	4	41	
9	Base + tetrazole to column	19	7	Coupling
10	Block flush	1	4	
11	No. 18 to waste	10	10	
12	Wait	4	31	
13	Reverse flush	2	3	
14	Bottle function	81	10	
15	Block flush	1	1	
16	No. 8 to column	26	20	Sulphurize
17	Bottle function	81	10	
18	Block flush	1	5	
19	Wait	4	45	
20	Reverse flush	2	5	
21	No. 15 to column	13	30	
22	Reverse flush	2	5	
23	Block flush	1	4	
24	No. 17 to column	11	30	
25	Reverse flush	2	5	
26	No. 16 to column	12	20	
27	Reverse flush	2	5	
28	Block flush	1	4	
29	Cap prep	16	5	
30	Cap to col 1	22	7	Capping
31	Wait	4	30	
32	Reverse flush	2	5	
33	Block flush	1	4	
34	No. 16 to column	12	30	
35	Reverse flush	2	5	
36	Block flush	1	4	
37	Cycle entry	33	1	
38	No. 16 to column	12	10	
39	Reverse flush	2	5	
40	Block flush	1	4	
41	Waste-port	6	1	
42	No. 14 to column	14	30	Detritylation
43	No. 16 to column	12	20	
44	Reverse flush	2	5	
45	Block flush	1	4	
46	Waste-bottle	7	1	
47	Advance FC	5	1	
48	No. 18 to column	9	20	
49	Reverse flush	2	10	
50	Block flush	1	4	
51	Cycle end	34	1	

[a] refer to reagents in *Table 2*
Total time, ~ 9 min

mixture of monomer and tetrazole that remains in the lines is jettisoned to waste. This latter change is a precaution to minimize side reactions of the monomer with tetrazole (21) which might lead to the formation of unwanted linkages in the product DNA. Finally, carbon disulphide (step 21) is used to wash sulphur from the column due to the insolubility of sulphur in the other solvents. When this step is included, few problems due to sulphur blockage in the machine lines are encountered.

Coupling yields can be estimated from trityl cation release by standard methods. While this provides an indication of how well a synthesis is proceeding, the trityl yields may vary. Trityl yields (recorded in dry dichloromethane) of 96–99%, depending upon the synthesis, have been obtained with these monomers. A better estimation of synthesis yields can be obtained using HPLC or gel electrophoresis (see below). Purity of the modified linkage can be assessed by ^{31}P NMR of the crude synthesis obtained after deprotection. Usually this is greater than 98%.

5. Deprotection and isolation of oligodeoxynucleotides containing the phosphorodithioate linkage

Deprotection of the product DNA is a two-step process. The first step involves treatment of the oligonucleotide bound to the solid support with triethylammonium thiophenolate to remove the 2,4-dichlorobenzyl group, the protecting group for the dithioate linkage. The second step is a standard treatment with aqueous ammonia which ensures total deprotection at phosphorus (should phosphotriester links be present in the molecule), cleavage of the DNA product from the solid support and removal of the base protecting groups. The deprotecting reagents are highly toxic. Work in a fume hood and use protective eyewear and clothing.

Protocol 6. Deprotection and isolation of oligodeoxyribonucleotides containing phosphorodithioate linkage

1. Flush the column containing the solid support with argon for 5 min.
2. Transfer the solid support to a 3.7 ml screw-capped deprotection vial. If the screw cap does not have a rubber seal, affix a Teflon seal.
3. Mix an appropriate volume of thiophenol, triethylamine and dioxane in the ratio thiophenol/triethylamine/dioxane (1:2:2). Mix in the order dioxane, thiophenol, triethylamine. Use glass syringes. Destroy all thiophenol residues in freshly mixed ethanolic bleach (bleach/ethanol, 1:1).

Protocol 6. *Continued*

4. Treat the solid support with the solution from step 3 (1.5 ml). Allow to stand at room temperature for 2 h.
5. Remove the thiophenolate solution with a glass pipette and discard into ethanolic bleach.
6. Wash the support with three volumes (3.7 ml) of methanol. After each wash, discard the solvent into ethanolic bleach.
7. Wash the support with three volumes (3.7 ml) of ether. After each wash, discard the solvent into bleach solution. *Care*—do so slowly as this solution may now be hot from reaction with thiophenol.
8. Allow the support to dry in air (20 min).
9. Treat the support with aqueous ammonia (1.5 ml). Screw cap tightly and place in an oven 15 h at 55 °C. (Check vial after 20 min for ammonia leaks).
10. Remove vial from oven and place in an ice bath. Unscrew cap and pipette the aqueous solution of product into a 1.5 ml Eppendorf tube. Wash the support with water and transfer the wash to a second tube.
11. The two solutions of product can be concentrated using a vacuum centrifuge.
12. The residues obtained can be suspended in water (0.4 ml) for purification.

6. Analysis and purification of oligodeoxyribonucleotide products

The degree of difficulty in purification of deoxyoligonucleotide analogue prepared by these methods depends largely on the extent of dithioate modification within the molecule. Oligomers bearing $\leqslant$50% dithioate linkages are easily purified using PAGE to separate the product DNA from other components in the mixtures obtained from synthesis. Standard gel extraction procedures for conventional phosphodiesters give good results for the isolation of these analogue. Oligomers containing small percentages of dithioate linkages can also be routinely purified by conventional HPLC methods (reversed phase and ion exchange) with excellent results.

Problems using these standard procedures are encountered when phosphorodithioates become the predominant linkage. However, it is possible to optimize gel extraction and chromatographic techniques so that polyphosphorodithioate oligodeoxyribonucleotides can be purified to homogeneity.

6.1 Analysis and purification using PAGE

PAGE affords a simple method for analysis of the products from a given synthesis. This can be seen from the analytical slab gel shown in *Figure 5*. This gel shows the total reaction mixtures from several different syntheses of the same oligonucleotide sequence where the only variable is the number of phosphorodithioate backbone modifications. Comparison to a control (phosphodiester) synthesis allows a qualitative estimate of the amount of product oligodeoxyribonucleotide in relation to failure sequences in the mixture. This, together with an estimate of the total amount of UV-absorbing material relative to the control, gives a good indication of the relative yield of analogue.

Materials for PAGE analysis and purification should be ultra pure. Most materials are either commercially available with sufficient purity or can be obtained by recrystallization of available reagents. Acrylamide is a potent neurotoxin and cancer suspect agent. Use protective eyewear, clothing and a dust mask when handling solid acrylamide. Use gloves at all times when handling acrylamide solutions.

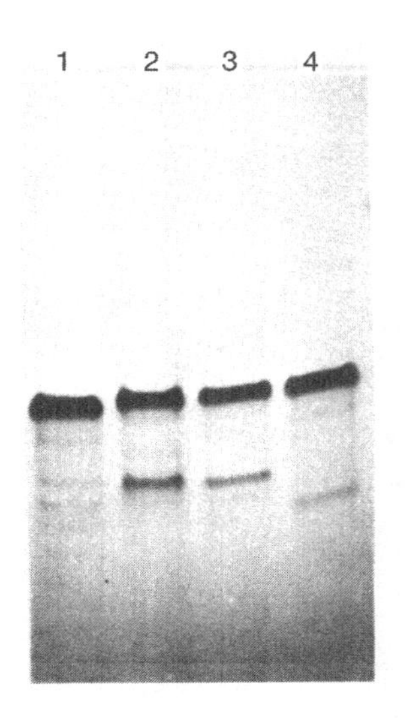

(1)GATTCAGCTAGTCCA
(2)GA*TT*CA*GC*TA*GT*CC*A
(3)GATTCAGC*T*A*G*T*C*C*A
(4)GATTCA*G*C*T*AGTCCA

Figure 5. 20% denaturing gel showing different extents of backbone modification in the same sequence. * Denotes the position of modification.

Protocol 7. Analysis and purification by PAGE of oligonucleotides containing phosphorodithioate linkages

1. Prepare a 40% (w/v) acrylamide solution with 19:1 cross-linking (380 g acrylamide, 20 g N,N′-methylene *bis*acrylamide per litre). Decolourize with Norit A and filter through a bed of celite 545.
2. Prepare 10 × *Tris*–borate EDTA (TBE). This solution should be
   - 0.9 M Tris base
   - 0.9 M boric acid
   - 0.1 mM Na_2 EDTA
   - pH 8.3

   Autoclaving this solution for 15 min prevents precipitation over time.
3. Prepare a 20% acrylamide, 8 M urea solution from the following components:
   - 48 g urea
   - 50 ml 40% acrylamide stock
   - 10 ml 10 × TBE
   - water up to 100 ml

   Make the solution 0.1% in ammonium persulphate.
4. Pour a slab gel (20 cm × 20 cm, analytical 0.75 mm thick; preparative 1.5 mm thick) by activating the gel solution with N,N,N′,N′-tetramethylethylenediamine (TEMED). TEMED should be added to approximately 0.1% just prior to pouring.
5. Prepare a sufficient volume of 1 × TBE in H_2O and pre-electrophorese gel for about 15 min before loading.
6. Take the aqueous suspension from the synthesis (0.4 ml, section 5, *Protocol 6*, step 12), determine total number of optical density units (1 OD unit = 1.0 A_{260} unit per ml). For gel wells of 2.5 cm width and 1.5 mm depth, up to 8 OD units can be loaded per lane and run with single base resolution.
7. Add deionized formamide (deionized over mixed bed resin) to the oligomer solution to a final concentration of 50% and reserve a gel lane for marker dyes (0.1% bromophenol blue (BPB), 0.1% xylene cyanol (XC) in 80% formamide, 1 × TBE). In 20% denaturing gels, BPB migrates at 6–8 nucleotides and XC at approximately 30 nucleotides. For analytical gels (loading approximately 1 OD unit), prepare a marker sample containing an equal amount of phosphodiester oligodeoxynucleotide.
8. Heat samples at 90 °C for 5 min before loading gel.

Protocol 7. *Continued*

9. Run gel at 400–600 V (30–40 mA) at constant voltage to elicit desired separation based on dye markers.
10. After electrophoresis visualize oligomers by UV shadowing. A polydithioate will migrate more slowly than the corresponding unmodified oligonucleotide (approximately equivalent to a difference of one nucleotide in chain length). Excise desired band and transfer to a sterile tube.

Research into the extraction of the oligomers from gel slices is an ongoing process. Currently, an active extraction method involving electroelution of the oligomer onto DEAE cellulose followed by elution with ionic denaturants (sodium thiocyanate or sodium chloride/formamide) is showing promise as a high-yielding procedure. The 'freeze–thaw' procedure described below is effective in the recovery of these analogue and conventional oligodeoxynucleotides.

Protocol 8. Gel extraction

1. Crush gel slice with a sterile instrument and tap gel material into the bottom of the tube.
2. Immerse crushed gel in sterile water or Tris–EDTA solution (10 mM Tris, 1 mM EDTA, pH 8). For polydithioates a 20 mM dithiothreitol solution is most effective in the extraction procedures. Add about 4 times the volume of crushed gel material and mix thoroughly.
3. Freeze gel slurry in dry ice until a white solid forms throughout the mixture (15–20 min).
4. Incubate tube at 37 °C with vigorous shaking for 3–4 h.
5. Repeat steps 3 and 4.
6. Filter water/gel slurry through siliconized glass wool and collect eluant.
7. Repeat steps 2–6 and pool eluants.
   (a) For oligomers of $\geqslant 8$ bases, ethanol precipitation is effective. Precipitate three times with 0.3 M sodium acetate, pH 5–7, and 4, 3.5, and 3 volumes of ethanol sequentially. Desalt on Sephadex G-25 if necessary.
   (b) For very short oligomers evaporate gel eluant to reduce total volume by approximately three-quarters. Desalt on a Sephadex G-25 column (60 cm × 1.5 cm diameter) in 50 mM triethylammonium bicarbonate. Follow oligomer by UV absorbance and evaporate desired fractions.

6.2 HPLC methods

Recent improvements in the synthesis of these analogue has facilitated attempts to purify them via HPLC. While ion exchange HPLC methods have not shown promise to date, conventional reversed phase chromatography (C_{18}) can be used to obtain purified polydithioate containing oligodeoxynucleotides. *Figure 6* shows traces of two crude syntheses. The first is a 5′-dimethoxytrityloligodeoxycytidine (dC_{12}) having phosphodiester linkages. The second shows the analogous 5′-dimethoxytritylpolydithioate dodecamer (S_2dC_{12}). Inspection of the traces indicates that the polydithioate analogue shows a much higher degree of lipophilicity compared to conventional phosphodiesters. Consequently, by changing standard HPLC protocols to account for this change in lipophilicity, it is possible to purify tritylated polydithioates on reverse phase chromatography. Moreover, these compounds can be detritylated without harm to the dithioate linkage. For these procedures, the following apparatus is required:

- standard HPLC equipment (pumps, controller, recorder, UV detector)
- 5 μm C_{18} reversed phase column

Protocol 9. Reversed phase purification of analogue

1. Synthesize the analogue of choice **Trityl ON**.
2. Deprotect as described previously (*Protocol 6*).
3. Take the aqueous suspension of product (*Protocol 6*, step 12) and remove solids (filter or centrifugation). Wash solids with water (0.2 ml). Concentrate in an Eppendorf tube to a small volume (approximately 0.1 ml).
4. Make the solution 30% in acetonitrile (or deionized formamide) and vortex tube. This precludes reversed phase effects of the oligomer product with the tube.
5. Determine the quantity of crude material by UV absorption.
6. Analyse the mixture on a 5 μm C_{18} reverse phase column using acetonitrile and 50 mM triethylammonium acetate solution (pH 7.0) as mobile phase.
7. Equilibrate column to 15% acetonitrile/85% triethylammonium acetate.
8. Inject 1 A_{260} unit into the HPLC system and elute the column with a gradient of 15–75% acetonitrile over a 66 min time period. Determine the retention time of the trityl on product from the trace obtained and the retention times of suitable markers (1 A_{260} unit of trityl on phosphodiester control and benzamide).†
9. Inject remainder of crude mixture and collect product as it elutes from column. Evaporate desired fractions.

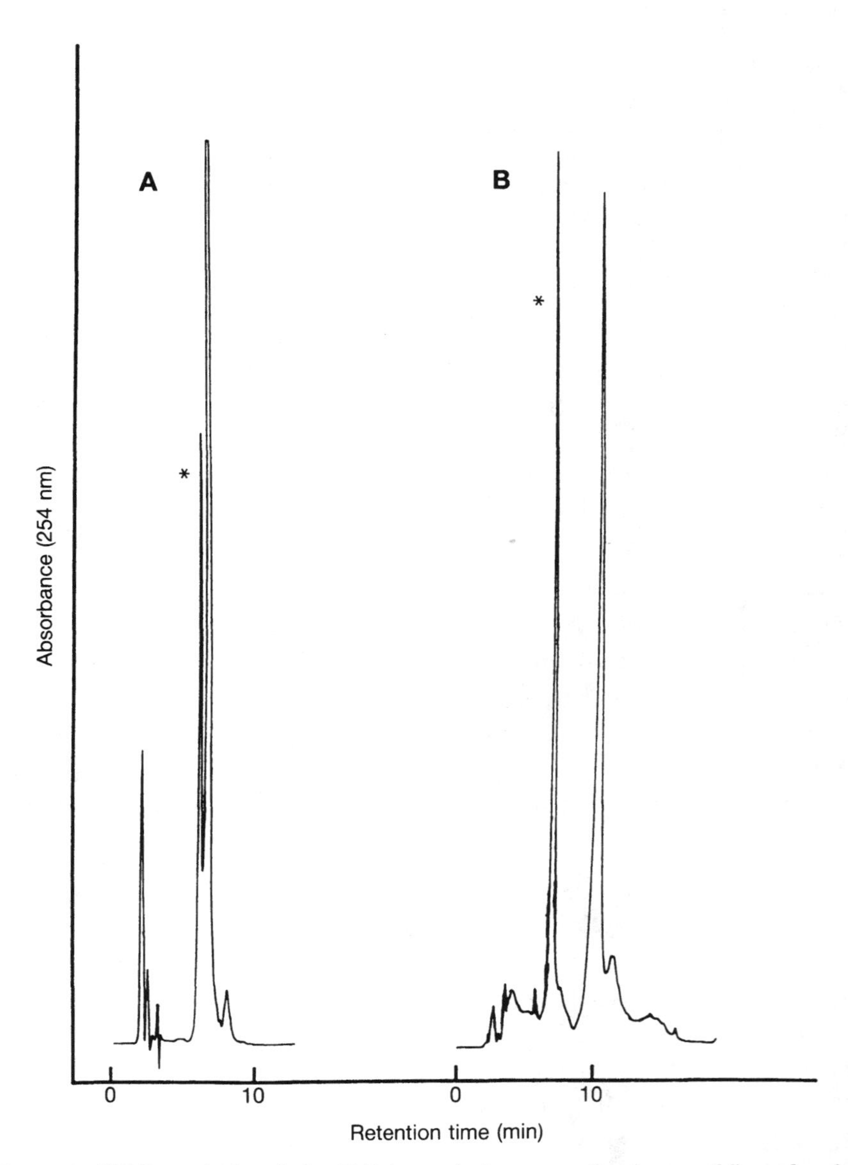

Figure 6. HPLC analysis of the **Trityl on** dodecamer oligodeoxycytidine. **A**, phosphodiester; **B**, polyphosphorodithioate. Gradient 25% acetonitrile/75% 50 mM triethylammonium acetate (TEAA, pH 7.0) to 75% acetonitrile/25% TEAA over 20 min. The peak labelled * was identified as benzamide.

Protocol 9. *Continued*

10. Resuspend residues in water. Make solutions 15–30% in acetonitrile to preclude reverse phase effects. Quantify product from UV absorption of solution.

† This gradient appears to give good separation of the trityl containing product even when substantial failure sequences are present.

Detritylation of the tritylated oligomer can be completed by a standard 80% acetic acid treatment and the product further purified by gel filtration chromatography. The method of desalting changes slightly due to the apparent reversed phase effects exhibited by polyphosphorodithioate oligodeoxyribonucleotides.

Protocol 10. Detritylation of phosphorodithioate-containing analogue

1. Swell LH-20 Sephadex in water by heating a slurry of the material in water at 100 °C (boiling water bath) for 30 min. Alternatively, autoclave the slurry.
2. When cooled, add acetonitrile (10 ml) to 20 ml of this mixture. Degas mixture on a water pump.
3. Pour a column (approximately 25 × 0.8 cm—disposable glass pipettes are very useful).
4. Equilibrate column with at least 5 volumes of 40 mM ammonium carbonate solution (30% in acetonitrile) or 30% aqueous acetonitrile.
5. Take the oligomer solution (~0.2 ml, 15–30% in acetonitrile). Chill on an ice bath.
6. Add glacial acetic acid (0.8 ml). Vortex tube and stand on ice for 15 min.
7. Extract mixture with ether (3 × 0.5 ml).
8. Take the small volume of aqueous phase (lower layer in each extraction—approximately 0.1 ml) and add water (0.3 ml).
9. Add concentrated aqueous ammonia (0.1 ml) dropwise. Care should be taken to insure that all the ether layer from step 7 has been removed.
10. Apply resulting solution to the LH-20 Sephadex column and elute column with buffer.
11. The product is eluted rapidly. Fractions containing product can be detected by measurement of their UV absorbance at 260 nm.
12. Pool and evaporate the desired fractions to obtain the purified oligodeoxyribonucleotide analogue.

While yields of dithioate-modified oligodeoxyribonucleotides are not as high as for their phosphodiester counterparts, this methodology easily allows for the synthesis and purification of these analogues for biological studies.

Currently, the use of deoxyribonucleoside phosphorothioamidites for the rapid synthesis of dithioate DNA is the method of choice. The approach requires only the four deoxynucleoside phosphorothioamidites for complete synthetic versatility and can be used in combination with deoxynucleoside phosphoramidites to generate polynucleotides of any sequence and having all combinations of phosphate and phosphorodithioate internucleotide linkages.

Access to these compounds have led us to investigate several potential biochemical, biological, and therapeutic applications for dithioate DNA (29). Because this derivative is completely resistant to degradation by snake venom and spleen phosphodiesterases, nuclease P1, the 3′-5′ exonuclease activity of T4 DNA polymerase, and the nucleases found in the HeLa cell nuclear and cytoplasmic extracts, we anticipate that dithioate DNA may be resistant to essentially all nucleases and therefore useful for many *in vivo* experiments. As a first step in this direction, we have examined the ability of dithioate DNA to stimulate RNase H activity in HeLa cell nuclear extracts. The results indicate that, indeed, deoxyoligonucleotides containing combinations of dithioate and normal linkages (also all dithioate links) are active in stimulating RNase H and, in contrast to normal DNA, are highly resistant toward endogenous nucleases. These results suggest that dithioate DNA may be useful as an antisense DNA therapeutic and for a large number of basic, biological experiments that need to be completed *in situ*. Other research has demonstrated that dithioate DNA can be used to probe protein–DNA interactions and as a very potent inhibitor of HIV reverse transcriptase. Because dithioate DNA can be readily tagged with heavy metal ions and, through alkylation, with various reporter groups such as fluorescein, we also anticipate that this derivative will find many additional uses in X-ray crystallography and other analytical biochemical applications. Because of all these potential uses plus many currently being explored, we predict that dithioate DNA will become an extremely valuable oligonucleotide analogue in the years to come.

Acknowledgements

This is paper 35 in a series on Nucleotide Chemistry. Paper 34 is Beaton, G., Brill, W. K.-D., Grandas, A., Ma, Y.-X., Nielsen, J., Yau, E. and Caruthers, M. H. (1991). *Tetrahedron*, **47**, 2377. This research was supported by the National Institutes of Health (Grant GM25680).

References

1. Matteucci, M. D. and Caruthers, M. H. (1981). *J. Am. Chem. Soc.*, **103**, 3185.
2. Beaucage, S. L. and Caruthers, M. H. (1981). *Tetrahedron Lett.*, **22**, 1859.
3. McBride, L. J. and Caruthers, M. H. (1983). *Tetrahedron Lett.*, **24**, 245.
4. Caruthers, M. H. (1985). *Science*, **230**, 281.
5. Groger, G., Ramalho-Ortigo, G., Steil, H., and Seliger, H. (1988). *Nucleic Acids Res.*, **16**, 7761.
6. Arnhelm, N. and Levenson, C. H. (1990). *Chem. Eng. News*, **68**, 36.
7. Leatherbarrow, R. J. and Fersht, A. R. (1986). *Protein Eng.*, **1**, 7.
8. Harrison, S. C. and Aggarwal, A. K. (1990). *Annu. Rev. Biochem.*, **59**, 933.
9. Kennard, O. and Hunter, W. N. (1989). *Q. R. Biophys.*, **22**, 327.
10. Caruthers, M. H. (1989). In *Oligonucleotides: Antisense Inhibitors of Gene Expression* (ed. J. S. Cohen), *Topics in Molecular and Structural Biology*, Vol. 12, pp. 7–24. Macmillan Press, London.
11. Nielsen, J. and Caruthers, M. H. (1988). *J. Am. Chem. Soc.*, **110**, 6275.
12. Eritja, R., Smirnov, V., and Caruthers, M. H. (1990). *Tetrahedron*, **46**, 721.
13. Stein, C. A. and Cohen, J. S. (1988). *Cancer Res.*, **48**, 2659.
14. Lesnikowski, Z. J., Jaworska, M., and Stec, W. J. (1988). *Nucleic Acids Res.*, **16**, 11675.
15. Nielsen, J., Brill, W. K.-D., and Caruthers, M. H. (1988). *Tetrahedron Lett.*, **29**, 2911.
16. Brill, W. K.-D. and Caruthers, M. H. (1988). *Tetrahedron Lett.*, **29**, 5517.
17. Farschtschi, N. and Gorenstein, D. G. (1988). *Tetrahedron Lett.*, **29**, 6843.
18. Grandas, A., Marshall, W. S., Nielsen, J., and Caruthers, M. H. (1989). *Tetrahedron Lett.*, **30**, 543.
19. Brill, W. K.-D., Tang, J. Y., Ma, Y. X., and Caruthers, M. H. (1989). *J. Am. Chem. Soc.*, **111**, 2321.
20. Stawinski, J., Thelin, M., and Zain, R. (1989). *Tetrahedron Lett.*, **30**, 2157.
21. Dahl, B. H., Bjergårde, K., Sommer, V. B., and Dahl, O. (1989). *Acta Chem. Scand.*, **43**, 896.
22. Dahl, B. H., Bjergårde, K., Sommer, V. B., and Dahl, O. (1989). *Nucleosides and Nucleotides*, **8**, 1023.
23. Porritt, G. M. and Reese, C. B. (1989). *Tetrahedron Lett.*, **30**, 4713.
24. Brill, W. K.-D., Yau, E. K., and Caruthers, M. H. (1989). *Tetrahedron Lett.*, **30**, 6621.
25. Porritt, G. M. and Reese, C. B. (1990). *Tetrahedron Lett.*, **31**, 1319.
26. Yau, E. K., Ma, Y.-X., and Caruthers, M. H. (1990). *Tetrahedron Lett.*, **31**, 1953.
27. Dahl, B. H., Bjergårde, K., Nielsen, J., and Dahl, O. (1990). *Tetrahedron Lett.*, **31**, 3489.
28. In order to facilitate ordering these compounds, protected deoxynucleosides are described using the nomenclature found in commercial catalogue.
29. Caruthers, M. H., Beaton, G., Cummins, L., Dellinger, D., Graff, D., Ma, Y.-X., Marshall, W. S., Sasmor, L. H., Norris, P., Wu, J.-V., and Yau, E. K. (1991). in *Host–Guest Molecular Interactions*, pp. 158–68 Ciba Foundation, John Wiley & Sons, Chichester, England.

Appendix: Chemical suppliers

Phosphine preparation

Pyrrolidine	Aldrich/P7,380–3
Phosphorus trichloride	Aldrich 31,011–5
Ether	Mallinckrodt/Aldrich

Thiophosphoramidite preparation

All solvents may be obtained from any general chemical supplier such as Aldrich or Mallinckrodt. 1-H tetrazole may be obtained from Aldrich (33,644–0, sublimed or 24,395–7). 2,4-Dichlorobenzylmercaptan can be purchased from Lancaster Synthesis (≥ 97%, catalogue no. 2795).

Protected deoxyribonucleosides may be obtained from Aldrich, Sigma, or Cruachem. Catalogue numbers for nucleosides sold by Cruachem are:

N-Benzoyl-5′-O-dimethoxytrityl-2′-deoxyadenosine	052120
N-Benzoyl-5′-O-dimethoxytrityl-2′-deoxycytidine	052130
N-Isobutyryl-5′-O-dimethoxytrityl-2′-deoxyguanosine	052110
5′-O-Dimethoxytritylthymidine	052100

DNA synthesis

Solvents and tetrazole may be obtained as described above. Acetonitrile, carbon disulphide and methanol should be HPLC grade (low water content). Other reagents may be obtained as follows:

Acetic anhydride	Aldrich/32,010–2
2,6-Lutidine	Aldrich/33,610–6
1-Methylimidazole	Aldrich/33,609–2
Trichloroacetic acid	Aldrich/25,139–9
Iodine	Aldrich/22,969–5
Sulphur	Aldrich/21,329–2

Deprotection

Thiophenol	Aldrich/24,024–9
1,4-Dioxane	Aldrich/29,630–9
Aqueous ammonia	Aldrich/22,122–8

Analysis and purification

Acrylamide	Boehringer Mannheim/100137
N,N′-Methylene bisacrylamide	Kodak/1198381
Norit A	Aldrich/26001–0

Celite	Aldrich/16743–6
Tris base	Boehringer Mannheim/604203
Boric acid	Sigma/B0394
Disodium EDTA	Aldrich/25,235,2
Urea	Aldrich/U270–9
Ammonium persulphate	Aldrich/24,861–4
N,N,N′,N′-Tetramethylethylenediamine	Aldrich/T2,250–0
Formamide	Aldrich/29,587–6
Bromophenol blue	Aldrich/11,439–1
Xylene cyanol	Sigma/X2751
Dithiothreitol	Boehringer Mannheim/100032
Sephadex G-25	Pharmacia/17–0033–01
Sephadex LH-20	Pharmacia/17-0090–01
Ammonium carbonate	Aldrich/20,786–1

Solvents should be HPLC grade.

General procedures
^{31}P NMR spectra were recorded on a JEOL FX90Q (phosphine preparations, analysis of phosphorothioamidite formations and products) using 85% phosphoric acid as external standard. ^{1}H NMR spectra were recorded on a Varian VXR-300S in deuterated chloroform with tetramethylsilane as internal standard. ^{31}P NMR of analogue products were recorded on a Varian VXR-300S in deuterium oxide referenced to 85% phosphoric acid. TLC was on aluminium backed sheets (silica gel 60F, 0.2 mM, E. Merck). DNA synthesis was performed on an Applied Biosystems 380A automated DNA synthesizer. Deprotection of products were manual as described. HPLC was performed using the following Waters equipment:

- two model 6000A pumps
- model U6K injection system
- model 440 absorbance detector (set to 254 nm)
- model 680 automated gradient controller
- model 745 data module
- Lichrosorb RP-18 5U column, 259 mm × 4.6 mm, B fittings (Alltech, catalogue no. C-600) fitted with guard column (Alltech)

6

Synthesis of oligo-2′-deoxyribonucleoside methylphosphonates

PAUL S. MILLER, CYNTHIA D. CUSHMAN, and JOEL T. LEVIS

1. Introduction

Oligo-2′-deoxyribonucleoside methylphosphonates contain nuclease resistant, non-ionic internucleotide bonds. These oligomers are taken up intact by mammalian cells in culture. In addition to their unique physical properties, these oligonucleotide analogues give sequence-specific inhibitory effects when targeted against specific messenger RNA or precursor messenger RNA in mammalian cells in culture. The use of oligonucleoside methylphosphonates as antisense reagents in cells has been extensively reviewed (1–3). This chapter will describe the automated synthesis, deprotection, and purification of these oligomers, methods used to characterize the oligomers, and a general description of procedures for preparing them for use in cell culture experiments.

2. Synthesis

The general structure of oligodeoxyribonucleoside methylphosphonates is shown in *Figure 1*. The oligomer contains a single phosphodiester linkage at the 5′-end. The presence of this linkage allows the oligomer to be purified by ion exchange chromatography and also enables the oligomer to be phosphorylated by the enzyme, polynucleotide kinase. The remaining linkages in the oligomer are methylphosphonate groups. This group is chiral and each linkage can exist in either an R_p or S_p configuration. Because the synthetic procedure does not result in stereospecific formation of the internucleoside linkages, each oligomer consists of 2^n diastereoisomers, where n is the number of methylphosphonate linkages in the oligomer. The oligomers are synthesized on controlled pore glass supports using an automated DNA synthesizer. Base protected 5′-dimethoxytrityl-2′-deoxyribonucleoside-3′-

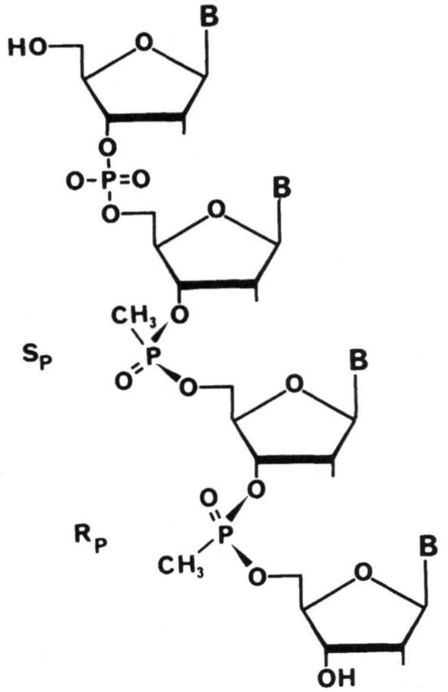

Figure 1. Structure of an oligonucleoside methylphosphonate where B is adenine, cytosine, guanine, or thymine. R_p and S_p indicate the configuration of the methylphosphonate linkage.

N,N-*bis*-(diisopropyl)aminomethyl-phosphonamidites are used as synthons (*Figure 2*). The synthetic cycle is analogous to that used to prepare oligodeoxyribonucleotides and consists of the following steps:

- detritylation
- coupling
- oxidation
- capping

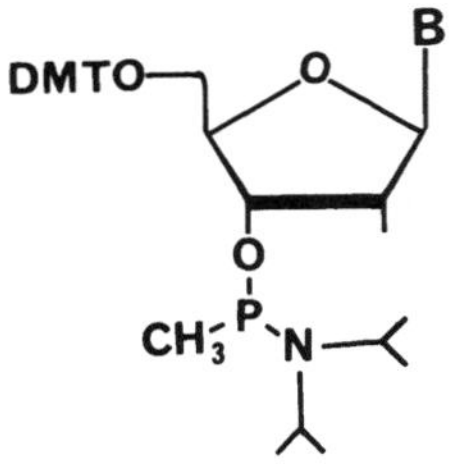

Figure 2. Methylphosphonamidite synthon used to prepare oligonucleoside methylphosphonates. DMT is a dimethoxytrityl protecting group and B is N^6-benzoyladenine, N^4-benzoylcytosine, N^2-isobutyrylguanine, or thymine.

2.1 Chemicals required

The appropriately protected nucleoside methylphosphonamidite synthons are available from American Bionetics Inc. or from Glen Research Inc.. The thymidine and deoxyadenosine synthons are dissolved in anhydrous acetonitrile whereas deoxycytidine and deoxyguanosine synthons are dissolved in a solution containing acetonitrile–methylene chloride (2:1 v/v). The final concentration of the synthons is 0.065 M. Anhydrous acetonitrile is prepared by refluxing reagent grade acetonitrile over calcium hydride, followed by distillation into 25 ml Erlenmeyer flasks which are sealed with Teflon-lined septa. Anhydrous methylene chloride is prepared by refluxing reagent grade methylene chloride over phosphorous pentoxide, followed by distillation into 25 ml Erlenmeyer flasks which are sealed with Teflon-lined septa. The other reagents used in the synthesis are:

- Deblock: dichloroacetic acid (32 ml) in methylene chloride (1 litre)
- 0.45 M tetrazole in anhydrous acetonitrile
- Cap A: dimethylaminopyridine (12.5 g) in dry pyridine (500 ml)
- Cap B: acetic anhydride (200 ml) in tetrahydrofuran (300 ml)
- Oxidizer: iodine (12.5 gm) in a solution of 2,6-lutidine (125 ml), water (125 ml) and tetrahydrofuran (250 ml)

Dry pyridine is prepared by refluxing reagent grade pyridine (1 litre) with chlorosulphonic acid (5 ml). **Care should be taken when adding the chlorosulphonic acid to the pyridine**. After refluxing for 3 h, the pyridine is distilled on to sodium hydroxide pellets. The pyridine is then refluxed with sodium hydroxide for 3 h and distilled onto 4 Å molecular sieves. The remaining reagents are all reagent grade or better. Premixed deblock, tetrazole, cap A, cap B, and oxidizer solution can be purchased from several of the manufacturers of DNA synthesizers (see Chapter 1).

2.2 Synthesis of oligonucleoside methylphosphonates in a DNA synthesizer

A Milligen/Biosearch Model 8700 DNA synthesizer is used to assemble the protected oligonucleoside methylphosphonate on the controlled pore glass supports. Other comparable instruments may also be used. The synthesis is carried out in 1 μmol scale synthesis columns which are filled with the appropriate controlled pore glass support. The loading of the support is usually between 25 and 35 μmol of nucleoside per gram. The fifth monomer bottle on the instrument contains a base protected 5′-dimethoxytrityl-2′-deoxyribonucleoside-3′-*β*-cyanoethyl-N,N-*bis*-(diisopropyl)amino phosphoramidite which is used to terminate the 5′-end of the oligomer. The standard synthesis program (see Chapter 1) is used except an additional capping cycle is inserted prior to the oxidation step. This capping cycle is identical to the capping cycle found in the standard synthesis program. The standard coupling program is used in the synthesis. The instrument is instructed to add the nucleoside phosphoramidite which is located in the 'U' reagent bottle, as the last nucleotide in the sequence. The dimethoxytrityl group is removed from the oligomer by the synthesizer. Upon completion of the synthesis, the support is washed with dry acetonitrile and then dried in a vacuum desiccator as described in *Protocol 1*, step 1 below. If desired, the reaction column and support can be stored in a desiccator at −20 °C until deprotection is carried out.

2.3 Deprotection of oligonucleoside methylphosphonates

The deprotection procedure consists of two reaction cycles. The first cycle, which is described in *Protocol 1* (steps 1–6) involves treatment of support-bound oligomer with a solution of hydrazine hydrate in a pyridine–acetic acid buffer. This reaction removes the base protecting groups from A and C residues. It is designed to prevent transamination of C residues during the subsequent ethylenediamine deprotection step. This step should be carried out for at least 24 h at room temperature and can be continued for up to 36 h. Under these conditions, the oligomer remains bound to the support and the G isobutyryl and phosphate β-cyanoethyl groups remain attached to the oligomer. The hydrazine treatment does not result in modification of C or T residues.

The second cycle, which is described in *Protocol 1* (steps 6–13) involves treatment of the support with ethylenediamine. This reaction cleaves the oligomer from the support, removes isobutyryl protecting groups from any G residues in the oligomer and removes the β-cyanoethyl protecting group form the 5′-terminal phosphotriester linkage. This reaction is most conveniently carried out after transfer of the support to a test-tube. The deprotected oligomer is removed from the support by sequential washing with a pyridine–ethanol solution followed by an aqueous acetonitrile wash and the solvents are evaporated with the aid of a vacuum pump. It is very important to ensure that all of the ethylenediamine is removed from the solution during the evaporation steps. This is achieved by repeated evaporations of the residue with 95% ethanol. The residue which remains is usually in the form of a white solid. This is dissolved in 50% aqueous ethanol and usually gives a solution whose pH is between 8 and 9. It is important to adjust the pH of this solution to between 6 and 7 by addition of a small amount of acetic acid. This will prevent any base-catalysed hydrolysis of the methylphosphonate linkages.

The solution obtained in step 13 may be stored at −20 °C. In most cases the oligomer is only partially soluble at this temperature and the solution will become cloudy. The oligomer can be redissolved by simply heating the solution with a hair dryer and vortexing the solution.

Protocol 1. Deprotection of oligonucleoside methylphosphonates

1. After the final synthetic coupling cycle, remove the synthesis column from the DNA synthesizer, place in a vacuum desiccator, and evacuate for 20 min using an oil pump.
2. Attach one end of the column to a 2.5 ml polypropylene syringe (Aldrich Chemical Co).
3. Fill the column using a second 2.5 ml syringe which contains the following solution:
   - pyridine 1.2 ml
   - glacial acetic acid 0.3 ml
   - 85% hydrazine hydrate 0.05 ml
4. Incubate the column at room temperature for 24 h. Occasionally flush the solution back and forth in the column using two syringes.
5. Flush the hydrazine solution from the column and wash the column with 5 ml of 95% ethanol followed by 5 ml of acetonitrile.
6. Dry the support in a vacuum desiccator as described in step 1.
7. Open the column and transfer the dried support to a glass screw-topped test-tube (15 × 100 mm).
8. Carefully add 0.5 ml of a solution containing 95% ethanol–ethylene-diamine (1:1 v/v) to the support. Make sure the support is completely covered and wetted by the solution, but do not allow the support to adhere to the wall of the test-tube.
9. Incubate the solution for 6 h at room temperature.
10. Use a Pasteur pipette to transfer the solution and the support to a 5 ml polypropylene chromatography column fitted with a porous frit (Whatman Inc.) and collect the solution in a screw-topped test-tube (20 × 150 mm). Wash the support with three 0.5 ml portions of pyridine–95% ethanol (1:1 v/v).
11. Evaporate the solution using an oil pump. The solution is vortexed during this operation and kept at room temperature using a warm stream of air from a hair dryer. Evaporate the residue with four 0.5 ml portions of 95% ethanol.
12. Wash the support with three 0.5 ml portions of acetonitrile–water (1:1 v/v) and add the combined washings to the residue obtained in step 11. Evaporate this solution and evaporate the residue with three 0.5 ml portions of 95% ethanol.
13. Dissolve the residue in 3 ml of 50% aqueous ethanol and adjust the pH to approximately 6 by addition of glacical acetic acid. This should only require 1–5 µl of acetic acid.

At this stage, the crude oligomer may be analysed by high performance liquid chromatography (HPLC) using a reversed phase C_{18} column. For example, a Whatman ODS-3 RAC II column is used with a linear gradient of 1.0–30% acetonitrile in 0.1 M sodium phosphate buffer (30 ml total volume) at a flow rate of 1.5 ml/min. Dodecamers generally elute between 12 and 16 min under these conditions. Longer oligomers may require higher acetonitrile concentrations.

2.4 Purification of oligonucleoside methylphosphonates

Because the desired oligomer contains a single, negatively charged phosphodiester linkage, it may be purified from the shorter, uncharged failure sequences by DEAE cellulose chromatography. The procedure is outlined in *Protocol 2*.

Protocol 2. Purification of oligonucleoside methylphosphonates by DEAE cellulose chromatography

1. Prepare a DEAE cellulose column (1.8 × 4 cm) in the bicarbonate form. Wash the column with 10 ml of 1 M ammonium bicarbonate and then wash with water until the pH of the solution is neutral. The column should be monitored at 254 nm using a UV detector and chart recorder.
2. Add 17 ml of 50% aqueous ethanol to the solution obtained in step 13 of *Protocol 1*.
3. Load this solution on to the column and wash the column with water until the recorder pen returns to the base line. This fraction contains the failure sequences.
4. Elute the oligomer with 0.1 M sodium phosphate buffer (pH 5.8). The oligomer usually elutes in a volume between 10 and 30 ml of buffer.
5. Add sufficient acetonitrile to this solution to give a final concentration of 5%. This will prevent biological contamination if the solution is stored for any length of time.

The oligomer obtained in step 5 of *Protocol 2* may be analysed by reversed phase HPLC as described above for the crude oligomer. In many cases the oligomer is sufficiently pure at this stage that further purification is not required. If this is not the case, the oligomer is further purified by preparative reversed phase HPLC. The oligomer solution is pumped directly onto a preparative Whatman ODS3 RAC column (0.9 × 10 cm). The column is eluted with a 50 ml linear gradient of 1–30% acetonitrile in 0.1 M sodium phosphate (pH 5.8) at a flow rate of 2.5 ml/min. The oligomer will usually elute near the end of this gradient. The acetonitrile is removed from the oligomer solution by evaporation.

2.5 Desalting oligonucleoside methylphosphonates

The oligomer solution obtained in step 5 of *Protocol 2* or after HPLC purification as described above is desalted on a SEP PAK C_{18} reversed phase cartridge. The procedure is described in *Protocol 3*.

Protocol 3. Desalting oligonucleoside methylphosphonates

1. Attach a SEP PAK C_{18} reversed phase cartridge (Waters Inc.) to a three-way nylon valve (BioRad Inc) which is fitted to the end of a 10 ml syringe.
2. Pre-equilibrate the SEP PAK with the following solutions:
   - acetonitrile 10 ml
   - 50% acetonitrile in 100 mM triethylammonium bicarbonate 5 ml
   - 25 mM triethylammonium bicarbonate 10 ml
3. Load the oligomer solution onto the SEP PAK and wash the SEP PAK with 20 ml of 25 mM triethylammonium bicarbonate.
4. Elute the oligomer with 4 ml of 50% acetonitrile in 100 mM triethylammonium bicarbonate. Keep this solution on ice.
5. Evaporate 1 ml portions of the solution in a screw-topped test-tube (20 × 150 mm) using a vacuum pump and vortexer as described in *Protocol 1*, step 11. Add 0.2 ml of water to the residue and evaporate. Repeat this operation until all the triethylammonium bicarbonate has been removed.
6. Lyophilize the oligomer from water.

The oligomer may be stored indefinitely as a solid. Alternatively, the oligomer may be dissolved in acetonitrile–water (1:1 v/v) and the solution stored at −20 °C. The oligomer is stable under these conditions for at least 6 months. A typical reversed phase HPLC of a dodecamer which was purified by DEAE cellulose chromatography followed by preparative reversed phase HPLC is shown in *Figure 3*.

3. Characterization of oligonucleoside methylphosphonates

The purity of oligonucleoside methylphosphonates may be determined by reversed phase HPLC as shown in *Figure 3* and by polyacrylamide gel electrophoresis as described below. The chain length of the oligomer may be determined by hydrolysis of the oligomer with aqueous piperidine and the

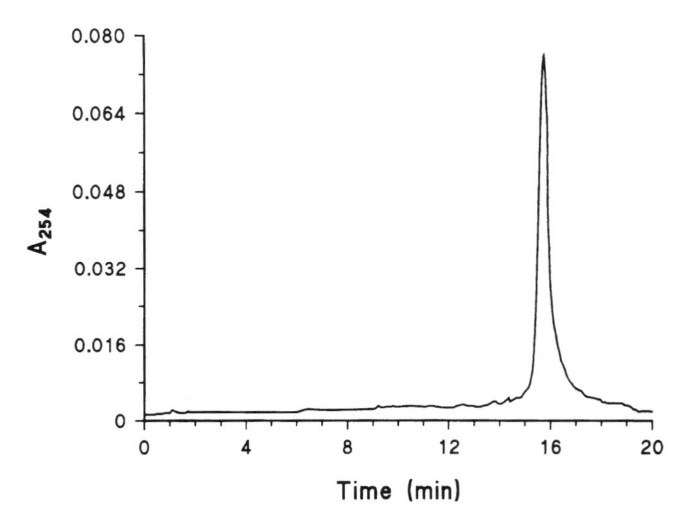

Figure 3. Reversed phase HPLC or d-ApTAGGATTTGTC (the underline indicates the positions of the methylphosphonate linkages) run on a Whatman C-18 ODS-3 RAC column (0.45 × 10 cm). The column was eluted with a linear gradient of 1–30% acetonitrile in 0.1 M sodium phosphate buffer (pH 5.8) at a flow rate of 1.5 ml/min.

positions of purine and pyrimidine residues within the oligomer may be verified by a chemical sequencing procedure.

3.1 Phosphorylation and gel electrophoresis of oligonucleoside methylphosphonates

Oligonucleoside methylphosphonates which contain a single phosphodiester internucleotide bond linking the 5'-terminal nucleoside with the rest of the oligomer may be phosphorylated enzymatically using polynucleotide kinase. Oligomers which lack this terminal diester linkage and which contain only methylphosphonate internucleoside linkages are not phosphorylated. The procedure outlined in *Protocol 4* is used to phosphorylate the oligomer with [^{32}P]phosphate.

Protocol 4. Phosphorylation of oligonucleoside methylphosphonates

1. Place a solution containing 0.5–1.0 nmol of the oligomer in a 500 μl Eppendorf tube and evaporate to dryness in a SpeedVac.
2. Dissolve the oligomer in a solution containing the following components:
   - water — 5.5 μl
   - 0.5 M Tris (pH 7.6), 50 mM magnesium chloride — 1.0 μl

Protocol 4. *Continued*

- 50 mM dithiothreotol 1.0 μl
- 100 μM adenosine triphosphate 1.0 μl
- [γ^{32}P]adenosine triphosphate (10 μCi/μl) 1.0 μl
- polynucleotide kinase (30 units/μl) 0.5 μl

3. Incubate the reaction mixture at 37 °C for 45 min.
4. Dilute the reaction mixture with 5 ml of 25 mM triethylammonium bicarbonate (pH 7.5).
5. Load the solution onto a SEP PAK reversed phase cartridge as described in *Protocol 3*.
6. Wash the SEP PAK with 10 ml of 25 mM triethylammonium bicarbonate and 10 ml of 5% acetonitrile in 25 mM triethylammonium bicarbonate. This step removes salts and unreacted adenosine triphosphate from the oligomer.
7. Elute the oligomer with 3 ml of 50% acetonitrile in 100 mM triethylammonium bicarbonate.
8. Remove the solvents by evaporation and remove the residual buffer by repeated evaporation with 0.5 ml aliquots of water.

Phosphorylated oligomers may be electrophoresed on denaturing 15% polyacrylamide gels following the general procedure described by Maniatis *et al.* for gel electrophoresis of oligodeoxyribonucleotides (4). The oligomer is visualized by autoradiography of the wet gel. A typical autoradiogram of a gel containing a dodecamer is shown in *Figure 4*.

The oligomer may be recovered from the gel by using the extraction procedure outlined in *Protocol 5*. The recovery of the oligomer using this procedure is generally 50% or better.

Protocol 5. Extraction of an oligonucleoside methylphosphonate from a polyacrylamide gel

1. Wrap the wet gel in plastic wrap and place radioactive ink markers near three of the corners of the gel.
2. Autoradiograph the wet gel at room temperature.
3. Lay the wet gel over the autoradiogram using the radioactive ink markers to properly align the gel.
4. Using the autoradiogram as a guide, excise the area of the gel corresponding to the oligomer.
5. Place the gel slice in a 1 ml Eppendorf tube and crush the gel into a paste using a plastic pestle.

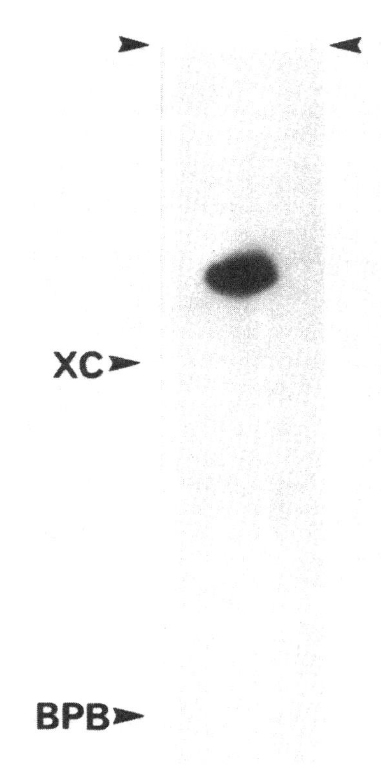

Figure 4. Polyacrylamide gel electrophoresis of d-[^{32}P]pApTAGGATTTGTC. The oligomer was electrophoresed on a 20 × 20 cm, 15% polyacrylamide gel under denaturing conditions using TBE buffer (4) at 800 V for 45 min. The arrowheads indicate the top of the gel. XC and BPB are xylene cyanol and bromophenol blue tracking dyes, respectively.

Protocol 5. *Continued*

6. Extract the crushed gel with four 0.5 ml aliquots of 1 M triethylammonium bicarbonate. For each extraction:
   - incubate the crushed gel with the 1 M triethylammonium bicarbonate for 15 min with occasional vortexing
   - briefly centrifuge
   - remove the supernatant
   - add fresh 1 M triethylammonium bicarbonate
7. Dilute the combined extracts with 5 ml of 25 mM triethylammonium bicarbonate.
8. Desalt the oligomer solution on a SEP PAK as described in *Protocol 3*.
9. Dissolve the oligomer in 500 µl of acetonitrile-water (1:1 v/v) and store the solution at −20 °C.

3.2 Hydrolysis of oligonucleoside methylphosphonates with piperidine

The methylphosphonate linkages of 5′-^{32}P-end labelled oligonucleoside methylphosphonates are hydrolysed in a random manner when the oligomer is treated with 1 M aqueous piperidine. This reaction may be used to verify the chain length of the oligomer as described in *Protocol 6*.

Protocol 6. Hydrolysis of oligonucleoside methylphosphonates with piperidine

1. Pipette aliquots which contain 20 000 d.p.m. of the ^{32}P-labelled oligomer obtained in step 10 of *Protocol 5* into 500 µl Eppendorf tubes.
2. Evaporate the solvent in a SpeedVac.
3. Dissolve the residue in 10 µl of 1 M aqueous piperidine.
4. Incubate the solutions at 37 °C for 10, 20, 40, or 60 min.
5. After the appropriate time, freeze the reaction in a dry ice–ethanol bath and lyophilize.
6. Dissolve the residue in 90% formamide loading buffer and electrophorese on a 15% polyacrylamide gel under denaturing conditions (4).

After gel electrophoresis and autoradiography, a set of bands is observed which correspond to two types of ^{32}P-labelled oligomers, those which terminate with a 3′-hydroxyl group and those which terminate with a 3′-methylphosphonate group. Two of the possible products of piperidine hydrolysis are shown in *Figure 5*. Lanes 2–4 show the products of the reaction after hydrolysis for 10, 20, and 40 min respectively. Since the phosphodiester linkage is not hydrolysed by this method, the limit digest consists of two dimers, d-pNpN and dpNpN*p*.

3.3 Determining the location of pyrimidines and purines in oligonucleoside methylphosphonates

Treatment of 5′-^{32}P-end labelled oligonucleoside methylphosphonates with acid or with hydrazine results in random formation of abasic sites at purine and pyrimidine residues respectively in the oligomer. These abasic sites undergo spontaneous chain cleavage at neutral pH to produce ^{32}P-labelled oligomer chains which terminate with a 3′-hydroxyl group as shown in *Figure 5*. Thus, in contrast to the chemical sequencing procedure used for oligodeoxyribonucleotides, a base catalysed *β*-elimination cleavage step is not required. These reactions serve to locate the position of purines and pyrimidines as shown in lanes 5 and 6 of *Figure 5* and can thus be used to

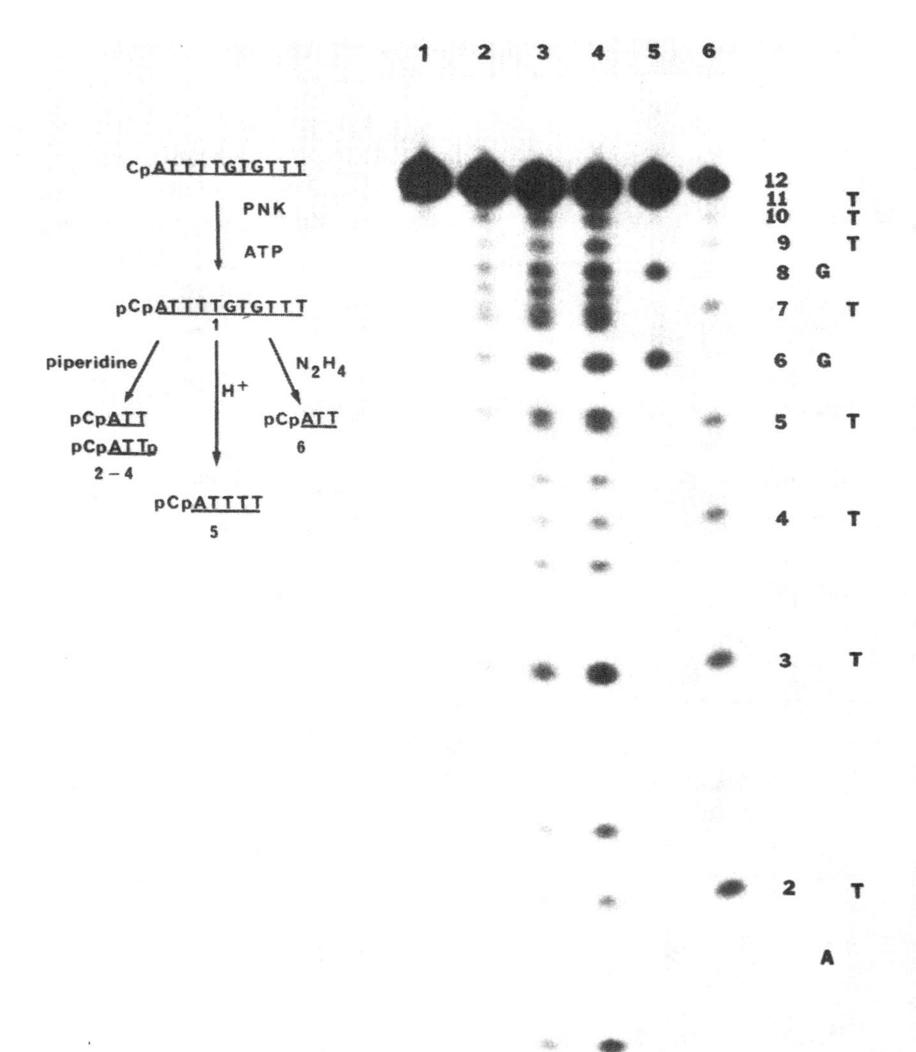

Figure 5. Hydrolysis of d-[^{32}P]pCpATTTGTGTTT. The oligomer (lane 1) was treated with 1 M aqueous piperidine at 37 °C for 10 min (lane 2), 20 min (lane 3), or 40 min (lane 4) or with 1 M hydrochloric acid at 37 °C for 30 min (lane 5), or with 85% hydrazine hydrate at 37 °C for 30 min (lane 6), and then subjected to polyacrylamide gel electrophoresis on an 18 × 30 cm, 15% polyacrylamide gel under denaturing conditions. The oligomer chain lengths and positions of the purine and pyrimidines are indicated at the right side of the gel. The nature of the cleavage reactions is indicated at the left side of the gel.

verify the sequence of the oligomer. The procedure for carrying out these reactions is given in *Protocol 7*.

Protocol 7. Purine and pyrimidine specific cleavage of oligonucleoside methylphosphonates

1. Place aliquots which contain 20 000 d.p.m. of the ^{32}P-labelled oligomer obtained from step 10 of *Protocol 5* in 500 μl Eppendorf tubes.
2. Evaporate the solvents in a SpeedVac.
3. Dissolve the residue in 10 μl of 1 M hydrochloric acid or 10 μl of 85% hydrazine hydrate.
4. Incubate the solutions at 37 °C for 30 min.
5. Freeze the solutions in dry ice and lyophilize.
6. Dissolve the residue in 10 μl of water, freeze the solution and lyophilize.
7. Dissolve the residue in 10 μl of water and incubate the solution at 37 °C for 30 min.
8. Freeze the solution and lyophilize.
9. Dissolve the residue in 10 μl of 90% formamide loading buffer and electrophorese on a 15% polyacrylamide gel under denaturing conditions (4).

4. Use of antisense oligonucleoside methylphosphonates in cell culture experiments

Antisense oligonucleoside methylphosphonates have been used to study gene expression in a variety of mammalian cell lines (5, 6) and in cells infected with viruses such as vesicular stomatitis virus (7), herpes simplex virus (8, 9) and human immunodeficiency virus (10, 11). Oligomers which are 12–17 nucleotides in length have sufficiently unique sequences to enable them to interact specifically with a single cellular or viral mRNA species. The target sites for the oligomers are usually the 5′-end or initiation codon region and to a lesser extent the coding region of mRNA, or the donor or acceptor splice sites of precursor mRNA.

Because the methylphosphonate linkages of oligonucleoside methylphosphonates oligomers are not hydrolysed by nucleases, it is not necessary to culture the cells in serum free medium, nor is it necessary to heat-inactivate the medium. Although the precise mechanism is unknown, it appears that oligonucleoside methylphosphonates are taken up intact by passive diffusion. Therefore the oligomer can be added directly to the cell culture medium. It is not necessary to use calcium phosphate precipitation, electroporation or other methods to effect uptake of the oligomer. A convenient method for preparing the oligomer for use in cell culture experiments is given in *Protocol 8*.

Protocol 8. Preparation of oligonucleoside methylphosphonates for use in cell culture experiments

1. Dissolve the lyophilized oligomer obtained in step 6 of *Protocol 3* in acetonitrile–water (1:3 v/v) to give a final oligomer concentration of approximately 1 mM.
2. Transfer the solution to a glass test tube which has been previously autoclaved. This transfer should be carried out in a laminar flow hood.
3. Fit a sterile rubber septum to the top of the test-tube.
4. Pierce the septum with a sterile 18-gauge syringe needle to which is fitted a sterile Millex 0.45 μm filter unit.
5. Freeze the solution in a dry ice–ethanol bath and lyophilize in a vacuum bottle. When lyophilization is complete air may be admitted to the bottle. The Millex unit will filter the air entering the test-tube thus keeping the sample sterile.
6. Return the test-tube to the laminar flow hood.
7. Remove the septum and dissolve the lyophilized oilgomer in cell culture medium to the desired concentration.

The kinetics of uptake of oligonucleoside methylphosphonates by mammalian cells growing in monolayer are not well understood, although it appears to be dependent upon the particular cell line being studied. Therefore in experiments involving viruses, it is often advantageous to pre-incubate the cells with the oligomer before infecting with virus. In cases where cellular gene expression is being studied, sufficient time should be allowed for uptake of the oligomer before assaying for inhibition.

4.1 Inhibition of virus protein synthesis in vesicular stomatitis virus infected cells

Antisense oligonucleoside methylphosphonates can be used to inhibit viral gene expression in virus infected cells. The exact conditions for growing the cells, and the assays employed, will vary with the system under investigation. As an example, the following describes a procedure for studying the effects of antisense oligonucleoside methylphosphonates on vesicular stomatitis virus

(VSV) protein synthesis in mouse L929 cells. The cells are cultured and treated with oligomer as described in *Protocol 9.* In these experiments the oligomers are used at concentrations between 50 and 150 μM. To reduce the amount of oligomer required, the cells are grown in 1.5 cm diameter culture chambers in a 24-well cell culture plate (Nunclon, Inc.). This enables the experiments to be carried out using 0.2 ml of cell culture medium per chamber. Three types of culture media, which are commercially available, are required in these experiments:

- 10% EMEM: Eagle's minimal essential medium supplemented with 10% fetal bovine serum
- 2% EMEM: Eagle's minimal essential medium supplemented with 2% fetal bovine serum
- 2% DMEM: Dulbecco's modified Eagle's medium which lacks L-glutamine and L-methionine and which is supplemented with 2% fetal bovine serum

---

Protocol 9. Treatment of VSV infected cells with antisense oligonucleoside methylphosphonates

1. Prepare a suspension of mouse L929 fibroblasts in 10% EMEM such that the density is 17 000 cells per ml of medium as determined using a Coulter counter.
2. Place 1 ml of the cell suspension into the culture well of a 24-well Nunclon cell culture plate. Three wells each are seeded with cells for each oligomer concentration to be tested. It is very important to seed the same number of cells in each well.
3. Incubate the cells at 37 °C in a 5% CO_2 atmosphere for three days. The cells should be subconfluent at this time.
4. Prepare the oligomer as described in steps 1–6 of *Protocol 8* and dissolve in 2% EMEM to give the desired concentration.
5. Remove the media from the cells and replace with 0.2 ml of media containing the oligomer.
6. Incubate the cells at 37 °C in a 5% CO_2 atmosphere for 8–10 h. To prevent the cells form drying out, sterile water is placed in unoccupied wells of the culture plate. The relative humidity in the incubator should be at least 95%.
7. Add VSV suspended in 10 μl of 2% EMEM such that the multiplicity of infection is between 5 and 10.
8. Return the cells to the incubator and shake the cell plate every 15 min for the first hour post-infection. Incubate the cells for an additional 5 h without shaking.

Protocol 9. *Continued*

9. Remove the medium from the cells 6 h post-infection. Wash the cells once with 0.5 ml of ice-cold Dulbecco's phosphate buffered saline and remove the wash.
10. Add 0.2 ml of 2% DMEM which contains 25 μCi of [^{35}S]methionine and incubate the cells at 37 °C for 30 min.
11. Remove the medium and wash the cells once with 0.25 ml of a solution which contains 0.14 M sodium chloride and 20 mM Tris–HCl, buffered at pH 7.4.
12. Remove the wash and incubate the cells for 5 min on ice with 0.1 ml of a solution which contains 0.14 M sodium chloride, 20 mM Tris–HCl, 0.75% NP40 buffered at pH 7.4. This treatment will lyse the cells.
13. Scrape the cell lysate from the bottom of the culture well. Remove 30 μl of lysate and aliquot into an Eppendorf tube which contains gel electrophoresis loading buffer composed of the following:
    - 0.125 M Tris–HCl pH 6.8
    - 0.4% sodium dodecyl sulphate
    - 20% glycerol
    - 10%β-mercaptoethanol
    - 2% bromophenol blue tracking dye
14. Briefly vortex the tube and place in a boiling water bath for 2 min. This solution can be stored at −20 °C.

The solution obtained in step 14 of *Protocol 9* is analysed by polyacrylamide gel electrophoresis according to the method of Laemmli (12). The gel consists of a 5% acrylamide stacking gel and a 10% acrylamide resolving gel. The stacking gel is prepared by mixing the following:

- 1.66 ml of 30% acrylamide (29.6 g acrylamide, 0.4 g bisacrylamide in 100 ml water)
- 2.5 ml 0.5 M Tris–HCl, pH 6.8
- 0.1 ml of 10% sodium dodecyl sulphate
- 5.7 ml of water
- 25 μl of 20% ammonium persulphate
- 12.5 μl of tetramethylenediamine

The resolving gel is prepared by mixing the following:

- 10 ml of 30% acrylamide
- 7.5 ml of 1.5 M Tris–HCl, pH 8.8

- 0.3 ml of 10% sodium dodecyl sulphate
- 12 ml of water
- 50 μl of 20% ammonium persulphate
- 25 μl of tetramethylenediamine

The solution is heated in a boiling water bath for 2 min and 15 μl aliquots are immediately loaded onto the gel. Electrophoresis is carried out using a buffer which contains 0.025 M Tris, 0.192 M glycine and 0.1% sodium dodecyl sulphate buffered at pH 8.3. The gel is run at 15 mA for 1.5 h or until the dye front reaches the resolving gel and then at 30 mA for 2.5 h or until the dye front is within 1 cm of the bottom of the gel. The gel is then stained for 1 h in a solution containing 0.125% Coomassie blue, 50% methanol and 10% acetic acid. The stained gel is then washed with 250 ml of 50% methanol, 10% acetic acid overnight at room temperature. The gel is dried and autoradiographed overnight at −80 °C.

The five VSV proteins are well separated by this procedure (7) and may be quantified by densitometry or by cutting out the protein bands from the gel and counting the bands in a liquid scintillation counter. The amounts of protein synthesized in the presence of various concentrations of antisense oligonucleoside methylphosphonate are compared with the amount of protein from VSV infected cells which have not been exposed to oligomer. The syntheses of VSV proteins are inhibited 57–98% by oligomers which are complementary to the initiation codon regions of VSV mRNAs at an oligomer concentration of 150 μM (7).

References

1. Miller, P. S. and Ts'o, P. O. P. (1988). *Annu. Rep. Med. Chem.*, **23**, 295.
2. Miller, P. S. (1990). In *Herpesviruses, the Immune System and AIDS.* (ed. L. Aurelian) pp. 343–360. Kluwer Academic Publishers, Boston, MA.
3. Uhlmann, E. and Peyman, A. (1990). *Chem. Rev.*, **90**, 544.
4. Maniatis, T., Fritsch, E. F., and Sambrook, J. (1982). *Molecular Cloning, A Laboratory Manual.* pp. 184–185. Cold Spring Harbor Laboratory Press, Cold Spring Harbor, NY.
5. Brown, D., Yu, Z., Miller, P., Blake, K., Wei, C., Kung, H.-F., Black, R., Ts'o, P., and Chang, E. (1989). *Oncogene Res.* **4**, 243.
6. Vasanthakumar, G. and Ahmed, N. K. (1989). *Cancer Commun.*, **1**, 225.
7. Agris, C. H., Blake, K. R., Miller, P. S., Reddy, M. P., and Ts'o, P. O. P. (1986). *Biochemistry*, **25**, 6268.
8. Smith, C. C., Aurelian, L., Reddy, M. P. Miller, P. S., and Ts'o. P. O. P. (1986). *Proc. Natl. Acad. Sci. USA*, **83**, 2787.
9. Kulka, M., Smith, C., Aurelian, L., Fishelevich, R., Meade, K., Miller, P., and Ts'o, P. O. P. (1989). *Proc. Natl. Acad. Sci. USA*, **86**, 6868.

10. Sarin, P. S., Agrawal, S., Civeira, M. P., Goodchild, J., Ikeuchi, T., and Zamecnik, P. C. (1988). *Proc. Natl. Acad. Sci. USA*, **85**, 7448.
11. Zaia, J. J., Rossi, J. J., Murakawa, G. J., Spallone, P. A., Stephens, D. A., Kaplan, R. E., Eritja, R., Wallace, R. B., and Cantin, E. M. (1988). *J. Virol*, **62**, 3914.
12. Laemmli, U.K. (1970). *Nature*, **227**, 680.

7

Oligodeoxynucleotides containing modified bases

BERNARD A. CONNOLLY

1. Introduction

Several hundred modified bases have been described in the literature and outlining their incorporation into oligonucleotides would clearly occupy several volumes. This chapter by necessity therefore only describes the preparation of a small number of bases in use in our laboratory. Each modified base has an exocyclic group replaced by either hydrogen or sulphur. We have mainly used oligodeoxynucleotides containing these altered bases to probe protein–nucleic acid interactions. However several of them also have potentially useful spectroscopic and chemical/photoreactivity properties suggesting further useful applications.

2. General methods

2.1 Safety precautions

Many of the solvents and reagents used in the syntheses in this chapter are toxic and flammable. All chemical syntheses should be performed in a well-ventilated fume hood. Wear safety glasses, rubber gloves, and a laboratory coat at all times. Some of the specific hazards associated with individual reagents are given in the protocols. In particular it should be noted that many base analogues are by their very nature toxic, mutagenic, and carcinogenic. It should as a matter of course be assumed that the base analogues and their intermediates mentioned in this chapter have these properties. Always handle them in a fume hood. Do not allow their solutions to come into contact with skin. When handling them in solid form, in which they often exist as fine powders, wear a disposable gauze face mask.

2.2 Analytical and preparative techniques

The two most important methods, enabling most of the chemistry outlined in the protocols to be successfully performed are thin layer chromatography

(TLC) and flash chromatography. TLC is an analytical technique, based on the separation of compounds on silica gel, used to monitor the progress of reactions. Flash chromatography is a preparative method that employs silica gel column chromatography to purify products after reactions. All the TLC mentioned in the protocols used silica gel $60F_{254}$ aluminium backed sheets (5×7.5 cm, layer thickness 0.2 mm) supplied by Merck (article 5549). Compounds were detected under short wavelength UV light. Flash chromatography used silica gel 60 (particle size 0.040–0.063 mm, 230–400 mesh ASTM) supplied by Merck (article 9385). Both techniques have been described in depth (1) and this reference should be consulted for practical details.

2.3 Necessity for anhydrous conditions

Many of the chemical reactions described in the protocols require dry conditions. For all the reactions outlined the following precautions are sufficient. Reactions should be carried out in stoppered quick fit glass flasks which have been dried in an oven at 80–90 °C for 2–3 h. Solid materials should be dried at room temperature *in vacuo* (use an oil pump) in a vacuum desiccator over silica gel. We recommend self-indicating silica gel (particle size 0.78–2.8 mm, 6–20 mesh) supplied by BDH. This gel is blue when dry but turns pink on absorbing H_2O. If after overnight drying of a solid the silica gel has turned pink it should be replaced and the drying continued. The sample is dry enough when the silica gel remains blue after overnight drying. Many of the intermediates of a synthesis are often produced as oils which do not require purification before the next step. These can be dried by co-evaporation from pyridine. This simply involves dissolving the material in 20–50 ml of dry pyridine and evaporation using an oil pump. The process should be repeated twice. In the course of evaporation any water is removed as an azeotrope with pyridine. Many solvents contain appreciable levels of H_2O and should be dried as outlined below. Providing these precautions are taken all the reactions should work well. It should be noted that super-anhydrous conditions are not necessary. Thus reaction vessels can be exposed to the atmosphere briefly for example to add reagents or remove small samples for TLC without significantly affecting yields.

Drying of solvents

- dichloromethane: distil from phosphorus pentoxide (10 g/litre) and store in a tightly stoppered bottle
- pyridine: reflux with barium oxide (about 15 g/litre) for 5 h and distil; store in tightly stoppered bottles
- dioxan/tetrahydrofuran: reflux with sodium metal (caution: highly reactive) and benzophenone until a deep blue colour is obtained. Distil and use at once

- triethylamine/N,N-diisopropylethylamine: reflux with calcium hydride (about 10 g/litre) for 5 h and distil. Store in tightly stoppered bottles
- methanol: reflux with iodine (0.5 g) and magnesium (5 g) per 500 ml for 2–3 h. Distil and store in tightly stoppered bottles
- hydrazine; use a freshly opened bottle of the 98% anhydrous grade from Aldrich (article 21, 515–5) (caution: extremely toxic). Do not attempt further purification or drying by any method especially distillation due to explosive risk

3. Synthesis of modified bases

We have used the phosphoramidite approach (see Chapter 1) to prepare oligodeoxynucleotides. Thus the modified base preparations described here are primarily designed to give a derivative suitable for the phosphoramidite method rather than the free base itself. This means preparing modified bases that contain a 5′-dimethoxytrityl group, a 3′-phosphoramidite function and base protection as appropriate. This is actually advantageous as in our experience, the free bases themselves are often non-crystalline and far too polar to purify by flash chromatography. The organic solvent-soluble 5′-dimethoxytrityl derivatives which can be purified by flash chromatography are much easier to obtain in a pure state. Many of the methods used in the preparation of modified bases are common to all the syntheses. These include the standard work-up procedure (used to remove water soluble materials from the organic solvent soluble product and to neutralize acidic species after a chemical reaction), the standard dimethoxytritylation procedure (used to add a 5′-dimethoxytrityl group to a modified deoxynucleoside) and the standard phosphitylation procedure (used to prepare the 3′-phosphoramidite). These are described in *Protocols 3*, *4*, and *11* respectively.

3.1 Preparation of a purine-9-β-D-2′-deoxyribofuranoside derivative suitable for oligodeoxynucleotide synthesis

This base is a derivative of 2′-deoxyadenosine (dA) and has the 6-NH_2 group replaced by hydrogen to give purine (dP) (*Figure 1A*). The modified base is prepared from dA as shown in outline in *Figure 1B* and described in detail in *Protocol 2*. The key step is the direct replacement of the 6-NH_2 group of 3′,5′-diacetyl-dA with pentylnitrile (2). Unfortunately the pentylnitrite must also be prepared prior to use as outlined in *Protocol 1*. The dP prepared is always contaminated with dA (about 10–15%). This arises because the 3′,5′-diacetyl-dA also contains 3′,5′-N-triacetyl-dA (in which the 6-NH_2 group is also acetylated) which does not react with the pentyl nitrite and on NH_3 treatment yields the starting dA. Additionally the reaction of 3′,5′-diacetyl-dA with pentylnitrite never proceeds in 100% yield and some starting material

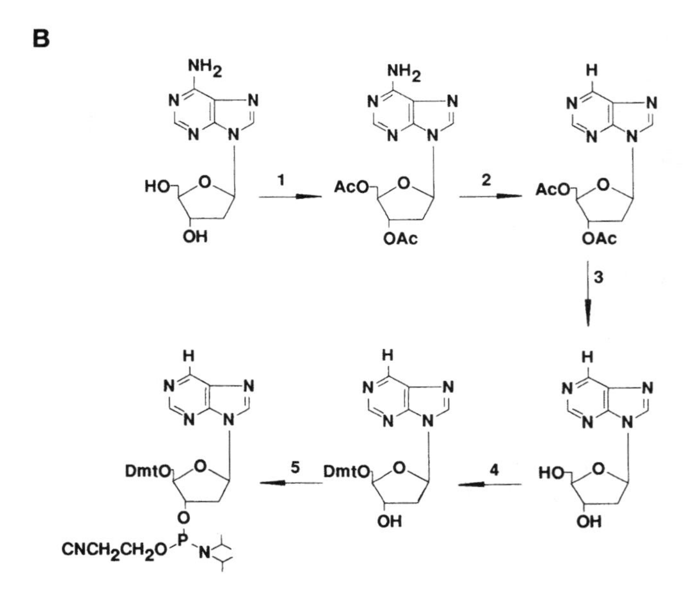

Figure 1. A. The structures of deoxyadenosine (dA) and deoxypurine (dP). **B.** The synthesis of the 5′-dimethoxytrityl,3′-cyanoethylphosphoramidite derivative of dP from dA. Reagents: (1) acetic anhydride, (2) 1-pentylnitrite, (3) NH_3, (4) dimethoxytritylchloride, (5) 2-cyanoethyl N,N-diisopropylchlorophosphoramidite. Ac = acetyl, Dmt = dimethoxytrityl.

remains which also gives dA with NH_3. Although the dA cannot easily be separated from the dP at this stage, after 5′-dimethoxytritylation the dmt-dP is easily separated from dmt-dA by flash chromatography. As dP contains no exocyclic functions on the base moiety no base protection is needed and the phosphoramidite is easily prepared by the standard approach (*Protocol 11*). It should be noted that after purification and solidification by precipitation this phosphoramidite is extremely sticky. However overnight drying *in vacuo* over silica gel gives an easy to handle free flowing powder.

Protocol 1. Preparation of 1-pentylnitrite[a]

1. Add 17 ml concentrated H_2SO_4 to 13 ml H_2O. **Take care—wear safety glasses!** Leave to cool. Add 70 ml of 1-pentanol and cool on ice.
2. Place 48 g (0.7 mol) of sodium nitrite in a 500 ml 2 neck round-bottomed flask. Add 190 ml H_2O and shake to dissolve.
3. Fit the flask with a thermometer, 200 ml dropping funnel and a magnetic stirring bar. Place in an ice–salt bath and cool to −5 °C.
4. Place the 1-pentanol/H_2O/H_2SO_4 mixture in the dropping funnel. Add dropwise to the sodium nitrite solution with vigorous stirring over 45 min. The temperature of the flask contents must not rise above 0 °C.
5. Leave the resulting yellow solution at room temperature for 90 min. It should separate into an upper yellow organic layer and a lower aqueous layer containing a dense precipitate.
6. Filter the mixture to remove the precipitate.
7. Place filtrate in a separating funnel and run off and discard the lower aqueous layer.
8. Add a solution of 1g $NaHCO_3$ and 12.5g NaCl in 50 ml H_2O to the yellow organic layer remaining in separating funnel. Shake and let the layers separate. Run off the lower aqueous layer and discard.
9. Transfer the organic layer to a 250 ml round-bottomed flask. Add about 3 g of anhydrous Na_2SO_4 and swirl to dry the organic layer. Remove Na_2SO_4 by filtration.
10. Purify the 1-pentylnitrite by distillation at atmospheric pressure under a stream of nitrogen. The pentylnitrite distils at 104 °C.
11. Store in the dark at 2 °C in a tightly stoppered flask. Use within 2 weeks of preparation.

[a] Pentylnitrite is toxic. It is a powerful heart stimulant that causes severe headaches. All manipulations involving this chemical should be performed in a well ventilated fume hood. Do not breathe its vapours or allow it to come into contact with the skin.

Protocol 2. Preparation of dmt-dP
(This protocol should be used in conjunction with *Figure 1B.*)

A. *Preparation of 3′,5′-diacetyl-dA*

1. Suspend 6.3 g (25 mmol) of dry dA in 50 ml of dry pyridine.
2. Add 5.2 ml (55 mmol) of acetic anhydride followed by 0.5 g of 4-dimethylaminopyridine. Protect from moisture and stir at room temperature for 3 h. During this time all the dA will dissolve.

Protocol 2. *Continued*

3. Test by TLC ($CHCl_3$ 90, CH_3OH 10). The product has an R_f of 0.5. Traces of higher and lower R_f material will be present. No starting material (R_f 0.05) should be visible.
4. Add 5 ml of CH_3OH to destroy remaining anhydride.
5. Evaporate to a sticky oil with an oil pump. Remove pyridine by co-evaporation from 2 × 50 ml toluene.
6. Purify the product by recrystallization from ethanol. Add 50 ml boiling ethanol, stir to dissolve and leave at 4 °C overnight.
7. Collect the crystals by filtration and wash with 50 ml ice-cold ethanol.
8. Yield should be about 80%. TLC shows one major spot (R_f 0.5) together with traces of triacetyl derivative (R_f 0.8).

B. *Preparation of 3′,5′diacetyl-dP*

1. Suspend 3.35 g of 3′,5′-diacetyl-dA in 50 ml of dry tetrahydrofuran. Use a thick walled glass vessel fitted with a Teflon screw cap. Add 10 ml of 1-pentylnitrite (*Protocol 1)* and heat at 50 °C for 48 h.
2. Add a further 10 ml of 1-pentylnitrite and heat at 50 °C for 48 h.
3. Examine by TLC ($CHCl_3$ 95, CH_3OH 5). Almost all of the starting material (R_f 0.3) should have been converted to product (R_f 0.45).
4. Perform standard work up (*Protocol 3*).
5. Dissolve the product in a small volume of $CHCl_3$ and purify by flash chromatography. Elute in turn with 500 ml $CHCl_3$ + 2% CH_3OH and 500 ml $CHCl_3$ + 4% CH_3OH. The product elutes with the last solvent. Check column eluate by TLC and pool pure fractions.
6. Evaporate to give an oil that will neither solidify nor crystallize.

C. *Preparation of dP*

1. Dissolve the 3′,5′-diacetyl-dP obtained as an oil above in 50 ml ethanol and 50 ml 35% aqueous nH_3. Use a thick walled glass vessel with a Teflon screw cap. Heat at 50 °C for 3 h.
2. Cool on ice and carefully open the vessel. Test by TLC ($CHCl_3$ 95, CH_3OH 5) which should show complete conversion of starting material (R_f 0.45) to product which remains at the origin.
3. Evaporate the solution and dissolve product in 100 ml H_2O.
4. Extract with 2 × 100 ml ethylacetate. Discard the organic (upper) layer at each stage and retain the aqueous (lower).
5. Evaporate the aqueous layer to dryness to give a crude sample of dP.

Protocol 2. *Continued*

D. *Preparation of dmt-dP*

1. Perform standard dimethoxytritylation (*Protocol 4*) on crude dP obtained above.
2. The yield of dmt-dP from 3′,5′-diacetyl-dA (3 stages) should be between 30 and 40%.

Protocol 3. Standard work-up procedure

1. Evaporate the reaction mixture to an oil using a water pump (all reaction solvents except pyridine). If pyridine is the reaction solvent evaporate to an oil using an oil pump. Co-evaporate the residue twice from 50 ml toluene to remove excess pyridine.
2. Dissolve residue in $CHCl_3$ (100–400 ml depending on scale).
3. Extract with an equal volume of 5% $NaHCO_3$ (twice) followed by an equal volume of saturated NaCl. Use a separating funnel. At each stage retain the organic (lower) layer and discard the aqueous (upper).
4. Dry the organic layer by addition of 5 g anhydrous Na_2SO_4 followed by manual swirling for 30 s.
5. Filter to remove the Na_2SO_4 and retain the organic filtrate.
6. Evaporate the solution using a water pump. This will yield either a dense syrup or a solid.

Protocol 4. Standard dimethoxytritylation procedure[a]

1. Dry the deoxynucleoside to be dimethoxytritylated by co-evaporation from 50 ml dry pyridine. Repeat twice.
2. Dissolve the dried deoxynucleoside in 50 ml dry pyridine. Protect from moisture (drying tube).
3. Add dimethoxytritylchloride (1.2 equivalents) and leave at room temperature for 3 h.
4. Examine by TLC. $CHCl_3$ 95, CH_3OH 5 is generally a useful solvent. The starting material (low R_f) should be completely converted to product (higher R_f). The product will stain bright orange if a drop of 3% trichloroacetic acid in CH_2Cl_2 is applied. Side products at the solvent front will also be visible.
5. If the reaction is not complete either leave overnight or add more (0.2–0.4 equivalents) dimethoxytritylchloride.

Protocol 4. *Continued*

6. On completion of the reaction add 25 ml CH_3OH and then perform the standard work-up (*Protocol 3*).
7. Dissolve the product formed in a small volume of $CHCl_3$ and purify by flash chromatography. A useful solvent is $CHCl_3$ containing 0.5% $(C_2H_5)_3N$ and between 0 and 5% CH_3OH.
8. Test column eluate by TLC and pool pure fractions.
9. Evaporate to dryness using a water pump. The product should turn solid. If not evaporate oil formed using an oil pump until solidification occurs.
10. Yields should be between 60 and 80%.

[a] This protocol is suitable for 1–10 g of deoxynucleoside.

3.2 Preparation of 2-aminopurine-9-β-D-2′-deoxyribofuranoside and 6-thiodeoxyguanosine derivatives suitable for oligonucleotide synthesis

In these related derivatives the 6-keto oxygen atom of dG is replaced by a hydrogen atom (to give the 2-aminopurine derivative, $d^{2am}P$) or a sulphur to give the 6-thiodeoxyguanosine analogue ($d^{6S}G$) (see *Figure 2A*). The initial steps in the syntheses of both the $d^{2am}P$ and $d^{6S}G$ derivatives are common resulting in an efficient overall synthesis. These common steps are shown in *Figure 2B* and described in *Protocol 5* and give initially a 3′,5′,N-tribenzoyl-dG derivative (by perbenzoylation of dG with benzoic anhydride) which is converted to the 6-(2,4,6-triisopropylbenzenesulphonyl) derivative by reaction with the arylsulphonylchloride. After this the pathways to $d^{2am}P$ and $d^{6S}G$ derivatives diverge (described in *Protocols 6* and *7* respectively) but are based on the well documented displacement of the benzenesulphonyl group from the 6-position of the dG with nucleophiles (3,4). The $d^{2am}P$ compound is accessible by displacement of the benzenesulphonyl group with hydrazine followed by an oxidative elimination of the hydrazino group with Ag_2O to yield the tribenzoylated-$d^{2am}P$ (5). The $d^{6S}G$ compound can be prepared by

A

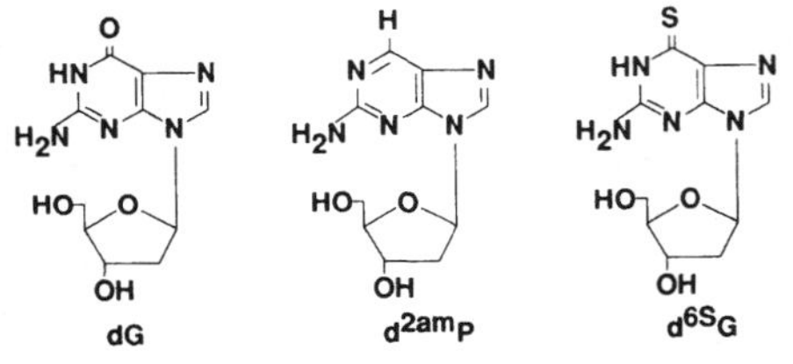

dG $d^{2am}P$ $d^{6S}G$

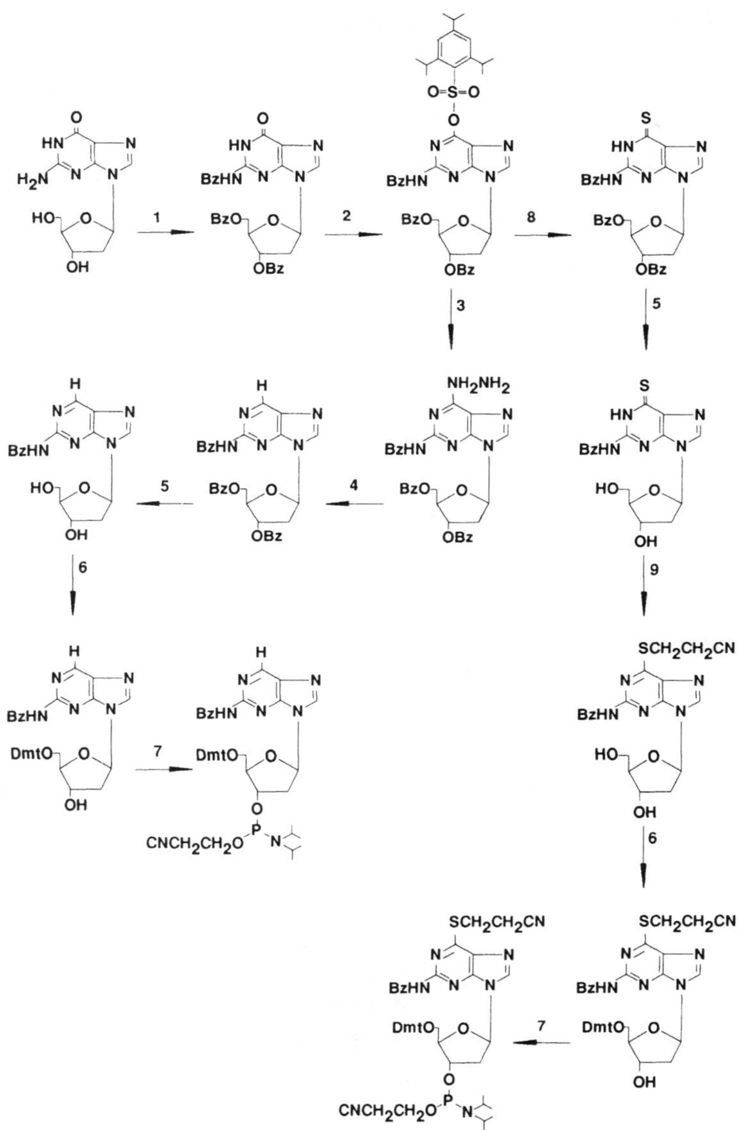

Figure 2. A. The structures deoxyguanosine (dG), 2-aminopurine-9-ß-D-2′-deoxyribofuranoside ($d^{2am}P$) and 6-thiodeoxyguanosine ($d^{6S}G$). **B.** The syntheses of the 5′-dimethoxytrityl 3′-cyanoethylphosphoramidite derivatives of N-benzoyl-2-aminopurine-9-ß-D-2′-deoxyribofuranoside and N-benzoyl-S-cyanoethyl-6-thiodeoxyguanosine from dG. Reagents: (1) benzoic anhydride, (2) 2,4,6-triisopropylbenzenesulphonylchloride, (3) hydrazine, (4) Ag_2O, (5) sodium methoxide, (6) dimethoxytritylchloride, (7) 2-cyanoethyl N,N-diisopropylchlorophosphoramidite, (8) lithium sulphide, (9) bromopropionitrile. Bz = benzoyl, Dmt = dimethoxytrityl.

direct displacement of the arylsulphonyl group with lithium sulphide to give tribenzoyl-$d^{6S}G$ (6). The two benzoyl ester groups can be removed from the 3′ and 5′-hydroxyl groups by mild alkaline hydrolysis leaving the benzoyl amide group in place on the 2-NH_2 function. This gives a suitable synthon for oligodeoxynucleotide synthesis with the exocyclic 2-NH_2 group protected. The N-benzoyl-$d^{2am}P$ requires no further base protection and can be converted to the 5′-dimethoxytrityl, 3′-phosphoramidite derivative in the usual fashion. This N-benzoyl group is completely removed by an overnight treatment with 35% aqueous NH_3 at 50 °C (standard deblocking for oligonucleotides). In contrast the sulphur atom of N-benzoyl-$d^{6S}G$ is expected to be reactive towards the conditions of oligodeoxynucleotide synthesis and must be protected. This can be achieved using the cyanoethyl group by reaction with bromopropionitrile. Both the N-benzoyl and the S-cyanoethyl groups are removed completely from this base by reaction with 35% aqueous NH_3 at 50 °C for 4 h. This lowered reaction time is needed with d^{6S} G derivatives to prevent production of 2,6-diaminopurine during NH_3 deblocking (see Section 4.1) after oligodeoxynucleotide synthesis.

Protocol 5. Preparation of 3′,5′,N-tribenzoyl-6-(2,4,6-triisopropylbenzenesulphonyl)-dG
(This protocol should be used in conjunction with *Figure 2B*.)

A. *Preparation of 3′,5′,N-tribenzoyl-dG*

1. Dry 2.85 g (10 mmol) of dG (monohydrate) by co-evaporation twice from 50 ml dry pyridine.

2. Suspend the dG in 40 ml dry pyridine.

3. Add 9.05 g (40 mmol) of benzoic anhydride and 0.35 g of dimethylaminopyridine.

4. Reflux for 18 h with exclusion of moisture.

5. Cool and check by TLC ($CHCl_3$ 90, CH_3OH 10). The product has an R_f of 0.6. No dG (R_f 0) or 3′,5′-dibenzoyl-dG (R_f 0.2) should be visible.

6. Add 20 ml of 5% $NaHCO_3$ solution and leave for 1 h.

7. Evaporate the contents to dryness.

8. Perform standard work-up procedure (*Protocol 3*).

10. Purify the product by flash chromatography eluting with 800 ml $CHCl_3$ containing 2% CH_3OH followed by 800 ml $CHCl_3$ containing 4% CH_3OH. The product elutes with the second solvent. Check column eluate by TLC and pool pure fractions.

11. Evaporate to dryness.

12. The yield should be 90%.

Protocol 5. *Continued*

B. *Preparation of 3′,5′,N-tribenzoyl-6-(2,4,6-triisopropylbenzenesulphonyl)-dG*

1. Dissolve 4.6 g (8 mmol) of dry 3′,5′,N-tribenzoyl-dG in 50 ml of dry CH_2Cl_2.
2. Add 0.5 g of dimethylaminopyridine and 5.6 ml of dry diisopropylethylamine.
3. Add 4.8 g (16 mmol) of 2,4,6-triisopropylbenzenesulphonylchloride.
4. Protect from moisture and stir at room temperature for 1 h.
5. Examine by TLC ($CHCl_3$ 97, CH_3OH 3). The starting material (R_f 0.15) should have been completely converted to product (R_f 0.6).
6. Transfer to a 1 lite round-bottomed flask and evaporate to remove solvents (caution: this mixture froths).
7. Disolve crude product in 20 ml $CHCl_3$ and purify by flash chromatography. Elute with $CHCl_3$ (500 ml) followed by $CHCl_3$ containing 2% CH_3OH. The product elutes with the second solvent. Check column eluate by TLC and pool pure fractions.
8. Evaporate to give product.
9. Yield should be 60%.

Protocol 6. Preparation of dmt-(N-benzoyl)-$d^{2am}P$
(This protocol should be used in conjunction with *Figure 2B*.)

A. *Preparation of 3′,5′,N-tribenzoyl-6-hydrazino-dG*

1. Dissolve 3.14 g (3.7 mmol) of 3′,5′,N-tribenzoyl-6-(2,4,6-triisopropylbenzenesulphonyl-dG (*Protocol 5*) in 70 ml of dry tetrahydrofuran.
2. Cool to between 0–5 °C on ice bath.
3. Add 0.45 ml (14.5 mmol) of anhydrous hydrazine (caution: extremely toxic). Protect from moisture and leave on ice for 2 h.
4. Examine by TLC ($CHCl_3$ 92, CH_3OH 8). The starting material (R_f 0.9) should have been completely converted to product (R_f 0.1–0.2 streaks).
5. Evaporate to dryness.
6. Perform standard work-up (*Protocol 3*) but use ethyl acetate instead of $CHCl_3$ as organic solvent. You need to keep the ethyl acetate layer. Remember it is the upper layer in this case.
7. Evaporate to give about 3 g of crude product which can be used directly in the next stage.

Protocol 6. *Continued*

B. *Preparation of 3′,5′,N-tribenzoyl-d^{2am}P*

1. Dissolve the product obtained above in 150 ml tetrahydrofuran containing 5% H_2O.
2. Add 1 g of Ag_2O and reflux for 30 min.
3. Add a further 1 g of Ag_2O and reflux for 30 min more.
4. Add a final 1 g of Ag_2O and reflux for a final 1 h.
5. Evaporate the mixture to dryness.
6. Suspend in 100 ml ethyl acetate and add 100 ml of 10% (w/v) potassium iodide solution. A black precipitate will be visible.
7. Filter through celite to remove insoluble material.
8. Place filtrate in a separating funnel and run off and discard the lower aqueous layer.
9. Extract with a further 100 ml of 10% potassium iodide followed by 2 × 100 ml of 10% sodium thiosulphate and 2 × 100 ml of H_2O. Discard the lower aqueous layer each time.
10. Dry the ethyl acetate layer by swirling with 5 g of anhydrous Na_2SO_4 and evaporate to dryness.
11. Purify the product by flash chromatography eluting with $CHCl_3$ containing 1% CH_3OH (500 ml) followed by $CHCl_3$ containing 2.5% CH_3OH (500 ml). The product elutes with the second sovlent. Check column eluate by TLC and pool pure fractions.
12. Evaporate to dryness to give 1.5 g of pure product (72% from 3′,5′,N-tribenzoyl-6-(2,4,6-triisopropyl)benzenesulphonyl-dG, 2 steps).

C. *Preparation of N-benzoyl-d^{2am}P*

1. Dissolve the 1.5 g of product obtained above in 30 ml of pyridine and add 4 ml of CH_3OH.
2. Cool to −20 °C using a salt ice-bath.
3. Add 3.6 ml of ice-cold 2 M NaOH.
4. Leave at −20 °C for 30 min.
5. Add Dowex 50W-X8 (pyridinium salt) in small batches until the pH reaches 7.
6. Filter off the Dowex and wash the Dowex with 50 ml of H_2O, pyridine mixture (4,1).
7. Pool filtrates and evaporate to dryness.

Protocol 6. *Continued*

D. *Preparation of dmt-N-benzoyl-$d^{2am}P$*

1. Perform the standard dimethoxytritylation reaction (*Protocol 4*) on the product obtained above.
2. After flash chromatography the product will probably not be pure by TLC. Final purification can be achieved by recrystallization from ethyl acetate (20 ml) containing a few drops of $(C_2H_5)_3N$.
3. The yield from 3′,5′,N-tribenzoyl-$d^{2am}P$ (two steps) should be 35%.

Protocol 7. Preparation of dmt-N-benzoyl-S-cyanoethyl-$d^{6S}G$
(This protocol should be used in conjunction with *Figure 2B.*)

A. *Preparation of 3′,5′,N-tribenzoyl-$d^{6S}G$*

1. Dissolve 4 g (4.7 mmol) of 3′,5′,N-tribenzoyl-6(2,4,6-triisopropylbenzenesulphonyl)-dG in 60 ml of dry tetrahydrofuran.
2. Add 0.25 g (5.4 mmol) of Li_2S (caution: stench—perform in fume hood).
3. Stir at room temperature for 2.5 h.
4. Examine by TLC ($CHCl_3$ 97, CH_3OH 3). The starting material (R_f 0.6) should have been completely converted to product (R_f 0.1).
5. If some starting material remains add a further 0.1 g of Li_2S and leave for 1 h more.
6. Evaporate to dryness.
7. Perform standard work-up (*Protocol 3*).
8. Purify the product by flash chromatography using $CHCl_3$ containing 1% CH_3OH as eluent. Monitor fractions by TLC and pool those containing pure product.
9. Evaporate to dryness. Yields of pure product are about 90%.

B. *Preparation of N-benzoyl-$d^{6S}G$*

1. Prepare this compound from the tribenzoyl derivative using limited NaOH hydrolysis exactly as for the conversion of 3′,5′-N-tribenzoyl-$d^{2am}P$ to the N-benzoyl derivative (*Protocol 6*).

C. *Preparation of N-benzoyl-S-cyanoethyl-$d^{6S}G$*

1. Dissolve the N-benzoyl-$d^{6S}G$ obtained above in 40 ml dimethylformamide (not everything will dissolve as this product contains insoluble impurities).
2. Place 100 ml of dimethylformamide, 5 ml of 3-bromopropionitrile and 6 g of anhydrous K_2CO_3 in a round-bottomed flask. Stir vigorously.

Protocol 7. *Continued*

3. Add the solution of N-benzoyl-$d^{6S}G$ dropwise to the mixture prepared in step 2 over about 10 min. Stir vigorously.

4. Stir overnight at room temperature.

5. Examine by TLC ($CHCl_3$ 85, CH_3OH 15). The starting material (R_f 0.4) should have been completely converted to product (R_f 0.5).

6. Filter to remove solid material. Wash the precipitate with 100 ml of $CHCl_3$ (85), CH_3OH (15) mixture and pool filtrates.

7. Evaporate to a yellow oil using an oil pump.

8. Dissolve the oil in about 200 ml of $CHCl_3$ (80), CH_3OH (20) mixture. Filter and discard the precipitate.

9. Evaporate the filtrate to give product in yields of about 80%. This product is virtually pure by TLC and can be used directly in the next stage.

D. *Preparation of dmt-N-benzoyl-S-cyanoethyl-$d^{6S}G$*

1. Perform the standard dimethoxytritylation reaction (*Protocol 4*) on the product produced above.

2. Yield should be about 60% (three steps from 3′,5′,N-tribenzoyl-$d^{6S}G$).

3.3 Preparation of 4-thiothymidine and 5-methyl-2-pyrimidinone-1-β-D-2′-deoxyribofuranoside derivatives suitable for oligonucleotide synthesis

In these two derivatives of thymidine the 4-keto oxygen is either replaced by sulphur (giving 4-thiothymidine, ^{4S}T) or a hydrogen atom (giving the pyrimidinone derivative, ^{4H}T) (*Figure 3A*). These two compounds are therefore conceptually related to $d^{2am}P$ and $d^{6S}G$ where the keto oxygen of dG is replaced with either hydrogen or sulphur. Once again much of the preparation of ^{4S}T and ^{4H}T compounds is via a common pathway leading to 5′-dimethoxytrityl-^{4S}T (*Protocol 8*). As shown in *Figure 3B* this is a four-step preparation from T consisting of an initial benzoylation of the 3′-and 5′-hydroxyl groups. Traces of a tribenzoyl derivative (presumably benzoylated at the 4-keto oxygen) are formed but this is also converted to the 3′,5′-dibenzoyl-^{4S}T in the next step. The 4-keto oxygen atom of 3′,5′-dibenzoyl-T is directly converted to sulphur using Lawesson's reagent. Previous approaches have used P_4S_{10} (7,8) but we have found Lawesson's reagent much easier to handle as it does not give brown tarry deposits during the synthesis (9). After removal of the benzoyl groups with sodium methoxide, the crude, sticky, and difficult to handle ^{4S}T can be directly converted to its 5′-dimethoxytrityl derivative. This is easily purified by flash chromatography to give a solid,

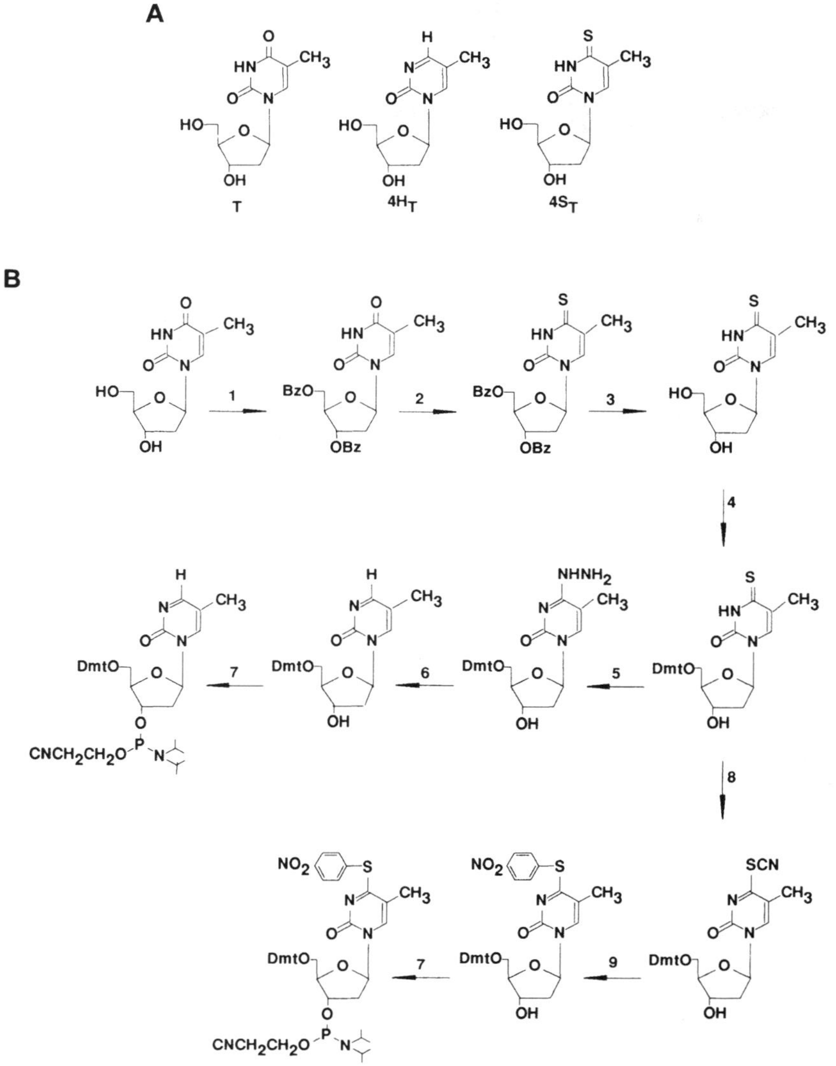

Figure 3. **A**. The structures of thymidine (T), 5-methyl-2-pyrimidinone-1-ß-D-2′-deoxyribofuranoside (^{4H}T) and 4-thiothymidine (^{4S}T). **B**. The synthesis of the 5′-dimethoxytrityl-3′-phosphoramidite derivatives of 5-methyl-2-pyrimidinone-1-ß-D-2′-deoxyribofuranoside and 4-p-(nitrophenylthio)-thymidine from T. Reagents: (1) benzoic anhydride, (2) Lawesson's reagent, (3) sodium methoxide, (4) dimethoxytritylchloride, (5) hydrazine, (6) Ag_2O, (7) 2-cyanoethyl N,N-diisopropylchlorophosphoramidite, (8) CNBr/KCN/18-crown-6, (9) p-nitrothiophenolate. Bz = benzoyl, Dmt = dimethoxytrityl.

free-flowing derivative. The pathways to ^{4S}T and ^{4H}T diverge here. The sulphur atom in ^{4S}T is very reactive (much more so than the sulphur in $d^{6S}G$) and requires protection during oligodeoxynucleotide synthesis. Previously we have used the sulphenylmethyl group ($-SCH_3$) and so achieved protection via a disulphide (9). Despite its ease of introduction (with methylmethanethio-sulphonate) and removal (with dithiothreitol) the sulphenylmethyl protecting group is far from perfect. The main problem arises as it is not 100% stable to the iodine and especially the acid steps that occur during oligodeoxy-nucleotide synthesis. This eventually leads to an oligodeoxynucleotide product that contains not only ^{4S}T at the desired site but also rougly equal amounts of T (i.e. thioketo to keto conversion) and of 5-methyldeoxycytidine (i.e. thioketo to amino conversion). Recently we have introduced the p-nitrophenyl derivative of ^{4S}T as a superior synthon for the preparation of oligodeoxynucleotides containing ^{4S}T (10). The required derivative is prepared from 5′-dimethoxytrityl-^{4S}T in two steps via the thiocynanato derivative (dmt-^{4SCN}T) as outlined in *Protocol 9*. The reaction of 4-thiopyrimidines with cyanogen bromide was first reported by Walker (11). Formation of the thiocyanato derivative (12) activates the base towards nucleophilic displacement at the 4-position enabling the facile production of 5′-dimethyloxytrityl-4-(p-nitrophenylthio)-thymidine with p- nitrothiophenol. After oligonucleotide synthesis the sulphur atom in the 4-p-(nitrophenyl)-thio-thymidine does not furnish the sulphur atom in the completely deblocked product. A three step deblocking procedure (outlined in Section 4.1) is required (rather than a simple one step deblocking with NH_3) that includes a treatment with the potassium salt of thioacetic acid. The sulphur atom in thioacetate acts as a nucleophile displacing the entire p-nitrothiophenolate with production of the unstable acetate ester of 4-thiothymidine. This is converted to 4-thiothymidine by the subsequent NH_3 treatment. Thus the sulphur atom in oligonucleotides containing 4-thiothymidine prepared by this method actually comes from thioacetate. This procedure gives much purer oligonucleotides than those prepared using the sulphenylmethyl protecting group. This is illustrated in *Figure 4* for a dodecamer containing two ^{4S}T residues.

The dmt-^{4S}T produced in *Protocol 8* can also be converted to dmt-^{4H}T as outlined in *Protocol 10*. This involves displacement of the sulphur atom with hydrazine followed by oxidative removal of the hydrazino function with Ag_2O in a very similar fashion to $d^{2am}P$ preparations. This method was developed by Cech and Holy (8) to produce 2-pyrimidinone derivatives and has also been used by McLaughlin and his co-workers (13). Our method given in *Protocol 10* is described in more detail in reference 9. It should be noted that the approach used to synthesize 5-methyl-2-pyrimidinones (i.e. containing a 5-methyl group and beginning from thymidine) can be used in an identical manner to prepare 2-pyrimidinones ($d^{4H}U$ derivatives having hydrogen rather than CH_3 at the 5 position) by commencing from 2-deoxyuridine rather than thymidine.

Protocol 8. Preparation of dmt-^{4S}T
(This protocol should be used in conjunction with *Figure 3B.*)

A. *Preparation of 3′,5′-dibenzoyl-T*

1. Dry thymidine (12.1 g, 50 mmol) by co-evaporation from 2 × 50 ml dry pyridine.
2. Suspend the dried thymidine in 250 ml pyridine.
3. Carefully add benzoylchloride (12.77 ml, 110 mmol). Protect from moisture and stir at room temperature for 3 h.
4. Examine by TLC ($CHCl_3$ 95, CH_3OH 5). The starting material (R_f 0.1) should have disappeared and there should be no traces of monobenzoyl derivatives (R_f 0.3–0.4). The product has an R_f of 0.8. Traces of a higher R_f product (a tribenzoyl derivative) will be seen. This is not detrimental.
5. Pour the mixture into 4 litres of stirred ice-cold water. The product precipitates as a copious white solid.
6. Collect the product by filtration and wash with 500 ml of ice cold water.
7. Dry in a vacuum desiccator over silica gel.
8. Yields should be near quantitative.

B. *Preparation of 3′,5′-dibenzoyl-^{4S}T*

1. Suspend 20 g (44 mmol) of dried 3′,5′-dibenzoyl-^{4S}T in 300 ml of dry dioxan. Use a 500 ml round-bottomed flask fitted with a reflux condenser.
2. Add 24 g (60 mmol) Lawesson's reagent (2,4-bis(4-methoxyphenyl)-1,3-dithia-2,4-diphosphetane-2,4-disulphide) (caution: stench, perform in fume cupboard).
3. Reflux for 2 h. Protect from moisture.
4. Test by TLC ($CHCl_3$ 98, CH_3OH 2). The starting material (R_f 0.25) should be completely converted to a yellow product (R_f 0.65).
5. Cool and add 5 ml H_2O.
6. Perform standard work-up (*Protocol 3*).
7. The crude product (yield > 90%) is suitable for the next step. Crystallization from ethanol (2 litres) is possible if a purer product is needed.

C. *Preparation of ^{4S}T*

1. Suspend 18.7 g (40 mmol) crude 3′,5′-dibenzoyl-^{4S}T in 300 ml dry CH_3OH.
2. Add sodium methoxide (27.6 ml of a 25% (w/v) solution) (120 mmol).

Protocol 8. *Continued*

3. Stir at room temperature for 30 min.
4. Examine by TLC ($CHCl_3$ 98, CH_3OH 2). The starting material (R_f 0.65) should have disappeared and been replaced by material at the origin.
5. Add 7 ml of glacial acetic acid.
6. Evaporate to dryness.
7. Dissolve in 200 ml H_2O and extract with 3 × 200 ml $CHCl_3$. Discard the organic (bottom) layer at each step and retain the aqueous (top). Evaporate to dryness.
8. Extract the solid product formed with 300 ml acetone. The yellow ^{4S}T will dissolve on stirring (30 min) leaving inorganic salts behind as an insoluble solid.
9. Filter to remove salts. Retain filtrate.
10. Evaporate to dryness.
11. This method yields a very hygroscopic product that should be stored *in vacuo* over silica gel.

D. *Preparation of dmt-^{4S}T*

1. Perform the standard dimethoxytritylation reaction (*Protocol 4*) on 5 g of ^{4S}T.
2. The yields of dmt-^{4S}T (four stps from thymidine) should be 40–60%.

Protocol 9. Preparation of dmt-4-p(nitrophenylthio)-thymidine

A. *Preparation of dmt-^{4SCN}T*

1. Dissolve 4.2 g (7.5 mmol) of dmt-^{4S}T in 100 ml dry CH_2Cl_2.
2. Add a solution of 2.7 g KCN (37 mmol) (CAUTION: TOXIC) in 70 ml 5% $NaHCO_3$, followed by a few crystals of 18-crown-6.
3. Stir the mixture vigorously at room temperature and add slowly a solution of 1.2 g CNBr (11 mmol) (poison, use a fume hood) in 20 ml CH_2Cl_2.
4. Stir 10 min at room temperature, then separate the phases, and wash the organic layer 3 × with H_2O.
5. Dry the organic layer with Na_2SO_4 and concentrate to dryness.
6. Purify by flash chromatography on 110 g of silica gel. Elute first with CH_2Cl_2/ethylacetate 3:1, then 2:1.
7. Monitor the fractions by TLC using CH_2Cl_2/ethylacetate 1:1. The product has an R_f of 0.7 and a characteristically blue fluorescence when viewed under long-wave UV-light.

Protocol 9. *Continued*

8. After chromatography, dissolve the product in a minimum amount of CH_2Cl_2 and precipitate into 250 ml of stirred petroleum ether at room temperature.
9. Yield: 3.80 g (86%), off white solid.

B. *Preparation of dmt-4-(p-nitrophenylthio)-thymidine*

1. Dissolve 1.9 g (3.2 mmol) dmt^{4SCN}T in 60 ml dry CH_2Cl_2.
2. Add 0.9 g (7.6 mmol) p-nitrothiophenol (stench; technical grade, 80+ %, obtained from Aldrich and used without further purification).
3. Cool to 4 °C, add slowly with stirring 1 ml dry $(C_2H_5)_3N$.
4. Remove the cooling bath, stir at room temperature for 30 min.
5. Wash the reaction mixture with 5% $NaHCO_3$ (5 times; the highly lipophilic p-nitrothiophenol is difficult to remove completely).
6. Purify by flash chromatography on 40 g of silica gel and elute with CH_2Cl_2-ethylacetate 3:1 to 1:1.
7. Monitor the fractions by TLC using CH_2Cl_2/ethylacetate 1:1. The product has an R_f of 0.75, i.e. it moves slighly faster than the starting material. It does not show any blue fluorescence.
8. After chromatography, dissolve in a minimum amount of CH_2Cl_2 and precipitate into 150 ml of petroleum ether at room temperature.
9. Yield: 1.50 g (68%) of a very pale yellow solid.

Protocol 10. Preparation of dmt-^{4H}T
(This protocol should be used in conjunction with *Figure 3B.*)

A. *Preparation of dmt-(4-hydrazino)T*

1. Dissolve 2.7 g (5 mmol) of dmt-^{4S}T (*Protocol 8*) in 100 ml of ethanol containing 5 ml H_2O.
2. Add 1.4 ml (25 mmol) of hydrazine hydrate (containing 55% hydrazine) (caution: extremely toxic).
3. Reflux for one hour.
4. Examine by TLC ($CHCl_3$ 90, CH_3OH 10). The starting material (R_f 0.85) should be completely converted to product (R_f 0.1–0.3 streaks).
5. Perform standard work-up (*Protocol 3*).
6. The product obtained can be used in the next stage without the need for further purification.

Protocol 10. *Continued*

B. *Preparation of dmt-^{4H}T*

1. Dissolve the dmt-(4-hydrazino)T (usually 2.6 g, 4.8 mmol) obtained above in 50 ml ethanol containing 0.5 ml triethylamine.
2. Add Ag_2O (2.3 g, 10 mmol).
3. Reflux for 2 h.
4. Examine by TLC ($CHCl_3$ 90, CH_3OH 10). The starting material (R_f 0.1–0.3) should have disappeared to give product (R_f 0.7). This product shows bright blue fluorescence under long wavelength (360 nm) UV light. Several other spots will be seen.
5. Filter through celite to remove most of the Ag_2O. Retain the brown filtrate.
6. Dissolve in 200 ml $CHCl_3$ and extract with 3 × 100 ml 10% (w/v) NaI. A black precipitate will form which should be removed by filtration and discarded. At each stage discard the aqueous (upper) layer and retain the organic (lower). At the end of these extractions the $CHCl_3$ layer should be clear and straw coloured.
7. Extract the $CHCl_3$ layer with 2 × 100 ml 10% (w/v) $Na_2S_2O_3$ followed by 100 ml saturated NaCl. Discard the upper aqueous layer at each stage and keep the lower organic layer.
8. Dry the organic layer by addition and swirling of 5 g anhydrous Na_2SO_4. Remove Na_2SO_4 by filtration.
9. Evaporate the $CHCl_3$ giving an oil.
10. Purify the product by flash chromatography using 500 ml each of ethylacetate containing 0.5% triethylamine followed by this mixture containing in turn 1, 3, and 5% CH_3OH. The product should elute with 5% CH_3OH. Test the column eluate by TLC and pool pure fractions.
11. Evaporate to give a white solid.
12. This reaction gives several side products and is never very clean. As a result yields are usually only about 50%.

3.4 Preparation of 3′-phosphoramidites

The final stage in the preparation of modified base derivatives is addition of the 3′-phosphoramidite group. The appropriate 3′-cyanoethylphosphoramidite can be prepared by reaction of the suitably protected base with 2-cyanoethyl N,N-diisopropylchlorophosphoramidite (14) as described in *Protocol 11*. The method given in this protocol works very well in our hands providing that the tetrahydrofuran is dried immediately prior to use by refluxing from sodium/benzophenone (Section 2.4). Although methods have been published for the preparation of 2-cyanoethyl N,N-diisopropylchlorophosphoramidite (14,15) we always use the product supplied by Aldrich (catalogue number 30, 230–9). Providing either freshly open bottles are used or this product is stored correctly (at −20 °C in a desiccator over desiccant) good results are obtained.

Protocol 11. Preparation of deoxyribonucleoside-3′-O(N,N-diisopropylamino)cyanoethylphosphoramidites

1. Dissolve 0.5–2 g of the appropriate deoxyribonucleoside in 20 ml of dry tetrahydrofuran.
2. Add 5 equivalents of dry diisopropylethylamine.
3. Protect from moisture with a rubber septum and cool on ice.
4. Add 1.5 equivalents of cyanoethyl N,N-diisopropylchlorophosphoramidite using a syringe.
5. Leave at room temperature for 30 min.
6. Examine by TLC ($CHCl_3$ 45, ethyl acetate 45, $(C_2H_5)_3N$ 10). The starting material (R_f usually 0.2) should have disappeared and been replaced by product (R_f usually 0.6–0.8). As the product exists as a pair of diastereomers two closely running product spots may be seen. Traces of deoxyribonucleoside-3′-phosphonate derivatives (a side product, R_f usually between 0.3–0.5) will also be seen.
7. Add 1 ml of CH_3OH.
8. Evaporate to give an oil.
9. Perform the standard work-up procedure (*Protocol 3*) but use ethyl acetate rather than $CHCl_3$ as the organic solvent. The upper organic layer should be retained at each stage.
10. Purify the product by flash chromatography using $CHCl_3$/ethyl acetate/$(C_2H_5)_3N$ (45:45:10 by volume) as solvent[a]. The products elute very rapidly. Monitor column eluate by TLC and pool pure fractions.
11. Evaporate to give an oil.
12. Obtain the products in solid form by dissolving in 10 ml of ethyl acetate and pipetting dropwise into vigorously stirred petroleum ether (500 ml) at −40 °C (dry ice–acetone bath). The products precipitate and can be collected by filtration.
13. Dry the products in a vacuum desiccator over silica gel. Store desiccated at −20 °C. Yields vary between 60 and 90%.

[a] In the case of phosphoramidite of 5′-dimethoxytrityl-4-p(nitrophenylthio)-thymidine the column should be the first eluted with petroleum ether 9, $(C_2H_5)_3N$ 1 followed by petroleum ether 45, CH_2Cl_2 45, $(C_2H_5)N$ 10. The product elutes with the second solvent.

3.5 Commercially available modified bases suitable for oligonucleotides synthesis

Three modified bases relevant to the study of protein–nucleic acid interactions can be purchased in a form directly suitable for oligodeoxynucleotide synthesis (i.e. having a 5′-dmt, 3′-phosphoramidite and appropriate base protection). These are deoxyinosine (dI) in which the 2-NH_2 group of dG is replaced with hydrogen (available from Applied Biosystems, article 400402 and Cruachem, article 20–8150–17), deoxyuridine (dU) in which the 5-methyl group of T is replaced with hydrogen (available from Cruachem, article 20–8140–17) and 5-methyl deoxycytidine ($d^{5Me}C$) which is a dC derivative having a CH_3 group instead of hydrogen at position 5 (available from Cruachem, article 20–8170–17). Very recently a large number of modified base phosphoramidites have become available from Glen Research. These include dP and also 7-deaza dA and 7-deaza dG.

4. Synthesis, purification, and characterization of oligonucleotides containing modified bases

All the methods used in this section are based on the standard approaches used to prepare oligodeoxynucleotides by the phosphoramidite approach. The reader should consult Chapter 1. We have emphasized any changes necessitated by the modified base.

4.1 Synthesis

We prepare all oligonucleotides containing modified bases using an Applied Biosystems 381 A DNA synthesizer on a 1 μmol scale using the standard cycle. Modified base phosphoramidites should be dried *in vacuo* over silica gel overnight (Section 2.4) and dissolved in DNA synthesis grade anhydrous acetonitrile (Applied Biosystems, article 400060) at the usual concentration of 0.1 M. As automated DNA synthesisers are very sensitive to particulate material, this solution must be filtered. We use 25 mm 0.5 μm PTFE Millex-SR filters (Millipore article SL5R 025 NB). The filtered solution should be introduced at the fifth base position (base X) on the machine. The efficiency of each base addition can be monitored by the orange dimethoxytrityl cation released at each cycle and this is especially useful to check that the modified base has coupled in good yield. All the modified base phosphoramidites mentioned in this article should couple with efficiencies of about 100% as monitored by trityl cation release. All syntheses are performed '**Trityl-ON**' to facilitate subsequent purification. After synthesis it is necessary to remove most of the protecting groups with NH_3. This is classically performed using 35% aqueous NH_3 at 50 °C overnight. For modified bases that are not chemically reactive (i.e. dP and $d^{2am}P$ and also the commercially available dI, dU, and $d^{5Me}C$), these deblocking conditions can be used unchanged. The other modified bases (i.e. ^{4S}T, ^{4H}T, $d^{4H}U$, and $d^{6S}G$) mentioned in this article are more reactive than the usual bases and therefore the NH_3 treatment should be shortened. We have found that with short oligodeoxynucleotides (up to 20 bases in length) 3 h in NH_3 at 50 °C is sufficient to remove all the protecting groups. With $d^{6S}G$ we use a 4 h treatment at 50 °C which completely removes base amino protecting groups and also the β-cyanoethyl function used to protect the thiol. With ^{4H}T a 3 h NH_3 reaction at 50 °C is best. With ^{4S}T containing oligonucleotides prepared by the 4-(p-nitrophenylthio)-thymidine method a three step deblocking is required consisting of: 1) treatment of the crude, 5′-tritylated, CPG-bound oligodeoxynucleotide with 0.3 ml t-butylamine in 2 ml CH_3CN for 45 min at room temperature, 2) incubation of the still CPG-bound product from 1 with a 0.3 M CH_3COSK solution in ethanol (1.5–2 h at 55 °C), 3) treatment of the CPG-bound product from 2 with concentrated aqueous NH_3 (3 h at 55 °C) or better 1 h at 55 °C, if very ammonia labile protected normal bases have been used. We have observed that $d^{4H}U$ is extremely sensitive to NH_3 and have only obtained satisfactory oligodeoxynucleotide syntheses using the normal bases protected with very NH_3 labile groups (16,17) and performing the deblocking with 35% NH_3 at 20 °C overnight. Very NH_3 labile bases have just become commercially available from Applied Biosystems (sold under the trade name FOD-fast oligonucleotide deblock). In view of the instability of the reactive bases to NH_3 we would recommend their use as standard.

4.2 Purification

Following NH_3 deblocking the desired full-length oligodeoxynucleotide which contains a dmt group can be easily separated from truncated failure sequences lacking the dmt function. This method is based on the strong retention of dmt-containing oligodeoxynucleotides by C_{18} reversed phase HPLC columns. It is a standard method used with normal oligonucleotides (18,19) and so is not discussed here. All the dmt oligomers that contain modified bases can be purified in this manner. Following purification, the dmt group can be removed by treatment with 80% acetic acid for one hour at room temperature. After dmt-specific purification and detritylatyion, oligonucleotides that contain the four normal bases are usually pure enough for all applications. A similarly high level of purity is observed for oligodeoxynucleotides that contain unreactive modified bases (dP, $d^{2am}P$, dI, dU, and $d^{5Me}C$). This is illustrated in *Figure 4* which shows an analytical reversed phase HPLC trace of d(GACPATATCGTC). Unfortunately oligodeoxynucleotides containing the reactive bases ^{4S}T, ^{4H}T, and $d^{6S}G$ are not quite as pure as this after only dmt-specific purification. This is also illustrated in *Figure 4* which shows analytical reversed phase HPLC traces of d(GACGA[^{4S}T]A[^{4S}T]CGTC) and d(GACGA[^{4H}]ATCGTC). With the

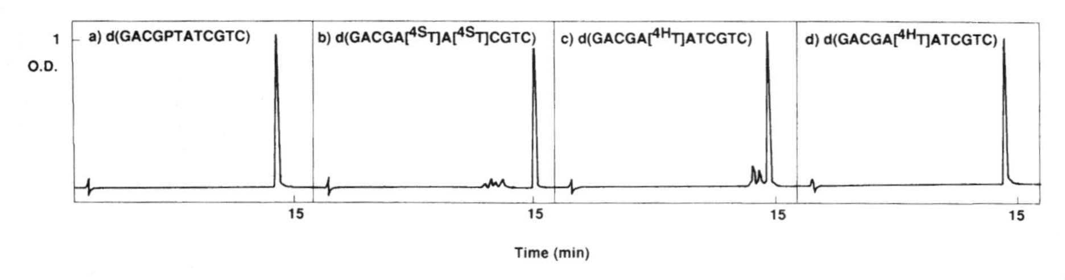

Figure 4. Reversed phase HPLC traces of oligodeoxynucleotides containing modified bases as indicated. (**a**) d(GACGPTATCGTC) after dmt-specific purification and detritylation; (**b**) d(GACGA[^{4S}T]A[^{4S}T]CGTC after dmt-specific purification and detritylation; (c) d(GACGA[^{4H}T]ATCGTC) after dmt-specific purification and detritylation; (d) d(GACGA[^{4S}T]ATCGTC) after a second HPLC purification. Column Apex-1 octadecylsilyl (25 × 0.45 cm, 5 μm particle size) (Jones Chromatography, Llanbradach, Glamorgan). Buffer A, 0.1 M triethylammonium acetate pH 6.5 containing 5% CH_3CN; Buffer B, 0.1 M triethylammonium acetate pH 6.5 containing 65% CH_3CN. t = 0, 5% B; t = 20 min, 25% B; linear gradient at 1 ml min^{-1}. Operating temperature: 50 °C.

oligonucleotide containing two ^{4S}T residues a small amount of early eluting contaminants are seen. However this level of purity should be adequate for most applications. If a purer product is desired it can be obtained by reverse-phase HPLC using shallow acetonitrile gradients (9). It should be noted that the 4-p-(nitrophenylthio)-thymidine method gives very much purer products than those obtained with sulphenylmethyl protection. Here even oligomers containing only one ^{4S}T residue were produced as roughly equimolar mixtures of the desired product (^{4S}T), a side product containing thymidine (T) in this position and a further side product containing 5′-methyldeoxycytidine (d^{5Me}C) (9). With the ^{4H}T containing oligomer small quantities of early eluting side products are also visible. These can be removed with shallow acetonitrile gradients on reverse-phase HPLC (9) as shown in *Figure 4*. (d^{5Me}C) (9). With the ^{4H}T containing oligomer small quantities of early eluting side products are also visible. These can be removed with shallow acetonitrile gradients on reverse-phase HPLC (9) as shown in *Figure 4*. d(GAC[^{6S}G]ATATCGTC) is of the same purity as normal oligonucleotide after dmt specific purification providing NH_3 deblocking and purification are carried out immediately post-synthesis. We have observed that this oligomer has a tendency to decompose to d(GACGATATCGTC) (i.e. d^{6S}G to dG conversion). This does not occur during synthesis or NH_3 deblocking but after for years. Oligonucleotides containing ^{4S}T and d^{6S}G have a tendency to slowly form disulphides due to air oxidation. This can be prevented by inclusion of 1 mM dithiothreitol. As mentioned above d^{6S}G oligonucleotides also decompose to their parents containing dG. However at −20 °C this is slow. Nevertheless for best results d^{6S}G oligomers should be used as soon as possible after purification.

4.3 Characterization

The simplest method of characterization involves base composition analysis by digesting the oligodeoxynucleotides to their monodeoxynucleoside constituents with snake venom phosphodiesterase and alkaline phosphatase. The digestion method is given in *Protocol 12.*

Protocol 12. Deoxynucleoside composition analysis of oligodexoynucleotides.

1. Dissolve about 0.5 of an absorbance unit (measured at 254 nm) of the oligodeoxynucleotide in 0.1 ml of 10 mM KH_2PO_4 pH 7 containing 10 mM $MgCl_2$.
2. Add 10 μl of snake venom phosphodiesterase (Boehringer article 108260, 1 mg/0.5 ml solution) and 10 μl of alkaline phosphatase (Boehringer article 713023, 1 unit/μl, special quality for molecular biology). It is important to use this alkaline phosphatase. Cheaper sources contain adenosine deaminase which converts dA to dI and interferes with the analysis.
3. Incubate at 37 °C for 5 h.
4. Inject an aliquot onto an analytical (25 × 0.45 cm) C_{18} reversed phase column. (See text and *Figure 5* for more details).

The monodeoxynucleosides produced can be analysed be reversed phase HPLC on analytical size C_{18} columns with 0.1 M triethylammonium acetate pH 6.5 containing 5% CH_3CN (buffer A) and 0.1 M triethylammonium acetate pH 6.5 containing 65% CH_3CN (buffer B) operated at 1 ml min^{-1} and room temperature. The most common system consists of isocratic elution with 95% HPLC buffer A and 5% HPLC buffer B with detection at 254 nm (however, this varies to a certain extent depending on the presence of the modified bases). Under these conditions the position of the four standard bases and all the modified bases described in this chapter are shown in the schematic *Figure 5*. We have only prepared oligodeoxynucleotides that contain dC, dG, T, dA, and one modified base and so have not determined the exact elution times of the modified bases relative to each other e.g. the scheme shown in *Figure 5* indicates for example that dU, $d^{5CH3}C$, dI, and $d^{4H}U$ elute between dC and G. Their order of elution is however not specified. The following points should be taken into account with specific bases.

- ^{4S}T is very tightly retained by C_{18} columns and is not eluted with the 95% A/5% B isocratic buffer given above. When ^{4S}T is present the column

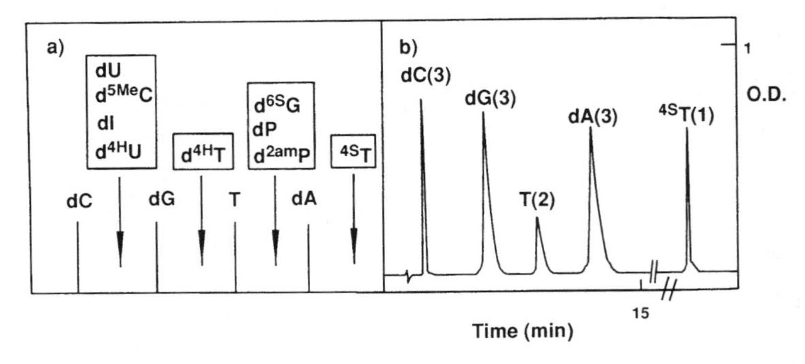

Figure 5. (**a**) Schematic diagram showing the relative elution positions on reversed phase HPLC columns of the modified bases mentioned in this article compared with the four normal bases dC, dG, T and dA. No time scale is given but dA elutes after about 13 min. (**b**) Base composition analysis of d(GACGATA[^{4S}T]CGTC). The monodeoxynucleoside products are shown together with their relative amounts (in brackets). Detection was at 254 nm for the four normal bases and 335 nm for ^{4S}T. More details of the elution buffers used, detection wavelengths and extinction coefficients of the bases are given in section 4.3 of the text.

should be run isocratically for 12 min and then developed with a gradient of 5–85% B over 10 min

- ^{4S}T, ^{4H}T, d^{4H}U, and d^{6S}G absorb weakly at 254 nm and are not easily detected here. With ^{4S}T the wavelength detector should be switched to 335 nm after dA elution. With d^{4H}U the detector should be changed to 305 nm after dC elution and back to 254 nm after d^{4H}U elution to allow detection of the remaining three normal bases (d^{4H}U elutes about midway between dC and dG allowing time for this change). ^{4H}T elutes very near and just before T and d^{6S}G very near and just after T. Here there is not enough time to alter the detector wavelength during a run due to peak closeness. With ^{4H}T two runs should be performed; one at 254 nm for normal base detection and a second at 315 for ^{4H}T. With d^{6S}G a first run at 254 nm will detect the normal bases and injecting an aliquot for a second analysis at 340 nm gives d^{6S}G

The amounts of each base can be determined by integration to give the areas under each peak. These must be divided by the following extinction coefficients to take into account the different absorbtions of the bases at the appropriate detection wavelength.

- 254 nm: dC (6×10^3), dG (13.5×10^3), T (7×10^3), dA (14.3×10^3), dP (6.5×10^3), dU (9.3×10^3), d^{5Me}C (5.1×10^3), d^{2am}P (3×10^3).
- 305 nm: d^{4H}U (6.47×10^3)
- 315 nm: ^{4H}T (5.3×10^3)
- 335 nm: ^{4S}T (22.3×10^3)

- 340 nm: $d^{6S}G$ (21.0×10^3)

  All extinction coefficients are in M^{-1} cm^{-1}.

As an example the base composition analysis of d(GACGATA[^{4S}T]CGTC) is given in *Figure 5*.

5. Applications of oligodeoxynucleotides containing modified bases

5.1 As probes of protein–DNA interactions

Many proteins that interact with double-stranded B-DNA show a very high specificity for a particular target sequence. Good examples are repressor proteins which usually bind to sequences 12–20 bases in length and type II restriction endonucleases which cut DNA at short recognition sequences, usually 4–8 base pairs in length. The mechanisms by which high selectivity is achieved are complex and probably due to several factors. Amongst these are: a protein making direct contacts to exposed portions of the bases in the major or minor groove, a protein recognizing a peculiar B-DNA conformation that depends on the cognate base sequence, and a protein being able to distort only the cognate sequence to a different conformation necessary for tight binding. It is possible to probe the first mechanism using base analogues. The parts of each of the bases in double-stranded B-DNA that are capable of interacting with proteins in the major and minor groove have been determined (20). Thus the 6-NH_2 group of dA can form hydrogen bonds with proteins via the major groove. This interaction can be probed using dP in which the 6-NH_2 group has been removed by replacement with hydrogen. As an example my group has been studying DNA recognition by the *Eco*RV restriction endonuclease and modification methylase (recognition sequence GATATC). The self-complementary dodecamer d(GACGATATCGTC) was prepared along with analogues in which the two dA residues within the recognition sequence were separately replaced with dP. Kinetic studies with the parent and the two analogues gave some ideas about the roles that these two 6-NH_2 groups played in the recognition of DNA (and also in the catalytic process) by the two enzymes (21, 22). Most of the analogues outlined here can be used to probe interactions of this kind and have been extensively used by many researchers (see references in above papers). Clearly the approach is not without criticism. The main one being that the modified base may change the overall oligodeoxynucleotide conformation and kinetic changes may be due to this rather than loss of a specific protein DNA contact.

5.2 As spectral probes

Some of the bases described here (^{4H}T, d^{4H}U, and d^{2am}P) are fluorescent, in contrast with the four normal bases which show no fluorescence. There has been one report (23) in which the fluorescence of ^{4H}T has been used as a probe for oligodeoxynucleotide dynamics and conformation. The sulphur-containing bases (^{4S}T, d^{6S}G) have a UV maximum at about 340–350 nm, well away from the 260–280 nm seen with the four normal bases. Oligodeoxynucleotides that contain these bases also show this long wavelength absorbance at 340 nm in addition to the maximum at 260 nm due to the normal bases. Perhaps more importantly, oligodeoxynucleotides that contain these thio-bases also have a circular dichroism transition at about 340 nm in addition to the usual features at 240–270 nm seen for B-DNA containing the four usual bases (6, 9). As circular dichroism spectroscopy is very sensitive to DNA conformation, it should be possible to monitor the environment of a single sulphur containing base in an oligonucleotide at 340 nm. This method may be extendable to oligonucleotides bound to proteins for the study of protein–DNA interactions.

5.3 As reactive probes

The bases ^{4H}T, d^{4H}U, ^{4S}T, and d^{6S}G have potentially interesting chemical and photochemical reactivities. All are reactive with nucleophiles (e.g. NH_3 and amines). With ^{4H}T and d^{4H}U nucleophiles appear to attack at carbon six giving a saturated derivative. In a further reaction, glycosidic bond cleavage followed by breaking of the oligonucleotide chain occurs (9). With ^{4S}T and d^{6S}G, nucleophiles can attack the carbon bearing the sulphur with the release of sulphide (6, 9). The reaction is reasonably rapid with ^{4S}T but much slower with d^{6S}G. There exists the potential of incorporating these bases into oligonucleotides that contain the recognition site of a protein of interest and using them in affinity labelling experiments. ^{4S}T and d^{6S}G are photoreactive at 340–350 nm (24–26) wavelengths removed from the usual absorbance maxima of proteins and nucleic acids. On photoactivation these bases generate reactive species possibly involving free radicles. Here again incorporating ^{4S}T and d^{6S}G into synthetic oligonucleotides may give probes suitable for the photoaffinity labelling of proteins. Finally the glycosidic bond of ^{4H}T and d^{4H}U is extremely acid labile. This has been exploited to prepare oligonucleotides containing a defined abasic site (27).

Acknowledgements

I would like to thank members of my research group; especially Dr P. C. Newman, Dr V. U. Nwosu, Mr T. R. Waters, Mrs H. Jones, and Dr. T. Nikiforov for their efforts in the syntheses and uses of the modified bases discussed in this article. The UK SERC, UK MRC and the EEC (Science plan) have supported much of the work financially. I am especially grateful to the Lister Institute of Preventine Medicine for a 5 year fellowship that has enabled me to pursue research unhampered by lecturing duties. Special thanks are due to Mrs S. Broomfield for typing this manuscript very quickly at short notice.

References

1. Sproat, B. . and Gait, M. J. (1984). In *Oligonucleotide Synthesis—a Practical Approach* (ed. M. J. Gait), pp. 199–202. IRL Press, Oxford.
2. Nair, V. and Chamberlain, S. D. (1984). *Synthesis*, 401.
3. Gaffney, B. L. and Jones, R. A. (1982). *Tetrahedron Lett.*, **23**, 2257.
4. Gaffney, B. L., Marky, L. A., and Jones, R. A. (1984). *Tetrahedron*, **40**, 3.
5. McLaughlin, L. W., Leong, T., Benseler, F., and Piel, N. (1988). *Nucleic Acids Res.*, **16**, 5631.
6. Waters, T. R. and Connolly, B. A. (1990). *Nucleosides and nucleotides.* In press.
7. Fox, J. J., van Pragg, D., Wempen, I., Doerr, I. L., Cheong, L., Knoll, J. E., Eidinoff, M. L., Bendich, A., and Brown, G. B. (1959). *J. Am. Chem. Soc.*, **81**, 178.
8. Cech, D. and Holy, A. (1977). *Coll. Czech. Chem. Comm.*, **42**, 2246.
9. Connolly, B. A. and Newman, P. C. (1989). *Nucleic Acids Res.*, **17**, 4957.
10. Nikiforov, T. T. and Connolly, B. A. (1991). *Tetrahedron Lett.*, in press.
11. Walker, R. T. (1971). *Tetrahedron Lett.*, **24**, 2145.
12. Nikiforov, T. T. and Connolly, B. A. (1991). *Tetrahedron Lett.*, **32**, 2505.
13. Gildea, B. and McLaughlin, L. W. (1989). *Nucleic Acids Res.*, **17**, 2261.
14. Sinha, N. D., Biernat, J., McManus, J., and Koster, H. (1984). *Nucleic Acids Res.*, **12**, 2261.
15. Claesen, C. A. A., Segers, R. P. A. M., and Tesser, G. I. (1985). *Recl. Trav. Chim. Pays-Bas.*, **104**, 639.
16. Vu, H., McCollum, C., Jacobson, K., Thiesen, P., Vinayak, R., Spiess, E., and Andrus, A. (1990). *Tetrahedron Lett.*, **31**, 7269.
17. Schulhof, J. C., Molko, D., and Teoule, R. (1987). *Nucleic Acids Res.*, **15**, 397.
18. Fritz, H. J., Belagaje, R., Brown, E. L., Fritz, R. H., Jones, R. A., Lees, R. F., and Khorana, H. G. (1987). *Biochemistry*, **17**, 1257.
19. McLaughlin, L. W. and Piel, N. (1984). In *Oligonucleotide Synthesis—a Practical Approach* (ed. M. J. Gait), pp. 117–33. IRL Press, Oxford.
20. Seeman, N. C., Rosenberg, J. M., and Rich, A. (1976). *Proc. Natl. Acad. Sci. USA*, **73**, 804.
21. Newman, P. C., Nwosu, V. U., Williams, D. M., Cosstick, R., Seela, F., and Connolly, B. A. (1990). *Biochemistry*, **29**, 9891.
22. Newman, P. C., Williams, D. M., Cosstick, R., Seela, F. and Connolly, B.A. (1990). *Biochemistry*, **29**, 9902.
23. Wu, P., Norlund, T. M., Gildea, B., McLaughlin, L. W. (1990). *Biochemistry*, 6508.
24. Rackwitz, H. R. and Scheit, K. H. (1974). *Chem. Ber.*, **107**, 2284.
25. Pleiss, M., Ochiai, H., and Cerutti, P. A. (1969). *Biochem. Biophys. Res. Commun.*, **34**, 70.
26. Pleiss, M. G. and Cerutti, P. A. (1971). *Biochemistry*, **10**, 3093.
27. Iocono, J. A., Gildea, B., and McLaughlin, L. W. (1990). *Tetrahedon Lett.*, **31**, 175.

8

Oligonucleotides with reporter groups attached to the 5′-terminus

NANDA D. SINHA and STEVE STRIEPEKE

1. Introduction

Traditionally, oligonucleotides or DNA have been labelled and detected via use of ^{32}P-labelled ATP and enzymatic phosphorylation. Although of high sensitivity, this technique has intrinsic hazards, expense, and problems associated with the short half-life of this isotope. Recent advances have shown that high sensitivity detection can be achieved by alternative markers based on fluorescent, chromophoric, or chemiluminescent detection. Incorporation of these labels can be achieved by enzymatic (1–8) or chemical (9–23) methods. One of the simplest and most useful methods involves introduction of primary amino or sulphyrdryl (thiol) groups to the 5′-terminus of the chemically assembled support-bound oligonucleotide (14–23). Subsequently fluorophores, chromophores, biotin, or alkaline phosphatase may be added. Further examples are discussed in Chapter 10. With this approach the position of the amino or thiol group is defined and unambiguous, and there is negligible interference in subsequent hybridizations. Additionally, the 3′-terminus of the oligonucleotide chain is free for enzymatic manipulations as required in Sanger's sequencing and PCR amplification applications (14, 24–26). The attachment of other groups to the 5′-terminus of oligonucleotides is described in Chapter 12.

A key advantage of the 5′-terminal modification approach is that it is based on the widely adopted method of solid phase synthesis (27) and can be performed routinely on a variety of commercially available DNA synthesizers. The 5′-modification reagents are either modified nucleosides (14, 19, 20) or derived from non-nucleoside molecules (15–18, 22–23). A lengthy spacer arm is desirable to separate the incorporated functionality from the nucleic acid sequence, especially when subsequent enzyme reactions are involved. For this reason, and because simple synthesis procedures can be used in their preparation, non-nucleoside reagents have gained general acceptance.

2. Selection, synthesis, and application of non-nucleosidic reagents

The most frequently used linkers for 5′-terminal modifications of oligonucleotides are aminohexyl and thiol-hexyl derivatives based on phosphoramidite (15–18) or H-phosphonate chemistry (22–23) (see *Figures 1* and *2*). After introduction on to oligonucleotide chains, primary amino groups can react with either iso-thiocyanate or N-hydroxysuccinimide ester derivatives of fluorophores, chromophores, or biotin. Similarly, free thiol groups can react with either maleimide, iodo-, or bromo-functionalized derivatives of fluorophores, chromophores, or proteins with free sulphydryl groups. Furthermore, thiol-linked oligonucleotides can also be immobilized on to solid supports, (e.g. CPG-SH; Pierce) via disulphide bond (28) or maleimide linkages.

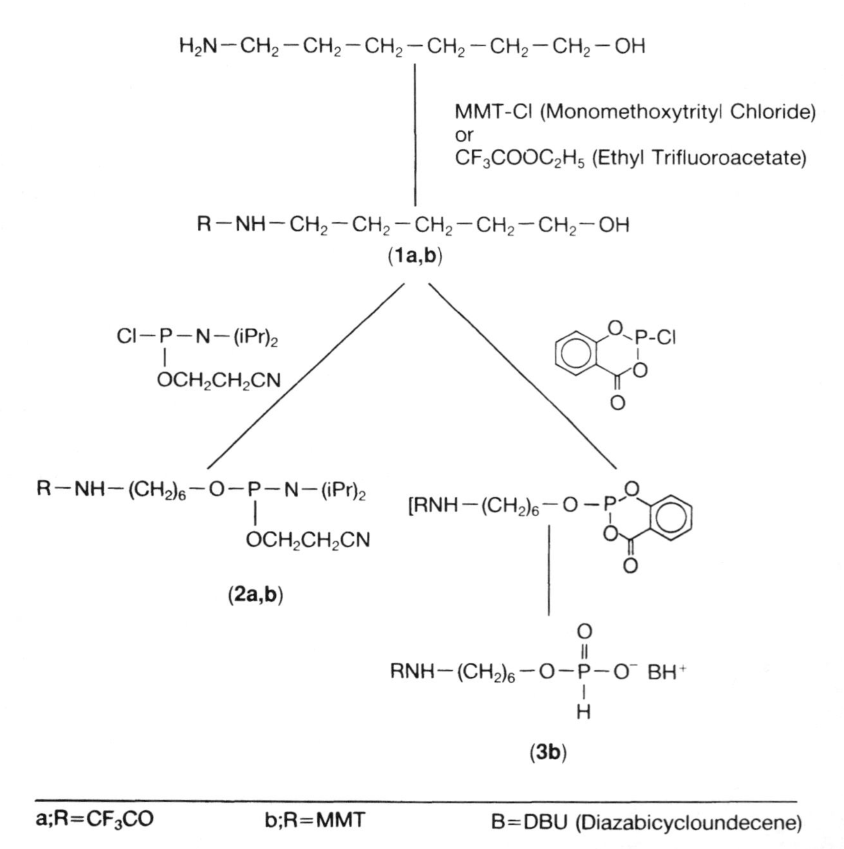

Figure 1. Preparation of protected aminohexyl phosphoramidites and H-phosphonate salts as amino linkers.

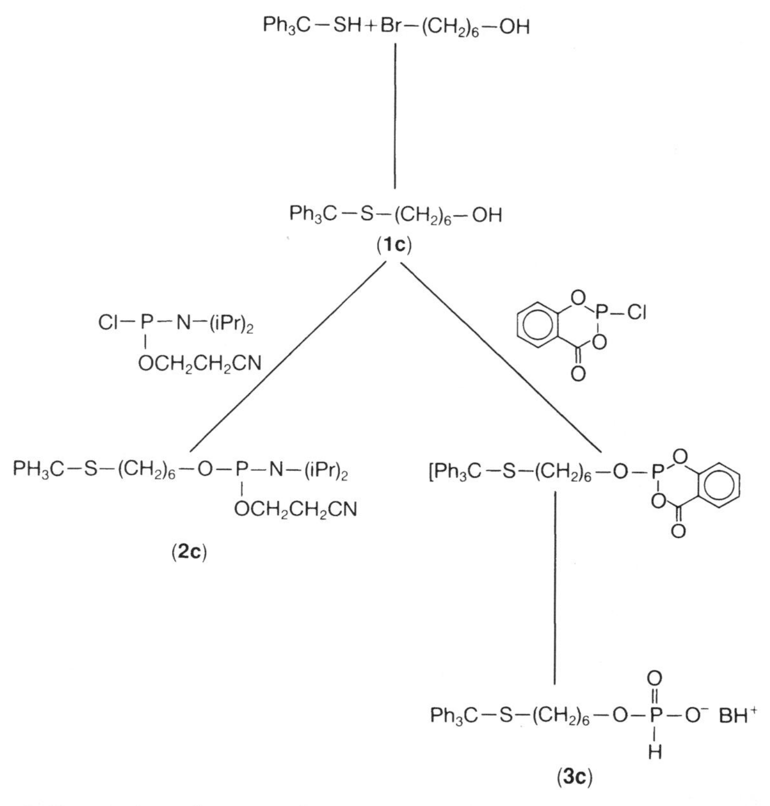

Figure 2. Preparation of protected mercapto or thio-hexyl phosphoramidites and H-phosphonate salts as thiol linkers.

The most commonly used amino-linkers for 5′-end modification are N-trifluoroacetyl (N-TFA) and N-monomethoxytrityl (N-MMT)-6-aminohexyl phosphoramidites (*Figures 1* and *2*). Standard tetrazole activation and coupling in acetonitrile incorporates these on to oligonucleotides (*Figure 3*). The removal of amino-protecting groups (TFA or MMT) differs: the trifluoroacetyl group is removed with ammonia during the final deprotection step, whereas the monomethoxytrityl group is removed by subsequent acid treatment.

Although both linkers introduce aminohexyl groups at the 5′-terminus of oligonucleotides, the monomethoxytrityl–aminohexyl linker provides advantages for purifying the amino-linked oligonucleotide from short and non-amino linked oligonucleotide using reversed phase HPLC or Oligo Pak column techniques. A MMT–amino-linked oligonucleotide can be labelled with a non-radioactive marker while still attached to the support.

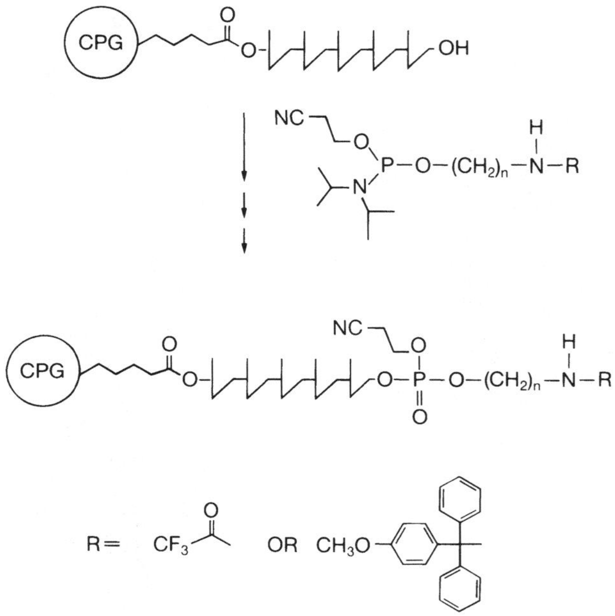

Figure 3. Coupling of amine-linker onto oligonucleotide chain attached to a solid support.

The incorporation of a thiol group is achieved with S-trityl-6-mercaptohexyl derivatives (15, 22). The trityl group also provides advantages for purification. Its presence facilitates the isolation of the desired product free from failure sequences, short sequences, and protecting groups by reversed phase HPLC or Oligo Pak column chromatography. Finally, the trityl group is removed by reaction with silver nitrate and the product is isolated by treatment with a DTT solution.

Amino- and thiol-linked oligonucleotides both have very similar application. The choice of modification, i.e. creation of a reactive free amino or thiol, depends on the particular application. Active esters or isothiocyanate derivatives are commonly used for tagging free amino-modified oligonucleotides. Maleimide, bromide, iodide, or sulphonyl derivatives are suitable for tagging thiol-linked oligonucleotides.

Recently, phosphoramidite derivatives of biotin (30–32) and fluorescein (33) have been introduced for incorporating these markers on to oligonucleotides on a solid support.

The first part of this chapter deals with the syntheses of various linkers, the second part with the methods used to incorporate these linkers, and the final part describes tagging with non-radioactive markers.

3. Synthesis and purification of linkers†

Protocol 1. Synthesis and purification of N-trifluoroacetyl-6-aminohexanol (*Figure 1* (**1a**))

A. *Synthesis*

1. Place 6-aminohexanol (23.4 g., 200 mmol) in a clean and dry round-bottomed flask (250 ml). Dissolve the solid in chloroform (25.0 ml).
2. Add dropwise under stirring a solution of ethyl trifluoroacetate (28.4 g, 200 mmol) in chloroform (25 ml) to a solution of 6-aminohexane under an inert atmosphere. Continue stirring for another 1 h. Monitor the reaction by TLC using ethyl acetate as solvent. The reaction is generally complete by this time. The protected aminohexal has a higher R_f (0.3) than aminohexanol and can be visualized by shortwave UV-light.
3. Evaporate the reaction mixture on a rotary evaporator to a viscous oil.

B. *Purification*

1. Pack a column (5 cm diameter or an appropriate size) with Merck silica gel-60 (silica to product ratio 15:1) in 4% methanol in methylene chloride.
2. Dissolve the viscous oil in a minimum volume of the above solvent and load the solution on to the column. Elute the column with 6% methanol in methylene chloride and collect 150 ml fractions.
3. Check fractions by TLC (ethyl acetate as solvent) and visualize by shortware UV light. When elution of the product is complete, combine all product-containing fractions.
4. Concentrate the combined fractions on a rotary evaporator and finally under high vacuum to give a solid (m.p. 45–46 °C), yield is typically 50–60%. ^{1}H NMR ($CDCl_3$): δ 6.5 (S, NH, 1H); 3.6 (m, O-CH_2, 2H); 3.3 (m, N-CH_2, 2H); 1.5 (m, 2 CH_2, 4H), and 1.35 (m, 2CH_2, 4H).

Protocol 2. Synthesis and purification of N-MMT-6-aminohexanol (*Figure 1* (**1b**))

A. *Synthesis*

1. Dry 6-aminohexanol (17.5 g, 150 mmol) by co-evaporation with pyridine (2 × 100 ml; HPLC grade) under reduced pressure. Remove residual pyridine by evacuation at 0.1 mm Hg for 2 h.

† The term linker is generally applied to the molecule which attaches a primary amine or sulphydryl group to an oligonucleotide.

Protocol 2. *Continued*

2. Dissolve the solid in methylene chloride (360 ml) and while stirring add freshly distilled diisopropylethyl amine (52.5 ml, 300 mmol).
3. To the above solution, add a solution of anisylchlorodiphenyl methane (monomethoxytritylchloride; 46.5 g, 150 mmol) in methylene chloride (150 ml) by addition through a dropping funnel under an inert atmosphere.
4. After 30 min of stirring, monitor the progress of the reaction by TLC (33% acetone in hexane as solvent). Visualize the TLC plate by shortwave UV and then by spraying with an aqueous 15% solution of sulphuric acid. (R_f of the product = 0.29). The reaction is usually complete in 1 h.
5. When the reaction is complete, wash the reaction mixture with 5% $NaHCO_3$ solution (3 × 250 ml) followed by saturated NaCl solution (250 ml). After drying the methylene chloride solution over anhydrous Na_2SO_4, evaporate the solvent on a rotary evaporator to a yellow oil.

B. *Purification*

1. Pack an appropriately sized glass column with 800 g of silica gel-60 using a mixture of methanol: ethyl acetate: methylene chloride: (0.5:6:93.5 v/v/v) containing 1% pyridine.
2. Dissolve the yellow oil in 100 ml of the above solvent mixture and load the solution on to the column. Elute with a mixture of methanol:ethyl acetate:dichloromethane (1:12:87 v/v/v) and collect 200 ml fractions.
3. Check each fraction for N–MMT–aminohexanol by TLC (33% acetone in hexane, R_f = 0.29) and pool all product-containing fractions.
4. Evaporate the pooled fractions on a rotary evaporator and finally dry the residue to constant weight on high vacuum. Usually the yield of purified material is between 65 and 70%. ^{1}H NMR ($CDCl_3$): δ 8.54 (s, NH, 1H); 7.4, 7.3, 7.2, 7.15, and 6.8 (14 aromatic H); 3.8 (s, OCH_3, 3H); 3.55 (t_1 O-CH_2-2H); 2.1 (t, N-CH_2-2H); 1.45 (m, -CH_2CH_2, 4H) and 1.22 (m, -CH_2CH_2, 4H).

Protocol 3. Synthesis of S-trityl-6-mercaptohexanol (*Figure 2* (**1c**))

1. Place triphenylmethyl mercaptan (13.8 g, 50 mmol) in a round-bottomed flask (100 ml) and add ethanol (20 ml) followed by aqueous sodium hydroxide solution (2.2 g, 55 mmol in 12.5 ml water).
2. To this light pink solution, add 6-bromo-hexanol (7.6 ml = 9.1 g, 55 mmol) dropwise over 30 min. Continue stirring for an additional 1 h

Protocol 3. *Continued*

Monitor the progress of the reaction by TLC ($R_f = 0.35$ in 33% acetone in hexane). The reaction is usually complete in 1–1.5 h.

3. Cool the reaction mixture in an ice-bath for 15 min and filter the reaction mixture through celite. Wash the residue with ethanol (2 × 15 ml). Combine the filtrate and washings and concentrate them under reduced pressure to a viscous oil.
4. Dissolve this viscous oil in methylene chloride (200 ml). Wash this solution with 5% aqueous sodium hydroxide (2 × 50 ml) and finally with water (2 × 100 ml).
5. Dry the methylene chloride solution with anhydrous sodium sulphate, filter, and concentrate the solution to a light yellow oil. Crystallize the product from ether–hexane (2:1 v/v) to give a colourless solid (65%), m.p. 70–72 °C. ^1H NMR ($CDCl_3$): δ (p.p.m.) 7.45 (m, aromatic, 6H); 7.2 (m, aromatic, 9H); 3.25 (t, $-0CH_2-$); 2.15 (t, $S-CH_2$); 1.55 (m, CH_2, 4H) and 1.35 (m, CH_2, 4H).

4. Synthesis of the 5′-terminal modifying agents (*Figure 1*, (**2**) and (**3**))

4.1 Synthesis of phosphoramidite derivatives as modifying agents

Protocol 4. Synthesis and purification of protected aminohexyl and thiol-hexyl phosphoramidites (*Figures 1* and *2* (**2a**), (**2b**), and (**2c**))

A. *Synthesis*

1. Dry protected hexanol [(**1a**), (**1b**), or (**1c**); 20 mmol] under high vacuum for 3 h and dissolve in freshly distilled THF (from sodium metal and benzophenone, 50 ml). Add diisopropyl ethyl amine (8.7 ml, 50 mmol) and stir at 0 °C for 10 min.
2. Add 'monochloridite'[a] (monochloro-β-cyanoethyl-N,N-diisopropylamino phosphoramidite, 5.5 ml, ~5.92 g, 25.0 mmol) dropwise through a syringe (27). The amine hydrochloride should precipitate within 5 min of addition. The reaction should be carried out under an argon atmosphere.
3. Stir for 30 min at 0 °C and allow to stir at room temp. for another 30 min to 1 h. Monitor the progress of reaction by TLC (50% ethyl acetate–hexane).

Protocol 4. *Continued*

4. When the reaction is complete, remove amine hydrochloride by filtering through a sintered glass funnel under argon and wash the solid with dry THF (2 × 20 ml).
5. Evaporate the combined filtrate to a viscous oil on a rotary evaporator. Release vacuum with argon to minimize the exposure to air.
6. Dissolve viscous oil in argon-purged ethyl acetate and wash the solution with ice-cold 5% aqueous sodium bicarbonate (2 × 50 ml) followed by saturated sodium chloride (50 ml).[b]
7. Dry the ethyl acetate solution over anhydrous sodium sulphate, filter, and concentrate the filtrate to a light yellow oil on a rotary evaporator.

[a] 'Monochloridite' is a pyrophoric compound. This reaction is only to be performed by personnel trained in handling hazardous chemicals.
[b] Sometimes aqueous work-up forms an emulsion; addition of solid sodium chloride removes this emulsion.

B. *Purification*

1. Pack a column (5 cm diameter or an appropriate size) with silica gel-60 (70–230 mesh) using 25% ethyl acetate in hexane containing 5% pyridine. The amount of silica gel should be 20 times the weight of the material to be purified.
2. Wash the silica packed column with one column volume of 25% ethyl acetate in hexane.
3. Load the sample dissolved in minimum volume of 50% ethyl acetate in hexane.

Protocol 5. Elution of protected aminohexyl and thiol-hexyl phosphoramidites

A. *N-TFA-aminohexyl phosphoramidite* **(2a)**

1. Elute the column loaded with the crude N-TFA-aminohexyl phosphoramidite with 30% ethyl acetate in hexane and collect 150 ml fractions in a screw-capped bottle or flask under argon. The product starts eluting in fraction 3.
2. Check the collected fractions by TLC (50% ethyl acetate–hexane). Detect the product by spraying with 5% $AgNO_3$ solution and heating. A grey to brown spot indicates the presence of the desired phosphoramidite (R_f = 0.5). Combine the product-containing fractions.
3. Evaporate the solvent under reduced pressure and final traces of solvent using a high vacuum pump. Check purity of the material by ^1H and ^{31}P

Protocol 5. *Continued*

NMR. Product purity is about 95–98% according to ^{31}P NMR. ^{1}H NMR (CD_3CN): δ 7.04 (b.s. NH, 1H); 3.8 (m, N-CH, 2H); 3.6 (m, $-CH_2$-OP, 4H); 3.35 (q, N-CH_2, 2H); 2.65 (t, $-CH_2$-CN, 2H); 1.6 (m, $-CH_2$-, 4H); ^{31}P NMR (CD_3CN) δ 144.5.

B. *N-MMT-6-aminohexyl phosphoramidite* (**2b**)

1. Elute the column loaded with the crude N-MMT-6-aminohexyl phosphoramidite (described in *Protocol 4* with 25% ethyl acetate in hexane. Collect 150 ml fractions. Fractions should be collected in a screw-capped bottle or flask under argon so that exposure to air can be avoided. The desired product starts eluting after the 2nd or 3rd fraction.
2. Check collected fractions by TLC (25% ethyl acetate in hexane). Visualize either with shortwave UV or acid spray. R_f value of the desired product is 0.38.
3. When elution is complete, combine the desired fractions and concentrate under reduced pressure using a rotary evaporator with controlled vacuum. Finally remove traces of solvents by evacuating under high vacuum to a constant weight. The yield of pure material as a viscous oil is about 70–75%.
4. Check the purity of the material by TLC, ^{1}H and ^{31}P NMR analysis: ^{1}H NMR (CD_3CN): δ 8.6 (d, NH, 1H); 7.5, 7.4, 7.2 and 6.8 (14 aromatic protons); 3.8 (m, N-CH, 2H); 3.75 (s, CH_3, 3H); 3.6 (m, $-OCH_2$-, 4H); 2.6 (t, CH_2-CN, 2H); 2.1 (q, N-CH_2, 2H); 1.65 (m, CH_2-4H); 1.35 (m, CH_2, 4H), and 1.15 (d,d, CH_3, 12H). ^{31}P NMR (CD_3CN) δ 147.95.

C. *S-trityl-6-mercaptohexyl phosphoramidite* (**2c**)

1. Elute the column loaded with the crude S-trityl-6-mercaptohexyl phosphoramidite with 30% ethyl acetate in hexane. Collect 150 ml fractions and monitor each fraction by TLC (acetone/hexane (1:2, v/v) and visualizing either with shortwave UV or acid spray). Combine pure fractions and evaporate down to a light yellow liquid.
2. Finally, remove traces of solvent by evacuating overnight under high vacuum (0.1 mm Hg).
3. Check the purity by TLC, ^{1}H and ^{31}P NMR. The product is usually 95% pure by ^{1}H and ^{31}P NMR. ^{1}H NMR (CD_3CN): δ 7.45 (m, aromatic 6H); 7.2 (m, aromatic 9H); 3.8 (m, N-CH, 2H); 3.6 (m, $-OCH_2$-4H); 2.6 (t, -CH_2-CN, 2H); 2.15 (t, S-CH_2, 2H); 1.55 (m, CH_2, 4H); 1.35 (m, CH_2, 4H), and 1.15 (d,d CH_3, 12H). ^{31}P NMR (CD_3CN): δ 148.07

4.2 Synthesis of H-phosphonate linker (*Figures 1* and *2*: (3b) and (3c))

Protocol 6. Synthesis of H-phosphonate salts of protected N-MMT-6-aminohexanol or S-trityl-6-mercaptohexanol [(**3b**) and (**3c**)]

A. *Synthesis*

1. Dry N-MMT-6-aminohexanol or S-trityl-6-mercaptohexanol (20 mmol) by evacuating under high vacuum (0.1 mm of Hg) for 6 h. Dissolve the dried alcohol in dry THF (50 ml), add diisopropylethyl amine (8.7 ml, 50 mmol). Cool this mixture in an ice-bath under argon or nitrogen with stirring.
2. To this mixture, add dropwise a solution of phosphonating reagent (22, 29) (2-chloro-5,6-benzo-1,3,2-phosphorin-4-one, 5.06 g, 25 mmol) in dry THF (25 ml) over 15 min.
3. Remove ice-bath after addition is complete and allow to stir for an additional 1 h at room temp. Usually the reaction is complete at this time which can be monitored by TLC using 10% methanol in methylene chloride. The desired product has an R_f value lower than that of the starting material.
4. When the reaction is complete, filter the insoluble amine hydrochloride and wash the solid with THF (2 × 10 ml). Combine the filtrate and washings and cool to 0 °C. Add water (5 ml) to this reaction mixture and stir for 10 min at 0 °C.
5. Evaporate the solvent on a rotary evaporator using a water aspirator. Dissolve the viscous oil in methylene chloride. Wash this solution with triethylammonium bicarbonate solution (2 × 150 ml; 0.1 M, pH 7.5) followed by water (1 × 100 ml).
6. Dry the methylene chloride solution over anhydrous sodium sulphate. Filter and concentrate the solution on a rotary evaporator to obtain a light-yellow viscous oil. The desired 'H-phosphonate linker' can be obtained by chromatographing this oil on a silica gel column.

B. *Purification*

1. Pack a column (5 cm diameter or an appropriate size) with silica gel-60 (70–230 mesh) in 5% methanol in methylene chloride containing 1% triethylamine. Amount of silica gel should be fifteen times the mass of crude material.
2. Dissolve the crude material in a minimum volume of 10% methanol in methylene chloride (25–30 ml). Apply this on to the silica gel column

Protocol 6. *Continued*

Elute the column with 10% methanol in methylene chloride containing 0.1% triethylamine.

3. Collect 150 ml fractions and check fractions by TLC (10% methanol in methylene chloride). Impurities and starting material elute before the desired material.
4. Once impurities are removed, elute the product with 15% methanol in methylene chloride. Combine all product-containing fractions.

C. *Further procedure for N-MMT-6-aminohexyl-O-H-phosphonate linker*

1. Wash the pooled fractions (obtained from step 4 of the purification procedure) with aqueous DBU bicarbonate solution (2 × 150 ml; 0.1 M, pH 7.5) followed by water (1 × 100 ml).
2. Dry this solution over anhydrous sodium sulphate, filter, and concentrate the filtrate to a clear viscous oil.
3. Finally, remove traces of solvent under high vacuum to obtain a white foam (65% yield, >95% pure). Characterize the product by ^{31}P NMR. ($CDCl_3$: δ 2.25.

D. *Further procedure for S-trityl-6-mercaptohexyl-O-H-phosphonate linker*

1. Wash the pooled fractions (obtained from step 4 of the purification procedure) with 0.1 M triethylammonium bicarbonate (2 × 150 ml).
2. Dry the organic phase over sodium sulphate, filter, and concentrate the filtrate to a viscous oil.
3. Finally, pump it down to a constant weight (70% yield, 95% purity). Characterize the product by ^{31}P NMR: ($CDCl_3$: δ 2.42 p.p.m.

5. Incorporation of linkers on to oligonucleotide chains and purification

5.1 Synthesis of 5′-end modified oligonucleotides using phosphoramidite chemistry

Protocol 7. Incorporation of phosphoramidite linkers on an automated DNA synthesizer (*Figures 3* and *4*)

1. Dissolve amino-linkers (**2a**), (**2b**), or (**2c**) in dry acetonitrile (100–250 mg in 2–5 ml) under an argon atmosphere.

Protocol 7. *Continued*

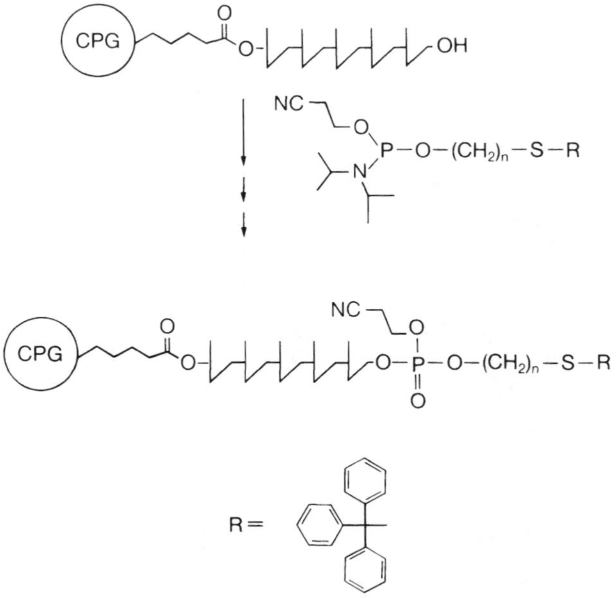

Figure 4. Coupling of thiol-linker onto an oligonucleotide chain attached to a solid support.

2. Place this solution into a clean extra reservoir (U-reservoir of the MilliGen Biosearch DNA synthesizer). Prime the line manually for a few seconds or use the priming program so that the delivery tube is filled with this reagent.
3. Write the desired sequence; the 5′-end having a base 'U' so that the modifying reagent is incorporated at the last step of synthesis by the instrument.
4. Print and verify the sequence.
5. Have 'DMT-on' option for synthesis program.
6. Start the synthesis using an appropriate scale (0.25 or 1.0 μmol) coupling program on the instrument.
7. At the end of synthesis, detach the column from the instrument and wash the support with methanol (3 × 5 ml) using a syringe to remove residual acid (from detritylation steps) and iodine (from oxidation steps)

   Complete removal of iodine is essential for thiol-linked oligonucleotides.

5.2 Synthesis of 5′-end modified oligonucleotides using H-phosphonate chemistry

Protocol 8. Incorporation of N-MMT-6-aminohexyl-O-H-phosphonate (**3b**) in an oligonucleotide on an automated synthesizer

1. Dissolve DBU salts of protected aminohexyl-O-H-phosphonate (**3b**) 50 mg/ml) in a mixture of anhydrous pyridine–acetonitrile (1:1 v/v) or commercially available H-phosphonate diluent.
2. Place this solution (3–4 ml) in an extra reservoir.
3. Rinse the line manually or using the priming program.
4. Write the desired sequence with the 5′-end bearing a residue 'U' on the instrument.
5. Print and verify the sequence so that linker is attached at the 5′-end.
6. Have 'DMT-on' option for synthesis program.
7. Start the synthesis using an appropriate synthesis scale coupling program.

Protocol 9. Incorporation of S-trityl-6-mercaptohexyl-O-H-phosphonate (**3c**) in an oligonucleotide

1. Dissolve the thio-linker (S-trityl-6-mercaptohexyl-O-H-phosphonate (**3c**) 50 mg/ml) in a mixture of anhydrous pyridine–acetonitrile (1:1) or commercially available H-phosphonate diluent.
2. Place this solution (3–4 ml) in an extra reservoir.
3. Rinse the line manually or using the prime program.
4. Write the desired sequence of oligonucleotide without linker at the 5′-end.
5. Start synthesis with 'DMT-off' option, using appropriate scale coupling program.
6. After chain assembly and oxidation with iodine solution, remove the synthesis column from the instrument.
7. Wash the support with acetonitrile–pyridine mixture to *remove the last traces of iodine.*
8. Place the synthesis column back on to the instrument.
9. Write a dimer synthesis, 'U' as the 5′-end base on the instrument.
10. Use a program which does not include oxidation with iodine or a program which has an option to stop the synthesis before oxidation.
11. When addition or coupling of linker from 'U'-reservoir is complete, after washing with wash solvent, stop the synthesis.

Protocol 9. *Continued*

12. Remove the column from the synthesizer and dry the support.

13. Pass 10 ml suspension of 10% water in carbon tetrachloride†, triethylamine, and N-methylimidazole (9: 0.5: 0.5) through the support using two 10 ml syringes for 10 min. *This oxidation prevents disulphide bond formation, generally obtained with iodine-solution oxidation.*

14. Remove this mixture, wash with acetonitrile. Discard the washing as a halogenated waste.

† Gloves must be used while handling carbon tetrachloride, a carcinogen. Perform this oxidation step in the fume hood.

5.3 Deprotection and isolation of 5′-modified oligonucleotides from solid support

This step can be performed following the protocol recommended by the instrument manufacturers or by the steps given below.

Protocol 10. Deprotection and removal of oligonucleotides from supports

1. Transfer the dried support into a screw-capped vial or screw-capped Eppendorf tube (1.5 ml).

2. Add 300 μl (0.25 μmol scale) or 1.0 ml (1.0 μmol scale) of conc. NH_4OH solution (30%).

3. Close the cap tightly and incubate the suspension at 55 °C for at least 5 h to overnight (*a longer period should be given for G-rich sequences*).

4. Cool to 0 °C and transfer the supernatant into another microfuge tube.

5. Rinse the support with same amount of water (HPLC grade or double distilled). Add this washing to the ammonia supernatant (step 4). *This ammonia solution contains full-length oligonucleotide with either a free aminohexyl group at the 5′-end (in the case of N-TFA-aminohexyl phosphoramidite incorporated) or protected amino- or thiolhexyl-linked oligonucleotide together with non-nucleosidic material and short sequences.*

Purification of oligonucleotides with free aminohexyl linkers can be achieved by anion exchange HPLC, ethanol precipitation (*Protocol 11*) or polyacrylamide gel electrophoresis.

Protocol 11. Purification by ethanol precipitation

1. Concentrate the ammonia solution (obtained from Protocol 10, steps 4 and 5) to dryness in a SpeedVac without heating.

2. Suspend the dried material in 200 μl of water and evaporate again.

Protocol 11. *Continued*

3. Dissolve the crude material in 200 μl of 0.5 M ammonium acetate solution, add 1.0 ml 95% ethanol. Mix the solution by vortexing. Cool the mixture at −20 °C for 30 min and then centrifuge for 3 min.
4. Remove the supernatant, wash the residue with cold ethanol (−20 °C). Shorter sequences, benzamide, and isobutyramide are removed by this step.
5. Dry the residue and resuspend the material in known volume of water (200 μl).
6. Measure the amount of oligonucleotide present in the solution at 260 nm. *This material is sufficiently pure for attachment of a non-radioactive marker molecule at the 5′-terminal primary amino group.*

If absolutely pure amino-linked oligonucleotide is required, the following standard gel electrophoresis procedure should be followed. *For sequences up to 30 bases long, use 20% polyacrylamide; for those up to 50 bases long, use a 15% polyacrylamide gel containing 7.0 M urea. The mobility of an amino-linked oligonucleotide is slightly slower than that of a non-modified oligonucleotide.* Follow any standard electrophoresis protocol—this is a routine procedure in molecular biology laboratories. The gel is then visualized by UV-shadowing at 260 nm and the product band is eluted from the gel.

Purification of oligonucleotides linked with protected primary amino- or thio-linkers can be achieved by reversed phase chromatography using either an Oligo-Pak column (MilliGen/Biosearch) (*Protocols 12* and *13*) reversed phase or HPLC (C_{18} Nova Pak or Bonda Pack, Waters).

Protocol 12. Oligo-Pak column purification for amino-linked oligonucleotides. Follow the protocol recommended by the suppliers.

1. First wash the Oligo-Pak column with 3 × 5 ml acetonitrile followed by 3 × 5 ml 1.0 M triethylammonium acetate solution (pH 7.0).
2. Using a syringe, load the entire sample (0.25 μmol scale) or one-third of a 1.0 μmol scale synthesis obtained from deprotection described in *Protocol 10*.
3. Wash the column 3 times as described in step 1.
4. Save the effluent: it may contain some full-length desired product.
5. Wash the support with 3 × 5 ml of 3% aqueous NH_4OH solution (10 times diluted concentrated ammonium hydroxide).
6. Subsequently wash with 3 × 5 ml of water.
7. Treat the Oligo-Pak column support with 1 × 5 ml of 2% trifluoroacetic

Protocol 12. *Continued*

acid solution over 2 min and wash the support immediately with water (3 × 5 ml) to remove residual trifluoroacetic acid.

8. Elute the product with 20% acetonitrile in water in two 1 ml fractions. Concentrate the eluant to dryness in a SpeedVac.
9. Dissolve the dried material in HPLC-grade water (200 μl) and dry the solution again.
10. Finally, dissolve the dried material in a known volume of water. Determine the amount of oligonucleotide in this solution by measuring the absorption at 260 nm.

Protocol 13. Oligo-Pak column purification for S-trityl-thiol-linked oligonucleotides

1. Load the entire or 1/3 volume of the sample obtained from *Protocol 10* on to the equilibrated Oligo-Pak column.
2. Wash with 3% aqueous ammonium hydroxide solution (3 × 5 ml) and then with water (3 × 5 ml).
3. Elute the product as the protected thiol-linked oligonucleotide with 40% acetonitrile in water (2 × 1 ml).
4. Concentrate the eluant to dryness in a SpeedVac. Dissolve the residue in a known amount of HPLC-grade water (200 μl) and determine the amount of oligonucleotide by measuring the absorption at 260 nm. *Removal of the trityl group should be carried out just prior to its use by treatment with silver nitrate solution followed by aqueous DTT solution. Use the free sulphydryl linked oligonucleotide immediately for non-radioactive marker incorporation.*

---

6. Incorporation of non-radioactive marker molecules on to free primary amine or sulphydryl-linked oligonucleotides

---

Protocol 14. Solution phase labelling of 5′-end amino-linked oligonucleotides with isothiocyanate or N-hydroxysuccinimide ester derivatives of fluorescein, rhodamine, Texas Red, biotin or NBD-fluoride.

A. *Marker incorporation*

1. Dissolve partially or fully purified amino-linked oligonucleotide (20–30 A_{260} units obtained from *Protocol 11*) in 250 μl of a mixture of 1.0 M $NaHCO_3/Na_2CO_3$ (pH 9.0).

Protocol 14. *Continued*

2. Check the pH of this solution using pH paper and *make sure it is basic.*
3. Add 500 µl of a solution of the dye derivative (10–15 mg) in a mixture of 1.0 M, $NaHCO_3/Na_2CO_3$ buffer pH = 9.0; DMF: Water (5:2:3 v/v).
4. Vortex the mixture and wrap the Eppendorf tube with aluminum foil to prevent light exposure.
5. After 20 h of incubation at room temperature, concentrate the mixture to a 250 µl volume.

B. *Removal of excess marker*

1. Pack a column (20 cm × 1.0 cm) with presuspended Sephadex G-25.
2. Equilibrate the column with HPLC-grade water (30–40 ml).
3. Apply the concentrated dye or biotin-labelled reaction mixture onto the column.
4. Elute the column with water. A faintly coloured band separates from the main coloured band, which contains the dye-tagged oligonucleotide.
5. Collect 1.0 ml fractions in Eppendorf tubes. *The desired product starts eluting after the void volume. Most of the desired product is eluted in fractions 3–9. Some fractions eluting after that are colourless followed by heavily coloured solution. The latter contains the unincorporated excess dye.*
6. Check the absorption of the faintly coloured solutions at 260 nm.
7. Concentrate the fractions which contain most of the material.
8. Combine these concentrated fractions into a single Eppendorf tube. *Usually 15–25 A_{260} units (80%) is obtained, which is free from excess dye. The purity of this material is generally better than 90–95%.*
   If necessary, the product can further be purified by electrophoresis (20% PAGE) or by HPLC.

Protocol 15. Labelling of 5′-end thiol linked oligonucleotide with maleimide derivative of biotin, 7-N,N-diethylamino-4-methyl-3-(4′-maleimido-phenyl)-coumarin, eosin, or monobromo bimane.

A. *Removal of trityl group*

1. Place a solution of purified S-trityl-thiol-linked oligonucleotide (30 A_{260} units; 100 µl) in an Eppendorf tube.
2. Add silver nitrate solution (25 µl, 1.0 M) and vortex the mixture.

Protocol 15. *Continued*

3. After one hour, add aqueous DTT solution (50 μl, 1.0 M), vortex again and centrifuge the mixture for 15 min.

4. Place the supernatant (containing free thiol-linked oligonucleotide and excess DTT) in a separate Eppendorf tube.

B. *Removal of excess DTT*

1. Add 500 μl of ethyl acetate to the supernatant, vortex, and centrifuge.

2. Remove ethyl acetate in the upper layer. Repeat this process once more. This process removes most of the excess DTT.

C. *Incorporation of marker molecules*

1. Immediately add a solution of maleimide or bimane derivative (10–15 mg) in DMF/THF (1:1 v/v) (500 μl) to the aqueous layer.

2. Vortex the reaction mixture, wrap the tube in aluminum foil and incubate at room temp. for 20 h or overnight.

3. Concentrate the mixture to a 200 μl volume.

D. *Removal of excess markers*

1. Place this reaction mixture onto a pre-equilibrated Sephadex column as described in *Protocol 14*.

2. Elute the product with HPLC-grade water. A faintly coloured band separates from the main coloured band. *The faintly coloured band contains most of the desired product.*

3. Collect 500 μl to 1.0 ml fractions in Eppendorf tubes. Most of the product elutes at the void volume of the column, usually in fractions 3–9.

4. Check the absorption of these fractions at 260 nm.

5. Concentrate the fractions which contain most of the material.

6. Combine these fractions in a single Eppendorf tube. *Usually 20–25 A_{260} units of purified material is obtained, which may contain up to 10% unincorporated dye and non-dye linked oligonucleotide.* If necessary, the product can be further purified by reverse-phase HPLC or 20% PAGE to result in homogeneous product.

Protocol 16. Labelling of 5′-end amino-linked oligonucleotide with a non-radioactive marker on a solid support (see *Figure 5*).

A. *Introduction of amino linker*

1. Introduce N-MMT–aminohexyl linker at the 5′-end of synthetic oligonucleotide as described in *Protocol 7* or *8*.

Protocol 16. *Continued*

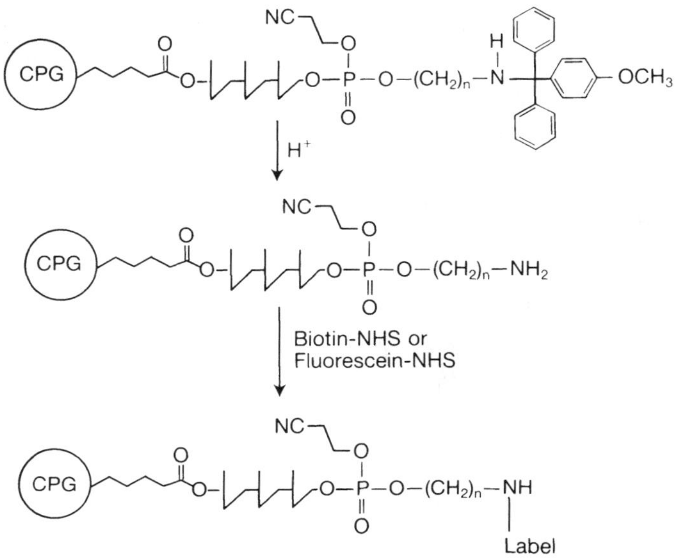

Figure 5. Introduction of non-radioactive reporter group at the 5′-end of an amine-modified oligonucleotide on a solid support. Biotin-NHS = N-hydroxysuccinicimide ester of biotin.

2. Remove MMT-group (monomethoxytrityl) from the oligonucleotide chain to generate a free amino group by passing a deblock solution until the support is free from yellow colour. *Rinse the support with wash solvent. These steps can be performed on the instrument by manual mode.*
3. Finally, remove the synthesis column and wash with a solution of 10% diisopropylethyl amine in dimethylformamide: (v/v; 2 × 5 ml). This removes the acid completely and generates a free amino group at the 5′-end of the oligonucleotide attached to the solid support.

B. *Introduction of marker molecules*

1. Incubate the supports present in the column between two 1 ml syringes with 500 μl solution of the isothiocyanate or the N-hydroxysuccinimide derivative of fluorescein, rhodamine, biotin, or ethylene diamine tetracetic acid anhydride (in 10% diisopropylethyl amine in DMF, 5 mg/ml).
2. After 1 h of incubation of the dark at room temp., wash the support with DMF to remove the excess dye using a 5 or 10 ml syringe [4 × 5 ml or 3 × 10 ml DMF washing, followed by acetonitrile (2 × 5 ml).
3. Dry the support by blowing air through with a syringe. Transfer the support into a screw cap Eppendorf tube and incubate with 1 ml of conc. NH_4OH either at 55 °C for 5 h or at room temp. for 24 h.

Protocol 16. *Continued*

4. Transfer the coloured supernatant into a separate Eppendorf tube and rinse the support with 2 × 250 μl of water. Combine the washings.
5. This solution contains a mixture of dye or non-radioactive marker tagged oligonucleotide, short DNA sequences, protecting groups and a small amount of free marker molecules (see *Figure 6*).

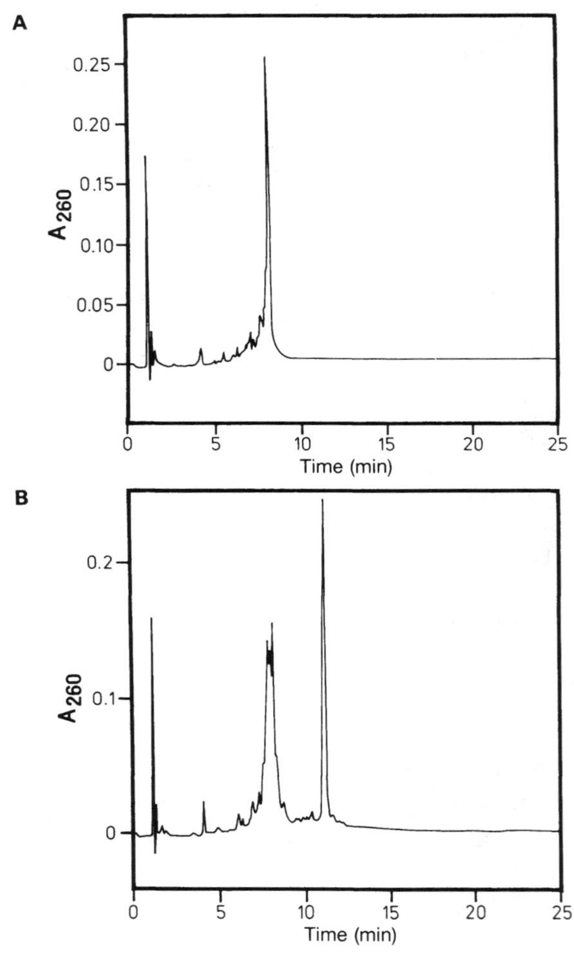

Figure 6. Reversed phased HPLC analysis of unpurified synthetic oligonucleotides. (**A**) Non-modified oligonucleotide (24mer) and (**B**) fluorescein isothiocyante attached to an amino-linked oligonucleotide (24mer) (peaks I = unreacted oligonucleotide and II = fluorescein-tagged oligonucleotide) on solid support. HPLC conditions: Column Waters C_{18} Delta Pak, 5 μm spherical particle; buffers: A = 0.1 M TEAA pH = 7.0, B = 100% acetonitrile; Gradient: linear gradient 0–1.0 min 95% A and 5% B then 1.0–25.0 min 40% B, Detection: 260 nm.

Protocol 16. *Continued*

C. *Purification by Oligo-Pak by a modified procedure*

1. Equilibrate Oligo-Pak column with acetonitrile (2 × 5 ml), water (2 × 5 ml) and 1.0 M triethylammonium acetate, pH 7.0 (2 × 5 ml).
2. Load the reaction mixture (1:1 diluted with water) on the column with a syringe in 3 steps.
3. Wash the column with 3% NH_4OH (2 × 5 ml) followed by water (2 × 5 ml).
4. Wash with 7% acetonitrile in water (1 × 5 ml) to remove the uncoupled amino-linked oligonucleotide.
5. Elute the dye-tagged oligonucleotide with 20% acetonitrile in two 1 ml fractions.
   Combine these fractions, lyophilize solution and check the purity of the product by HPLC (see *Figure 7*) or gel electrophoresis.

7. Applications of oligonucleotides carrying non-radioactive reporter molecules

7.1 DNA sequencing

Fluorescein, rhodamine, or other fluorophores linked to the 5′-end of oligonucleotides have been used as chain extension primers in automated DNA sequencing (14, 24–33). Recently, biotin-linked oligonucleotides have also been used in DNA sequencing followed by chemiluminescent (25) and colourimetric (26) detections.

7.2 Diagnostic probes

With the improvement of non-radioactive detection systems, the use of synthetic oligonucleotides with non-radioactive reporter groups as diagnostic probes is increasing. Examples are:

(a) A colourimetric method for visualizing biotin-labelled DNA probes hybridized to DNA or RNA immobilized on nitrocellulose (1b).
(b) A colourimetric method for DNA hybridization (34).
(c) Chemically modified nucleic acids as immuno-detectable probes in hybridization experiments (35)
(d) Comparison of non-radioisotopic hybridization assay methods using fluorescent, chemiluminescent, or enzyme-labelled synthetic oligonucleotides (36).

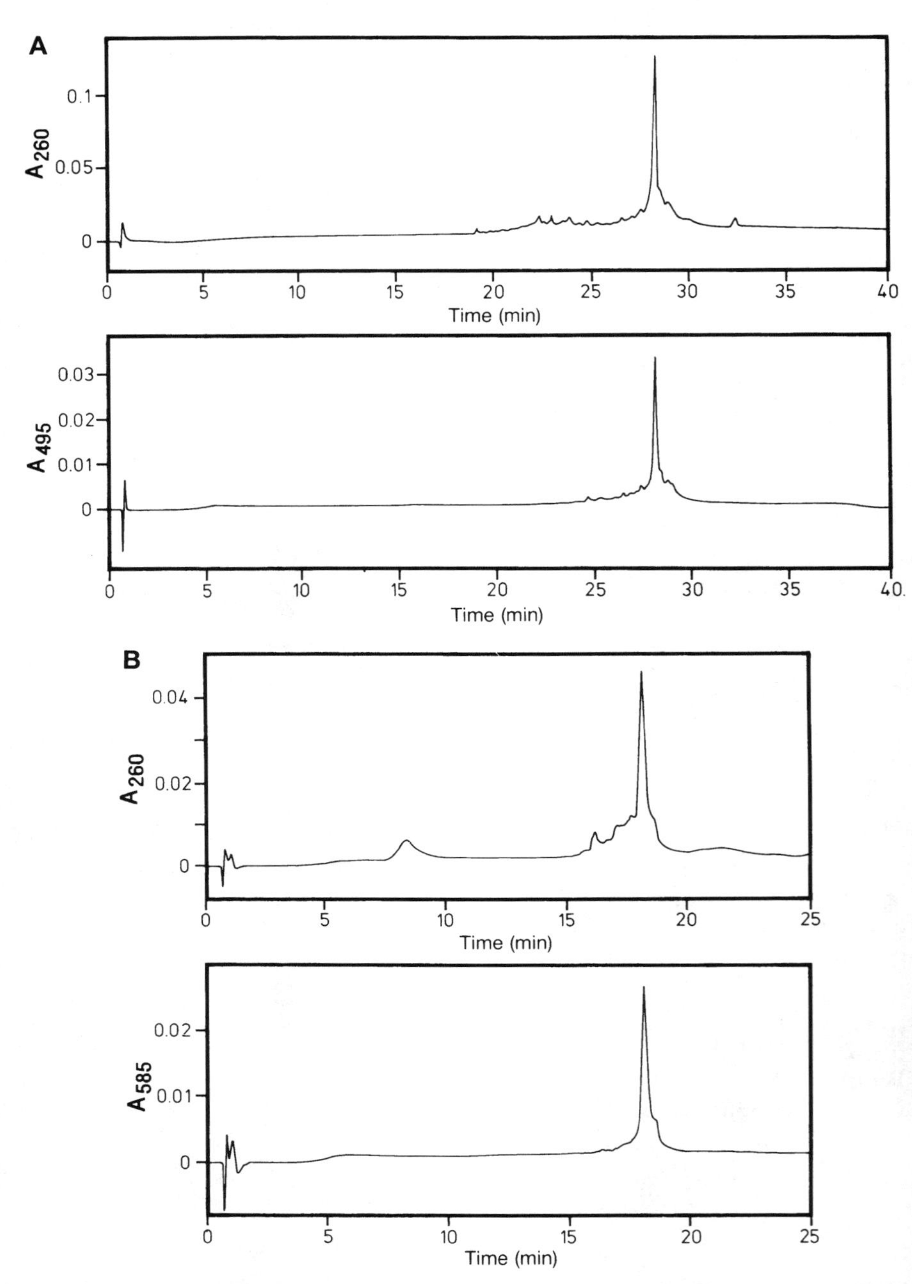

Figure 7. Anion exchange HPLC analysis of Oligo-Pak column-purified, fluorescein isothiocyanate-attached oligonucleotide (24mer) (**A**) and rhodamine isothiocyanate-attached oligonucleotide (15mer) (**B**). HPLC conditions: column: Water Gen-Pak Fax anion-exchange; buffers: A = 25 mM Tris–HCl in water pH = 7.5, B = 25 mM Tris–HCl/1.0 M NaCl in water pH = 7.5; gradient: linear gradient 0–1.0 min 100% A, then 1–35 min 0–60% B; detection: (A) at 260 and 495 nm, (B) at 260 and 585 nm.

(e) Rapid chemiluminescent nucleic acid assays for the detection of TEM-1 β-lactamase-mediated penicillin resistance in *Neisseria gonorrhoea* and other bacteria (37).

Acknowledgements

The authors sincerely thank Dr D. Hudson for his encouragement and help during the preparation of the manuscript and also Rosie Freeman and Sundee O'Neel for typing the manuscript.

References

1. a) Langer, P. R., Waldrop, A. A., and Ward, D. C. (1981). *Proc. Natl. Acad. Sci. USA*, **78**, 6633.
   b) Leary, J. J., Brigati, D. J., and Ward, D. C. (1983). *Proc. Natl. Acad. Sci. USA*, **80**, 4045.
2. Richardson, R. W. and Gumport, R. I. (1983). *Nucleic Acids Res.*, **11**, 6167.
3. Marasuji, A. and Wallace, R. B. (1986). *DNA*, **3**, 269.
4. Shimkus, M., Levy, J., and Herman, T. (1985). *Proc. Natl. Acad. Sci. USA*, **82**, 2593.
5. Gillan, I. C. and Tener, G. M. (1986). *Anal. Biochem.*, **157**, 199.
6. Kumar, A., Tchen, P., Roullet, F., and Cohen, J. (1988). *Anal. Biochem.*, **169**, 376.
7. Trainor, G. L. and Jenson, M. A. (1988). *Nucleic Acids Res.*, **16**, 11846.
8. Allen, D. J., Darke, P. L., and Benkovic, S. J. (1989). *Biochemistry*, **28**, 4601.
9. Ruth, J. L. (1984). *DNA*, **3**, 123.
10. Chou, B. C. F. and Orgel, L. E., (1985). *DNA*, **4**, 327.
11. Kempe, T., Sundquist, W. I., Chou, F., and Hu, S. L. (1985). *Nucleic Acids Res.*, **13**, 45.
12. Chollet, A. and Kawashima, E. H. (1985). *Nucleic Acids Res.*, **13**, 1529.
13. Forster, A. C., McInnes, J. L. Skingle, D. C., and Symmons, R. H. (1985). *Nucleic Acids Res.*, **13**, 745.
14. a) Smith, L. M., Fung, S., Hunkapillar, M. W., and Hood, L. E. (1985). *Nucleic Acids Res.*, **13**, 2399.
    b) Smith, L. M., Sanders, J. L., Heiner, C., Kent, S. G. H., and Hood, L. E. (1986). *Nature*, **321**, 674.
15. Connolly, B. A. and Rider, P. (1985). *Nucleic Acids Res.*, **13**, 4445.
16. Coull, J. M., Weith, H. L., and Bischoft, R. (1986). *Tetrahedron Lett.*, **27**, 3991.
17. Agrawal, S., Christodoulou, C., and Gait, M. J., (1986). *Nucleic Acids Res.*, **14**, 6227.
18. Connolly, B. A. (1987). *Nucleic Acids Res.*, **15**, 3131.
19. Sproat, B. Beijer, B., Rider, P., and Nuener, P. (1987). *Nucleic Acids Res.*, **15**, 4837.
20. Sproat, B., Beijer, B., and Rider, P. (1987). *Nucleic Acids Res..*, **15**, 6181.
21. Gibson, K. J. and Benkovic, S. J. (1987). *Nucleic Acids Res.*, **15**, 6455.
22. Sinha, N. D. and Cook, R. M. (1988). *Nucleic Acids Res.*, **16**, 2659.

23. Kansal, V. K., Huynh-Dinh, T., and Ingolen, J. (1988). *Tetrahedron Lett.*, **29**, 5537.
24. Kristensen, T., Voss, H., and Ansorge, W. (1987). *Nucleic Acids Res.*, **15**, 5507.
25. Beck, S., O'Keefe, T., Coull, J., and Koester, H. (1989). *Nucleic Acids Res.*, **17**, 5115.
26. Richterich, P. (1989). *Nucleic Acids Res.*, **17**, 2181.
27. Sinha, N. D., Biernat, J., McManus, J. P., and Koester, H. (1984). *Nucleic Acids Res.*, **12**, 4539.
28. Bischoff, R., Coull, J. M., and Regnier, F. E. (1987). *Anal. Biochem.*, **164**, 336.
29. Marugg, J. E., Tromp, M., Kuyl-Oyeheskiely, E., Vander-Marcel, G. A., and van Boom, J. H. (1986). *Tetrahedron Lett.*, **27**, 2661.
30. Alves, A. H., Holland, D., and Edge, M. D. (1989). *Tetrahedron Lett.*, **30**, 3089.
31. Cocuzza, A. J. (1989). *Tetrahedron Lett.*, **30**, 6290.
32. Misiura, K., Durrant, I., Evans, M. R., and Gait, M. J. (1990). *Nucleic Acids Res.*, **18**, 4345.
33. Schubert, F., Ahlert, K., Cech, D., and Rosenthal, A. (1990). *Nucleic Acids Res.*, **18**, 3427.
34. Reuz, M. and Jurtz, C. (1984). *Nucleic Acids Res.*, **12**, 3425.
35. Tchen, P., Fuchs, R. P. P., Sage, E., and Levy, M., (1984). *Proc. Natl. Acad. Sci. USA*, **81**. 3466.
36. Urdea, M. S., Warner, B. D., Running, J. A., Stempien, M. S., Clyne, J., and Horn, T., (1988). *Nucleic Acids Res.*, **16**, 4937.
37. Sanchez-Pescador, R., Stempien, M. S., Urdea, M. S., (1988). *J. Clin. Microbiol.*, **26**, 1934.

Appendix: Chemical suppliers

Protection of 6-aminohexanol

ethyl trifluoroacetate	Aldrich
6-aminohexanol	Aldrich
p-methoxyphenyldiphenylmethyl chloride (MMT-chloride)	Aldrich
N,N-diisopropylethyl amine	Aldrich
silica gel-60 (70–230 mesh)	Merck
precoated silica gel GF-254 plates	Merck
sodium sulphate	Merck

Protection of 6-mercaptohexanol

triphenylmethyl mercaptan	Aldrich
6-bromo-1-hexanol	Sigma
sodium hydroxide	Merck
sodium sulphate	Merck

Synthesis of amino- or thio-hexyl phosphoramidite linkers

monochloro-β-cyanoethoxy-N,N-diisopropyl-amino phosphine (phosphitylating agent)	Applied Biosystems Biosyntech American Biometrics Cruachem Glen Research
phosphoramidite linkers now also available from	Milligen-Biosearch Clonetech Glen Research

Synthesis of amino- or thio-hexyl phosphonate linkers

2-chloro-4H,1,3,2-benzodioxaphosphorin-4-on (phosphonylating agent)	Aldrich
diazobicyloundecen	Aldrich

Incorporation of linker on to oligonucleotide chain

Phosphoramidite linkers

deoxynucleoside phosphoramidites and ancillary DNA synthesis reagents	Milligen-Biosearch Applied Biosystems American Biometrics

H-phosphonate linkers

deoxynucleoside H-phosphonates and ancillary DNA synthesis reagents	Milligen-Biosearch Applied Biosystems American Biometrics
triethylamine	J. T. Baker Aldrich
carbon tetrachloride	Aldrich
N-methyl imidazole	Aldrich

Deprotection and purification

concentrated ammonium hydroxide solution (30%)	Aldrich
materials for polyacrylamide gels (acrylamide, bisacrylamide, ammonium persulphate)	Aldrich
Oligo-Pak	Milligen-Biosearch
triethylamine	J. T. Baker
glacial acetic acid	Aldrich
trifluoroacetic acid	Aldrich

Attachment of a non-radioactive markers

Amino-linked oligonucleotides	
dimethylformamide (analytical grade; free from primary and secondary amines)	
triethylamine, isothiocyanate, or N-hydroxy-succinimide ester derivatives of fluorescein,	Molecular Probes
rhodamine and Texas Red	Aldrich
N-hydroxysuccinimide ester of biotin	Molecular Probes Pierce
Thiol-linked oligonucleotides	
Silver nitrate	Aldrich
dithiothreitol (DTT)	Sigma
maleimide derivative of biotin	Sigma
eosin	Molecular Probes
7-diethylamino-4-methyl-3-(4'-maleimidlphenyl)-coumarin	Molecular Probes
monobromo bimane	Molecular Probes

Solvents

Analytical or HPLC grade solvents are commonly used. These are dichloromethane, ethyl acetate, pyridine, N,N-dimethylformamide(DMF), tetrahydrofuran (THF), methanol, hexane, and acetonitrile, which can be obtained from any reputable supplier.

9

Site-specific attachment of labels to the DNA backbone

NANCY E. CONWAY, JACQUELINE A. FIDANZA, MARYANNE J. O'DONNELL, NICOLE D. NAREKIAN, HIROAKI OZAKI, and LARRY W. McLAUGHLIN

1. Introduction

Sequence-specific attachment of reporter groups, drug derivatives, or chemically reactive species to DNA sequences has the potential to provide new materials for detailed spectroscopic and biophysical studies as well as a host of new DNA therapeutics and diagnostics. The covalent binding of a variety of such agents at specific locations within the nucleic acid sequence can be achieved by a number of procedures depending on whether the nucleobase, carbohydrate, or phosphate residue is employed as the site of attachment. These procedures often exploit the availability of specific functional groups (such as terminal phosphomonoesters, see Chapter 8) or the reactivity of selected sites on the purine or pyrimidine building blocks (such as the C5 position of pyrimidines, see Chapter 11) in order to attach an appropriate linker or the active agent directly. While the manner in which the nucleic acid is labelled may be dictated by the specific study involved, in general the principles of simplicity and versatility are best employed to guide the choice of labelling procedure.

Some consideration should also be given to the structural effects the label or agent will have the nucleic acid. For example, the addition of an agent or label to a terminal phosphomonoester would be unlikely to alter the structure or stability of a double-stranded or even triple-stranded complex. On the other hand, internal labelling may be more advantageous if the product complex involves an interstrand covalent cross-link or an intercalating agent. Labelling of a site within a sequence has generally relied upon the synthesis of a modified nucleoside building block in which a linker is incorporated for attachment of the label. However, in some cases this can destabilize the duplex structure, for example when the exocyclic amino group of the cytosine residue was used as a site of attachment, the labelled helices exhibited biphasic melting curves (1, 2) suggesting local or even global structural modulation. Therefore, the nature of the target nucleic acid species to be labelled, or the complex to be formed in the presence of the labelled DNA or RNA sequence, such as single- or double-stranded DNA or RNA, antisense complexes, hairpin structures or triple helix complexes should also influence the method chosen for labelling.

In the present chapter we will describe a series of experiments which allow for the labelling of the phosphate backbone of DNA sequences at single or multiple sites in a sequence-specific manner. These procedures provide versatile labelling at virtually any position within a sequence and are simple enough to be incorporated into standard solid phase based chemical synthesis procedures designed for nucleic acids.

1.1 Labelling techniques directed towards the DNA backbone

The attachment of labels or other agents to the DNA backbone offers a number of advantages over the modification of a terminal phosphomonoester or the modification of a base residue of a nucleoside building block. By using internal phosphodiesters instead of terminal phosphomonoesters to attach the label or agent, virtually any site within the sequence is amenable for introduction of the desired functionality. The phosphate residues are not involved in inter-strand base pairing so the attachment of a linker or label at such sites should not drastically alter the stability of the nucleic acid complex. Sites for backbone labelling can be incorporated into the nucleic acid sequence during the assembly of the sequence by standard chemical techniques and is not limited by the time consuming process involving the synthesis of a modified nucleoside or nucleoside analogue containing the desired linker or label. Multiple labels can also be incorporated without increasing the complexity of the procedure.

In addition, modification of the prochiral phosphodiester residue with a single moiety creates a chiral site of two phosphorus diasteroisomers, (R_p and S_p). Although for many experiments, stereochemically pure derivatives are not essential to obtain the desired information, in cases where stereochemistry is important, the preparation and isolation of each diastereoisomer can be achieved as described in this chapter. The ability to use a singly labelled phosphotriester diastereoisomer can also have certain advantages. For example, with double-stranded structures one phosphorus stereoisomer would direct the covalently attached derivative towards the major groove of an A or B form helix while the second diastereoisomer would direct the agent towards the minor groove. If the desired agent binds or reacts preferentially in one of the helix grooves, stereochemical labelling of the backbone can assist in enhancing the desired reactivity.

Presently two approaches have been employed to introduce labels into the backbone of DNA sequences. Letsinger (3) has described the oxidation of an

H-phosphonate to the N-substituted phosphoramidate with the desired agent tethered via the nitrogen and others have used his approach (4). Reaction with the substituted amine occurs, instead of other forms of oxidation, immediately after the introduction of the H-phosphonate linkage. The phosphoramidate is stable to the chemical DNA synthesis and deprotection conditions and yields the desired modified sequence. Our own recent work has involved the use of phosphorothioates (5–9) as sites for alkylation by the label of interest and this approach has also been used by others (10). In this case a phosphoramidite coupling is followed by oxidation with sulphur (13–15) to generate the phosphorothioate diester [H-phosphonate chemistry followed by sulphur oxidation (11,12) can also be employed if phosphothioate diesters are introduced at each site in the oligodeoxynucleotide]. After completion of the synthesis, deprotection, and isolation of the fragment, the phosphorothioate is amenable to alkylation (at sulphur) by a variety of functional groups. Both of these procedures result in DNA fragments carrying the agent of interest covalently bound to a specific internucleotide phosphate residue.

In principle either the H-phosphonate or the phosphoramidite procedure can be used to incorporate multiple labels into the nucleic acid fragment. By using DNA fragments containing multiple phosphorothioate diesters we have described a technique (6) which allows the incorporation of hundreds or more bimane fluorophores (one at each phosphorus residue) which results in highly sensitive fluorometric detection of nucleic acids in polyacrylamide gels.

The remainder of this chapter will describe the procedures necessary to modify the DNA backbone at single or multiple sites using one of the methods described above.

2. The synthesis of oligodeoxynucleotides containing phosphorothioate diesters

The synthesis of an oligodeoxynucleotide containing phosphorothioate diesters can be accomplished by both chemical (13–15) (see Chapter 4) and enzymatic (16) means. The methodology in this section employs the chemical synthesis of small oligomers with one or several specifically placed internucleotidic phosphorothioate diesters. Phosphorothioate triesters can be introduced by oxidizing the intermediate phosphite triester obtained during standard phosphoramidite chemistry with elemental sulphur (13–14) to generate a pentavalent phosphorus in which the phosphorothioate triester exists as a thione:

$$\mathrm{N{-}O{-}\underset{\displaystyle O{-}CH_2CH_2CN}{P}{-}O{-}N} \xrightarrow{S_8/CS_2/\text{Lutidine}} \mathrm{N{-}O{-}\overset{\displaystyle O{-}CH_2CH_2CN}{\underset{\displaystyle \| \atop S}{P}}{-}O{-}N}$$

Recent work (15, 17) has also indicated that oxidation to the phosphorothioate triester proceeds rapidly with 3H-1,2-benzodithiol-3-one 1,1-dioxide which exhibits greater solubility than elemental sulphur in organic solvents. However, at present this derivative is not readily available, so the procedures described here employ elemental sulphur. The thione formed in this manner is stable to the subsequent oxidation steps necessary to generate internucleotidic phosphodiesters (as judged from ^{31}P-NMR experiments) in the remainder of the oligodeoxynucleotide. Phosphorothioates can also be formed from the reaction of an internucleotidic H-phosphonate with elemental sulphur (11,12):

$$\text{N—O—}\overset{\text{O}}{\overset{\|}{\underset{\text{H}}{\underset{|}{\text{P}}}}}\text{—O—N} \xrightarrow{S_8/CS_2/\text{Lutidine}} \text{N—O—}\overset{\text{O}}{\overset{\|}{\underset{\text{S}^-}{\underset{|}{\text{P}}}}}\text{—O—N}$$

In this approach the H-phosphonate is oxidized to the phosphorothioate diester. This species is not compatible with subsequent oxidations employing H_2O/I_2/THF/lutidine. However, it is perfectly adequate for the synthesis of oligonucleotides with a uniform phosphorothioate backbone as described in Chapter 4.

2.1 Synthesis of an oligodeoxynucleotide containing a single internucleotidic phosphorothioate diester

The sequence of interest is prepared by automated DNA synthesis using standard phosphoramidite chemistry on controlled pore glass with chain elongation occurring in the 3′ to 5′ direction as described in Chapter 1 in this book. The program is interrupted at the appropriate step in order to introduce the phosphorothioate by oxidation with sulphur (see also Chapter 4). The solution containing the sulphur oxidant is added manually to the column. This is done because the sulphur will tend to precipitate out of the carbon disulphide/lutidine solution and when this occurs in the delivery lines of a DNA synthesis machine the lines become plugged. As noted above, the 3H-1,2-benzodithiol-3-one 1,1-dioxide derivative described by Iyer *et al.* (15,17) exhibits better solubility in organic solvents and can be used directly in the automated synthesis. The following procedure uses the simpler sulphur oxidant.

Protocol 1. Oxidation of the intermediate phosphite triester to the phosphorothioate triester

Materials

- Four 1 ml glass syringes
- Elemental sulphur, carbon disulphide, lutidine

Protocol 1. *Continued*

- Additional reagents necessary for automated phosphoramidite DNA synthesis.

Method

1. Assemble the sequence in the normal manner until the nucleoside residue 3′ to the desired phosphorothioate linkage has been added, capped and oxidized to the phosphate triester. Couple the nucleoside residue 5′ to the phosphorothioate and follow with a capping step. At this point the synthesis is interrupted and the column removed from the machine. With many DNA synthesizers the machine can be programmed to interrupt at this point in the cycle.
2. Fill a 1 ml syringe with a solution of 2.5 M elemental sulphur in carbon disulphide/lutidine (1:1) prepared shortly before use (to prevent extensive precipitation of the sulphur) in an oven dried flask (for a 1 μmol synthesis) according to the following:
   - 0.401 g elemental sulphur (Aldrich gold label: 99.999% pure).
   - 2.5 ml carbon disulphide.
   - 2.5 ml lutidine (previously distilled over Ninhydrin followed by distillation over KOH).

   Insert a second empty 1 ml syringe into the outlet of the synthesis column and insert the filled syringe into the inlet. Add the sulphur solution to the synthesis column such that any excess solution is collected in the second syringe. Allow the oxidation reaction to continue for 30–60 min. Every 5–10 min pass the solution back and forth through the column.
3. After 30–60 min wash the synthesis column with carbon disulphide/lutidine (1:1) using two syringes as described above. This washes the excess sulphur from the column and redisolves any precipitated sulphur.
4. Remove the excess solvent from the column prior to its attachment to the machine. Attach the column to the machine and wash the column manually with acetonitrile for 120 s and then re-initiate the synthesis program (excluding the first oxidation step).
5. After assembly of the oligodeoxynucleotide, deprotect, and purify the fragment as described for unmodified oligodeoxynucleotides. Use the isolation procedure in which the 5′-terminal DMT group is not removed from the sequence.

In most cases there is virtually no difference in retention time on a reversed phase column for the modified (phosphorothioate containing) or the unmodified DMT containing oligodeoxynucleotide. With increasing numbers of phosphorothioate diesters some peak broadening is observed presumably as a result of the presence of numerous diasteroisomers.

After isolation and desalting, the purity of the phosphorothioate containing oligomer can be analyzed by reversed phase chromatography using Buffer A: 0.02 M KH_2PO_4 pH 5.5, Buffer B: 0.02 M KH_2PO_4, pH 5.5 containing 70% methanol with a flow of 1.5 ml/min and a gradient of 0–100% Buffer B in 60 min as previously described. The oligomer containing a single pair of phosphorothioate diasteroisomers will elute as two peaks with only slight separation between isomers or as a broad peak with little or no separation depending upon the length of the fragment.

2.2 Synthesis of stereochemically pure oligodeoxynucleotides containing phosphorothioate diesters

The chemical oxidation of the intermediate phosphite triester with sulphur results in a roughly equal mixture of the R_p and S_p diastereoisomers. Reaction of this mixture with the desired label or agent will yield a corresponding isomeric mixture of labelled DNA fragments (6). With short fragments it is sometimes possible resolve the two diastereoisomers chromatographically but this is a less than ideal approach, and with longer fragments there is often no detectable difference in retention between the labelled R_p and labelled S_p isomer. In order to effectively generate stereochemically pure DNA labelled at the phosphorothioate, it is most efficient first to generate the isomerically pure phosphorothioate dinucleotide building block with the techniques first reported from the laboratory of Fritz Eckstein (18,19). In this procedure the desired nucleoside dimer containing a phosphorothioate diester is prepared in solution followed by separation of the two diastereoisomers. After conversion of each diastereoisomer into the appropriate phosphoramidite, the dimer of known chirality can be incorporated into the DNA fragment (see *Scheme*).

Protocol 2. Preparation and purification of the fully protected R_p and S_p isomers of the dimer Cp(s)C, and its conversion to the corresponding phosphoramidite derivative

Materials

- Methoxyacetic acid, DCC, anhydrous pyridine
- 4,4′-Dimethoxytrityl-N-benzoyl-2′-deoxycytidine
- Elemental sulphur, carbon disulphide, pyridine
- Acetic acid, concentrated aqueous ammonia, dioxane
- Silica gel 60 (0.40–0.63 μM or finer than 0.63 μM)
- Methyl-N,N-diisopropylchlorophosphoramidite
- Dichloromethane, methanol, hexane

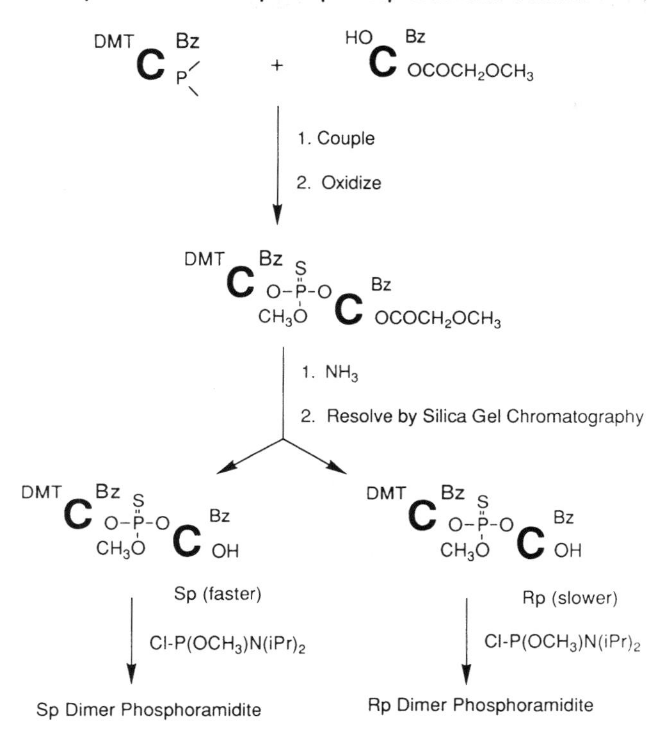

Protocol 2. *Continued*

A. *Synthesis of methoxyacetic anhydride*

Combine 1.217 ml (15.86 mmol) of methoxyacetic acid and 6 ml of dichloromethane. Add 1.6336 g (7.93 mmol) of DCC over a 10 min period. Stir the reaction for 1 h under anhydrous conditions. Filter the solution to remove the urea derivative.

B. *Synthesis of N-benzoyl-3′-O-(methoxyacetyl)-2′-deoxycytidine*

1. Dissolve 2.0080 g (3.172 mmol) of 4,4′-dimethoxytrityl-N-benzoyl-2′-deoxycytidine in anhydrous pyridine. Add the methoxyacetic anhydride (see above) to the pyridine solution under anhydrous conditions. Monitor the reaction by TLC (9:1 CH_2Cl_2:CH_3OH). If the reaction is not complete after approximately 4 h, add an additional 7.5 mmol of methoxyacetic anhydride and stir for an additional 2 h. Evaporate the reaction mixture to dryness under high vacuum and co-evaporate twice from toluene.

Protocol 2. *Continued*

2. Add 20 ml of 80% acetic acid to the residue and stir for 2.5 h. When the reaction is complete (TLC), add 4 ml of H_2O to quench the reaction and evaporate the mixture to dryness. Dissolve this residue in 50 ml of dichloromethane and wash it twice with 5% aqueous sodium bicarbonate and dry. Evaporate the organic phase to dryness and purify the residue on a 30 g column of silica gel. Elute the product using 100 ml portions of dichloromethane using an increasing methanol gradient.

C. *Synthesis of R_p and S_p isomers of the fully protected d[Cp(s)C] dimer*

1. Dry 0.4836 g (1.2 mmol) of N-benzoyl-3′-O-(methoxyacetyl)-2′-deoxy-cytidine, 1.4312 g (1.8 mmol) of 5′-O-(4,4′-dimethoxytrityl)-N-benzoyl-3′-O-(methoxy phosphoramidite)-2′-deoxycytidine, and 0.3362 g (4.8 mmol) of tetrazole (Aldrich, gold label) in vacuum oven at 50 °C for 12 h. After this time, add 20 ml of anhydrous acetonitrile to dissolve all reagents. Monitor the reaction with TLC. The reaction is generally complete within minutes but is often stirred for an hour or more.

2. Dissolve 0.3847 g of elemental sulphur (Aldrich, gold label) in 16 ml of anhydrous carbon disulphide/pyridine solution (1:1). Add this solution dropwise to the reaction mixture, (some sulphur precipitation results). Stir this reaction for 2 h and then filter the solution to remove the precipitate. Evaporate the yellow solution to dryness under high vacuum and co-evaporate twice from toluene. Dissolve the residue in 50 ml of dichloromethane and wash twice with 5% aqueous sodium bicarbonate, twice saturated sodium chloride, dry the organic phase (sodium sulphate) and evaporate it to dryness.

3. Dissolve the residue in 20 ml of dioxane and add 5 ml of 25% aqueous ammonia. Allow this to stir for approximately 2.5 h while monitoring by TLC. When the reaction is complete, evaporate the solution to dryness.

4. Purify the product by chromatography on a column of silica gel 60.

D. *Synthesis of the methyl phosphoramidite derivative of the fully protected d[Cp(s)C] dimer*

1. To 200 mg (200 mmol) of a dimer in 5 ml dry dichloromethane containing 150 µl of diisopropylethylamine add 75 µl (525 µmol) of N,N-diisopropyl-methylphosphonamidic chloride. Stir the reaction mixture for 2 h at ambient temperature, TLC analysis should indicate that the reaction is largely complete. Stop the reaction with 1 ml of methanol, evaporate to dryness, and chromatography the residue on a column of silica gel 60 using dichloromethane (containing 1% triethylamine) and a methanol gradient.

2. Pool the fractions containing product and evaporate to dryness. Dissolve

Protocol 2. *Continued*

the residue in a minimum amount of dichloromethane and precipitate into hexane. Filter the precipitate, dry it under vacuum, and store it in a desiccator (over KOH). Yields were typically between 50 and 70%.

3. Reactions of phosphorothioate diesters

Oligodeoxynucleotides containing a uniquely placed covalently bound reporter group tethered to the DNA backbone by alkylation of a phosphorothioate diester are easily obtained by incubation of a phosphorothioate-containing oligodeoxynucleotide with the probe of choice in aqueous or largely aqueous solution within a pH range of 5–8 and from 25 to 50 °C. Our preliminary experiments employed the simplest phosphorothioate diester Tp(s)T in order to establish the rates of reactivity with different functional groups under a variety of conditions. Working with the dimer is advantageous because alkylation of the sulphur residue can be easily monitored by HPLC with the large shift in retention time of the labelled versus unlabelled phosphorothioate diester. However, the protocols described in this section are designed for sequences from 10 to 24 residues. Three functional groups can be employed to covalently introduce a label or other agent to the DNA backbone by alkylation of the phosphorothioate diester. Reagents containing γ-bromo-α,β-unsaturated carbonyls, iodo or bromo acetamides, or aziridinylsulphonamides, function effectively to alkyate the sulphur residue and produce the corresponding phosphorothioate triester:

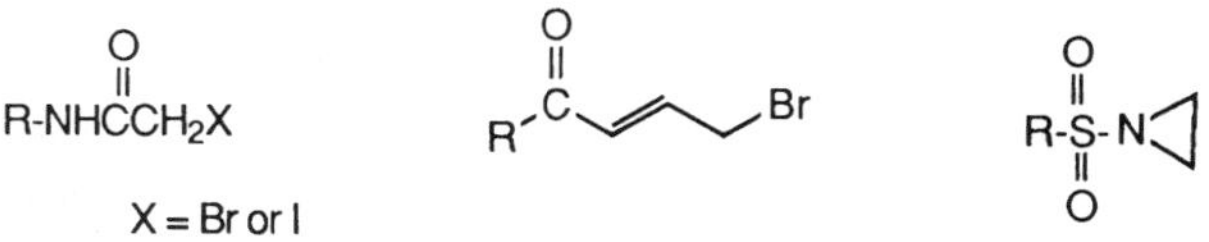

R-NHC(=O)CH_2X X = Br or I

R-C(=O)-CH=CH-CH_2Br

R-SO_2-N(aziridine)

Both the R_p and S_p diastereoisomers exhibit equal reactivity in all reactions monitored to date.

The concentration and amount of label required for rapid and efficient reaction with the phosphorothioate diester varies significantly depending upon the functional group employed for the alkylation since the rates of competing hydrolytic reactions also vary. Additionally the aqueous character of the reaction mixture will depend largely upon the solubility of the label as well as the oligodeoxynucleotide. Aqueous or dimethylformamide solutions can commonly be employed to dissolve both the oligodeoxynucleotide and the label or agent of interest. The amount of DMF required in a given reactions is dependent upon the solubility of the appropriate reagent. However, with DMF concentrations greater than 55%, precipitation of the oligodeoxynucleotide may become problematic. Three protocols are described in this section; the first employs a largely water soluble spin label to modify a 12mer, the second a hydrophobic fluorophore to modify a 24mer and the third uses a drug derivative to alkylate a 12mer.

Protocol 3. Attachment of the PROXYL spin label to a phosphorothioate containing 12mer

Materials

- 3-(2-Iodoacetamidomethyl)-PROXYL
- 100 mM potassium phosphate, pH 6.0
- d[CGCA(s)AAAAAGCG]

Method

1. Prepare the reaction mixture in a 0.5 ml Eppendorf tube as described below:

Reagent (mM)	Volume (μl)	Amount (μM)	Equivalents	Final concentration (mM)
PROXYL (25) in 30% aqueous DMF	100	2.5	31	12 14% DMF
12mer (1.35)	60	0.08	1	0.39
Buffer (100)	50			24
Total:	210 μl			

2. Seal the Eppendorf tube with Parafilm and completely submerge it in a water bath at 50 °C for 6 h.
3. At this point the reaction is 80–90% complete and product is isolated by HPLC as described in *Protocol 6*.

Protocol 4. Attachment of 5-iodoacetamidofluorescein to a phosphorothioate containing 24mer

Materials

- 5-iodoacetamidofluorescein (5-IAF)
- 100 mM Tris–HCl, pH 8.0
- d[CGAACT(s)AGTTAACTAGTACGCAAG]

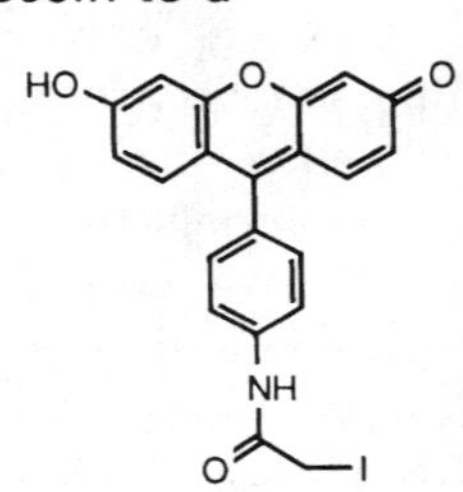

Protocol 4. *Continued*

Method

1. Prepare the reaction mixture in a 0.5 ml Eppendorf tube as describe below:

Reagent (mM)	Volume) (µl)	Amount (µM)	Equivalents	Final concentration (mM)
5-IAF (50) in DMF	30	1.5	16.7	15 30%
24mer (2.0)	45	0.09	1	0.9
Buffer (100)	25			25
Total:	100 µl			

2. Seal the Eppendorf tube the Parafilm, cover with aluminum foil and completely submerge it in a water bath at 50 °C for 24 h.
3. At this point the reaction is 80–90% complete and product is isolated by HPLC as described in *Protocol 6*.

Protocol 5. Attachment of a CC-1065 analogue to a phosphorothioate containing 12mer

Materials

- CDPI-Br
- 100 mM sodium acetate, pH 5.0
- d[CGCA(s)AAAAGCG]

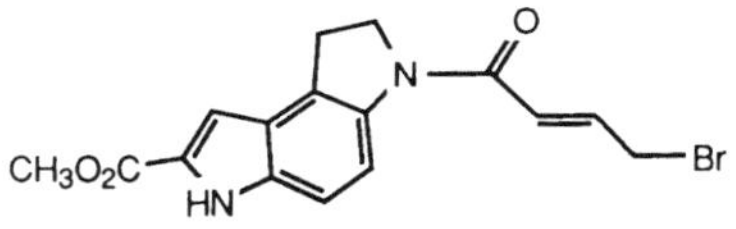

Method

1. Prepare the reaction mixture in a 1.0 ml Eppendoft tube as described below:

Reagent (mM)	Volume) (µl)	Amount (µM)	Equivalents	Final concentration (mM)
CDPI-Br (20) in DMF	156	3.1	20	3.7
12mer (0.3)	525	0.16	1	0.19
Buffer (100)	43			10
DMF	100			30% total
H_2O	26			
Total:	850 µl			

2. Seal the Eppendorf tube with Parafilm and completely submerge it in a water bath at 50 °C for 12 h.
3. At this point the reaction is 80–90% complete and product is isolated by HPLC as described in *Protocol 6*.

3.1 Analysis and purification of the labelled oligodeoxynucleotides

The reactions are typically monitored and the products isolated by HPLC using reversed-phase chromatography. This method is advantageous since alkylation of a phosphorothioate diester converts the charged diester to the neutral triester. Often the alkylating moiety is itself hydrophobic such that the product oligodeoxynucleotide has more hydrophobic character than the unlabelled starting material and consequently elutes from the column with an increased retention time. For example, resolution of the labelled and unlabelled dodecamers obtained after reaction of the single-stranded 12mer with 3-(2-iodoacetamido)-PROXYL is illustrated in *Figure 1*. This analysis additionally illustrates a common observation with diastereomeric mixtures, in that the R_p and S_p phosphorothioate diesters are not resolved by the column, but addition of a label at the chiral centre enhances separation such that resolution of the two labelled diastereoisomers occurs. In some cases, as

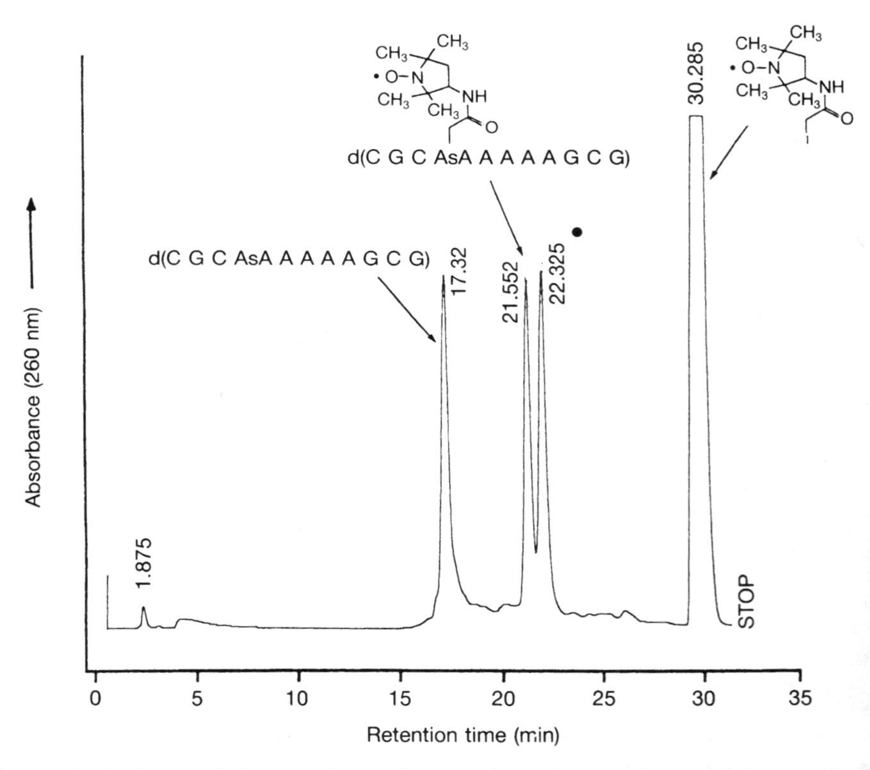

Figure 1. Analysis of the reaction mixture of a dodecamer containing a single phosphorothioate diester and the iodoacetamidoPROXYL spin label after two hours of reaction. Column conditions similar to that described in *Protocol 6*.

illustrated by the chromatogram of *Figure 1*, it should be possible to purify the two isomeric compounds by HPLC. However, this is unlikely to be an effective general method. Typically such isomer resolution decreases with increasing chain length (see also the analysis of the labelled 24mer, *Figure 3*). Isolation of chirally pure material is much more efficient if the chiral phosphorothioate dinucleotide building block is prepared for incorporation into the sequence as previously reported (18, 19) and described above in *Protocol 2*.

Effective resolution of the labelled and unlabelled material can often be achieved with a column equilibrated in 20 mM phosphate and employing a methanol gradient as described below:

Protocol 6. HPLC analysis and isolation of the labelled oligodeoxynucleotide derivatives

Materials

- 4.6 × 250 mm reversed phase column (ODS-Hypersil)
- 1.0 M potassium phosphate pH 5.5
- HPLC grade methanol
- Gradient HPLC system (Beckman)
- Membrane filtration set-up (Millipore)
- Sep-Pak or Nensorb cartridge

Method

1. Prepare two buffers by appropriate dilution of the stock solution and addition of methanol: Buffer A: 20 mM potassium phosphate pH 5.5. Buffer B: 20 mM potassium phosphate pH 5.5, 70% methanol.
2. Filter both buffers through a 0.45 μm membrane filter to remove particulates.
3. Wash the column with buffer B and then re-equilibrate with buffer A.
4. Programe the HPLC system to elute the column under the following conditions: Flow rate: 1.5 ml/min, Linear gradient: 0–100% buffer B in 60 min.
5. For isolation of the labelled fragment, collect the appropriate peak, evaporate it to dryness. Prepare a Sep-Pak (or Nensorb) column in the following manner: Attach a 10 ml plastic syringe cylinder to the Sep-Pak column as a solvent reservoir. Wash the column with 20 ml of methanol followed by 20 ml of double-distilled water. Dissolve the residue containing the labelled fragment in distilled water and add this solution of the Sep-Pak column.

Protocol 6. *Continued*

6. Wash the Sep-Pak (Nensorb) column with 10 ml of water to remove the phosphate buffer followed by 10 ml of 50% aqueous methanol (higher concentrations may be required in some instances depending upon the moiety bound) to elute the labelled DNA fragment.
7. Lyophilize to dryness and store at −20 °C.

HPLC resolution of the labelled and the unlabelled fragments has been very successful but ultimately may be limited by the length of the fragments and the hydrophobic character of the attached derivative. A 24mer labelled with a iodoacetamidofluorescein derivative can be easily resolved from the unlabelled 24mer (*Figure 2*). However, the iodoacetamidodansyl derivative of a dodecamer was more difficult to obtain in purified form. In the latter case, modification of the 12mer with iodoacetamidodansyl converts the negatively charged phosphorothioate diester to the neutral triester which should allow its separation from unlabelled 12mer. However, the dansyl group itself carries a negatively charged sulphonic acid residue which appears to obscure differences in polarity between the labelled and unlabelled

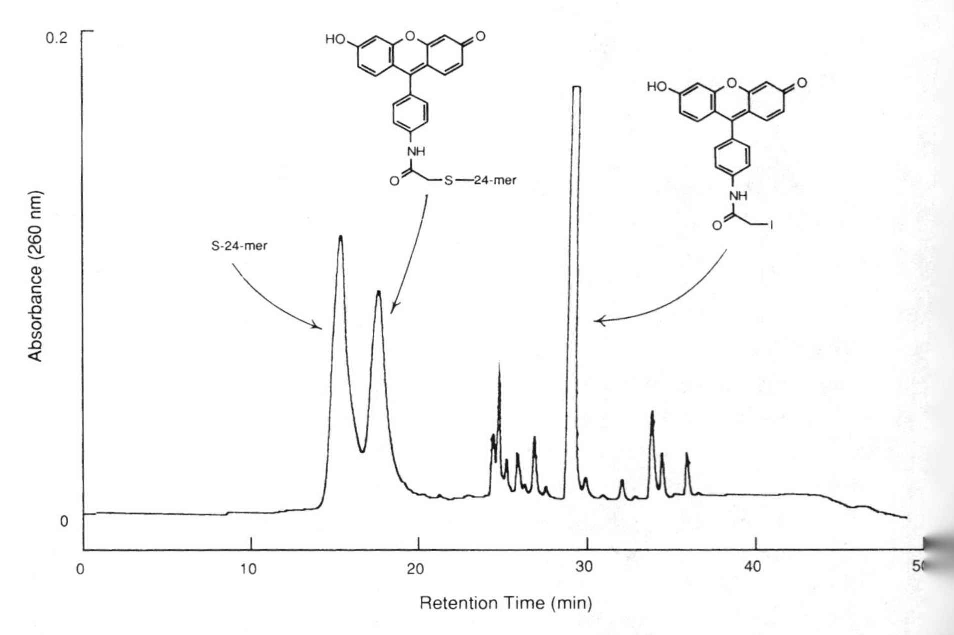

Figure 2. Analysis of the reaction mixture of a 24mer containing a single phosphorothioate diester and the iodoacetamidofluorescein fluorophore after six hours of reaction Column conditions were similar to that described in *Protocol 6* (small unlabelled peaks are present in the fluorophore preparation or are related hydrolysis products).

fragments. By using a shallower methanol gradient than the described in *Protocol 6* (0–30% methanol in 60 min), the dansyl labelled dodecamer could be purified (*Figure 3*), however similar success with this label and longer fragments may be more difficult to achieve.

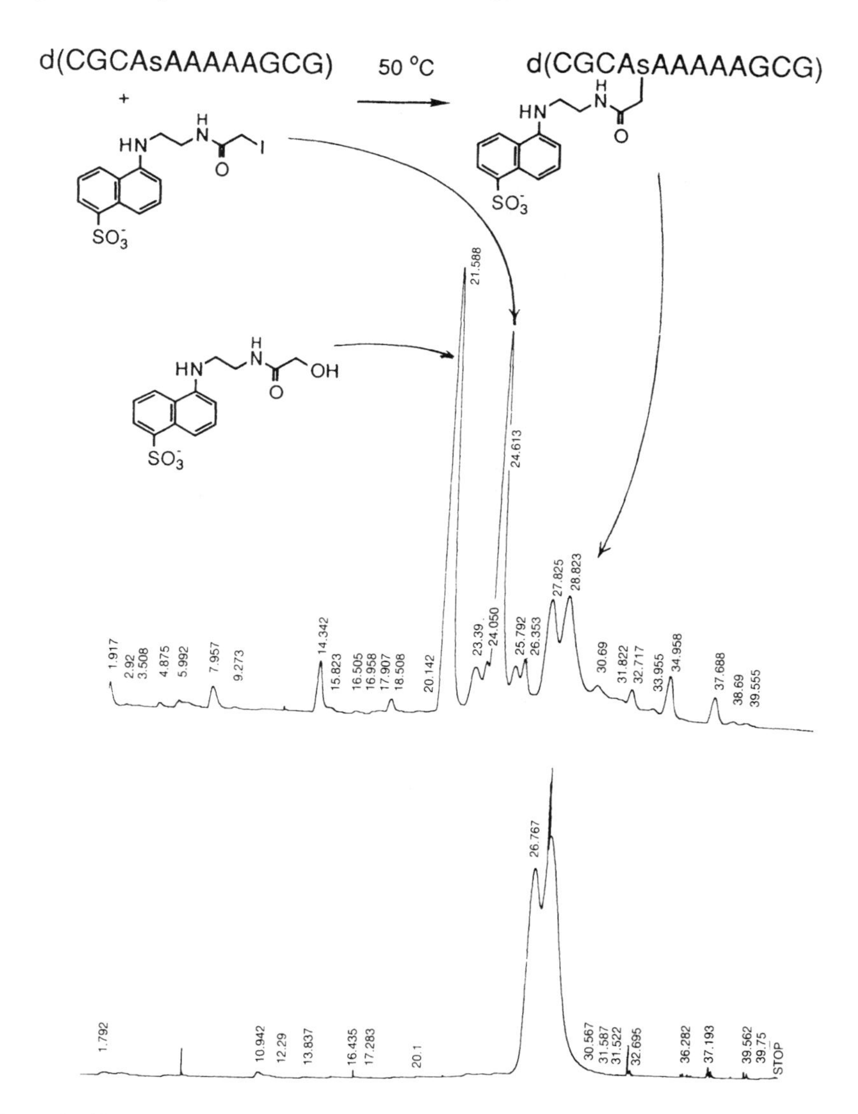

Figure 3. Above: analysis of the reaction mixture of a dodecamer containing a single phosphorothioate diester and the bromoacetamidodansyl fluorophore after four hours of reaction. Column conditions similar to that described in *Protocol 6* but incorporated a 0–35% methanol gradient in 60 min (small unlabelled peaks were present in the fluorophore preparation). Below: HPLC analysis of the isolated labelled product.

3.2 Variations in reactivity

Alkylation of the phosphorothioate diester is generally more efficient in a simple dimer or a single-stranded oligodeoxynucleotide, than for a self complementary or double-stranded (for example an eicosomer) fragment. This difference in reactivity can be dramatic at ambient temperature where a self-complementary eicosomer (20mer), or longer fragment, is largely double-stranded and suggests that nucleic acid secondary structure can reduce the accessibility of the internucleotidic phosphorus residue to the reporter group. This difficulty can be overcome by heating the reaction mixture to 50 °C and presumably disrupting the DNA helical structure without any noticeable side reactions. Since reaction conditions which incorporate elevated temperatures enhance reactivity as well as break down secondary structure, we routinely use incubation temperatures of 50 °C.

A large portion of the phosphorothioate diester is converted to the labelled phosphorothioate triester within the first few hours of reaction as illustrated for the 24mer in *Figure 4*. However, reaction mixtures are typically and conveniently incubated overnight. Reactions yields under such conditions are typically 80–90% as observed in the plot of *Figure 4* for a 24mer.

Under these conditions we cannot detect any side reactions when the haloacetamido or γ-bromo-α,β-unsaturated carbonyl derivatives are utilized for labelling. However, the HPLC analyses of reactions employing the aziridinyl sulphonamides indicates the presence of minor products in addition to the major product, the phosphorothioate triester. This is not surprising since aziridinyl derivatives are known to modify DNA bases. Using HPLC it

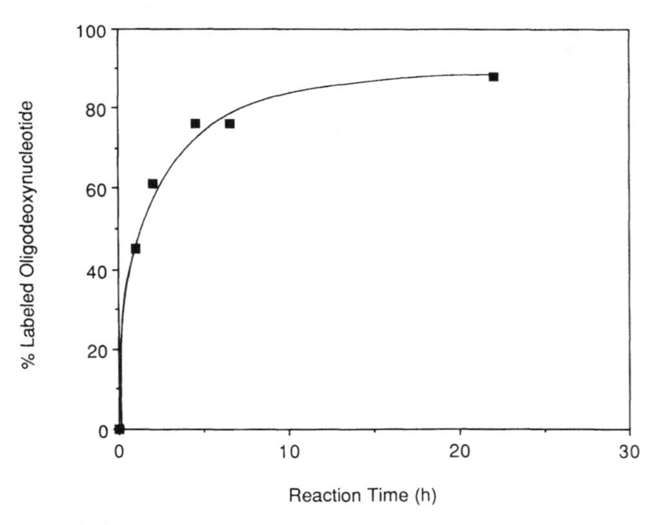

Figure 4. Graph of the extent of product formation for the fluorescein labelled 24mer after various reaction times (for conditions see *Protocol 5*).

is still possible to isolate fragments labelled with aziridinyl sulphonamides in high purity, however, due to the additional reactivity exhibited by these compounds we prefer to use haloacetamide or γ-bromo-α,β-unsaturated carbonyl derivatives to attach the label of interest to the DNA backbone.

3.3 Stability of the product phosphorothioate triesters

In order for the oligodeoxynucleotides labelled by these procedures to the valuable for further studies, including those of structure and dynamics, it is required that the phosphorothioate triesters exhibit reasonable stability in aqueous solutions. Hydrolysis of the phosphorothioate triester proceeds almost exclusively by P-S bond cleavage to yield the native oligodeoxynucleotide with an internucleotidic phosphodiester in place of the phosphorothioate triester and the corresponding sulphydryl derivative of the reporter group [minor amounts of chain cleavage also occur and this phenomenon has been the basis for phosphorothioate based DNA sequencing (20)]. We have not quantified the fate of the reporter group in these experiments but have monitored the conversion of phosphorothioate triester to phosphate diester.

Previous experiments had indicated that the triester formed from the dimer, d[Tp(s)T], was stable for long periods of time at pH values below 7 but was rapidly hydrolysed ($t_{1/2} < 30$ min) at pH 11. To compare the stability of a simple dimer with that of a dodecamer and self complementary eicosomer (20mer), the appropriate derivatives are incubated at pH values of 7.0, 8.0, and 10.0, at ambient temperature over a 24 h time period. Under these conditions the stability of the phosphorothioate triester increases in the order: dimer < dodecamer < self complementary (*Table 1*). The stability of the modified DNA will vary with the character of the label. The hydrolysis results reported in *Table 1* employed the iodoacetamido-dansyl group which is one of the least stable labels we have examined. In general, the triester resulting from reaction with the γ-bromo-α,β-unsaturated carbonyl exhibited stability similar to that of the haloacetamido labelled derivatives, while that resulting from reaction with the aziridinyl sulphonamide is significantly more stable.

Table 1. Stability of phosphorothioate labelled DNA

	Quantity of labelled oligodeoxynucleotide remaining after incubation at ambient temperature for 24 h		
pH	**Tp(s)T-Dansyl**	**12mer-danysl**[a]	**20mer-dansyl**[b]
7.0	90%	99%	> 99%
8.5	71%	69%	98%
10.0	0%	0%	37%

[a] Single-stranded
[b] Double-stranded

3.4 Thermal stability of labelled phosphorothioate triester-containing DNA

When a probe is introduced into a native biopolymer it is necessary to consider the structural implications and possible perturbation of the structure which might occur. Attachment of a label to the outer surface of the biomolecule, in this case to the phosphorus residue, may be of minimal structural consequence, but a large group may still interfere with proper Watson–Crick base pairing. As an example of the stability of backbone labelled oligodeoxynucleotide helices, a number of modified sequences have been investigated to determine if a backbone modification generally destabilizes the duplex structure. The thermal melting curves (helix-to-coil transitions) obtained for a dodecamer containing a single PROXYL spin label, a drug derivative or a self complementary eicosomer containing two labelled phosphorothioate triesters, indicate that incorporation of a reporter group does not generally impart any significant instability to the helical structure. Thermal melting points are measured in 1.0 M NaCl and 10 mM KH_2PO_4, pH 7.0. The unlabelled dodecamer exhibits a T_m (55 °C) which is indistinguishable from the T_m values obtained for the PROXYL labelled or drug labelled dodecamer helices (T_ms of 55 and 54 °C respectively) under the same conditions. The T_m value for the self complementary eicosomer with two labels shows a slight increase of 1.5 °C in comparison to the unlabelled fragment. The conversion of one or two negatively charged diesters to the neutral triesters within 10–20 base pairs does not appear to have any significant structural implications.

4. Applications for site-specifically modified DNA fragments

Site-specifically labelled DNA fragments can be valuable in a number of different biophysical studies including the study of nucleic acid structure and dynamics as well as protein–DNA or ligand–DNA complexes. The case of attachment of reporter groups such a fluorophores should simplify the incorporation of donor and acceptor groups for use in energy transfer studies on a variety of unusual structures and complexes. Additionally, the sequence-specific attachment of reactive agents such as bifunctional alkylating agents, incorporation of donor and acceptor groups for use in energy transfer studies series of modified antisense oligodeoxynucleotides as an approach to antiviral therapy. Furthermore, such modified DNA fragments may soon initiate a new class of potentially sensitive materials for use as what might be termed DNA diagnostics or therapeutics. In the present chapter we will describe the use of site-specifically labelled DNA to monitor the sequence-specific binding by a DNA repressor protein.

4.1 Detection of sequence-specific protein–DNA binding

One of the best understood mechanisms of genetic expression is that described by the operon model in which the expression of a particular enzyme or series of enzymes is controlled by the binding of a repressor protein to an operator sequence which in part overlaps the RNA polymerase promoter sequence. With the repressor protein bound, RNA polymerase binding is inhibited and no expression of the gene occurs. In the absence of the repressor protein, RNA polymerase can bind and the gene is expressed.

4.2 Detection of binding between the tryptophan repressor and operator

The tryptophan operon controls expression of the bacterial enzymes responsible for the synthesis of the amino acid tryptophan. Binding of the tryptophan repressor protein to the operator sequence is modulated by the tryptophan amino acid which functions as a co-repressor. Upon binding tryptophan the repressor protein undergoes a conformational change which dramatically enhances its affinity for the operator sequence. Unfortunately there are not presently many efficient assays for measuring solution binding between macromolecules such as proteins and nucleic acids. Most assays such as nitrocellulose filter binding or gel shift assays involve trapping the protein–nucleic acid complex. It would be desirable if the protein–nucleic acid complex could be detected directly in solution, ideally under equilibrium conditions. In order to assess such possibilities we have prepared a series of site-specifically labelled operator sequences. In this case the choice of site to label was dictated by the crystal structure of the tryptophan repressor–operator complex. We prepared three site-specifically labelled operator sequences by placing a phosphorothioate diester between residues −1 and +1 (site 1 below), between residues −1 and −2 (site 2 below) and between residues −2 and −3 (site 3 below):

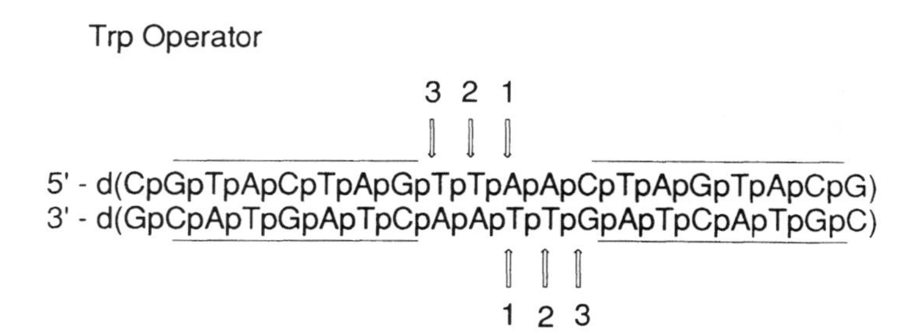

These sites were chosen since the crystal structure indicated that the repressor protein did not directly contact this area of the operator and we did not want the fluorophore to inhibit binding by the protein (the lines above mark the bases of the operator directly contacted by the repressor). However, it was anticipated that the fluorophore would be very close to the protein upon binding such that complex formation might alter the environment of the fluorophore. With a fluorophore sensitive to such change in the local environment, it was possible that protein binding would be reported by changes in the characteristics of the emitted fluorescence. The fluorophore chosen for this study was the iodoacetamido derivative of nitrobenzdiimidazole. Three sequences were prepared with the label linked to the internucleotidic phosphorothioate diester as typified by the example below:

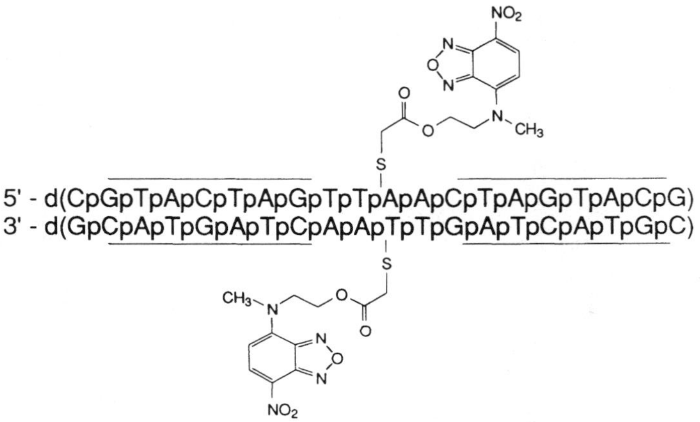

5' - d(CpGpTpApCpTpApGpTpTpApApCpTpApGpTpApCpG)
3' - d(GpCpApTpGpApTpCpApApTpTpGpApTpCpApTpGpC)

The labelled sequences were prepared by the procedure described in *Protocols 3*, *4*, and *5* of this chapter. After isolation and purification of the labelled fragments as described in *Protocol 6* each of the fragments was titrated with repressor protein under the conditions described below. To the operator solution (0.4 ml containing 10 mM Tris–HCl pH 7.4, 250 mM sodium chloride, 0.5 mM L-tryptophan, 0.1 mg/ml BSA and 5 μM NBD-labelled 20mer) was added 1.6 μl aliquots of the repressor (0.1 ml containing 265 μM tryptophan repressor dimer (in 10 mM phosphate). After addition and mixing, the fluorescence of the solution was measured. Additions were continued until the concentration of the repressor dimer had reached 10 μM.

As can be observed in *Figure 5*, the fluorescence intensity of the solution increased proportionately with increasing concentrations of repressor protein. When the concentration of the protein equalled that of the operator site the fluorescence increase reached its maximum. At that point, any further increase in repressor concentration did not yield a corresponding increase in fluorescence intensity. During this titration, no change in the excitation or emission maxima for the nitrobenzdiimidazole derivative was observed.

The same titration performed in the absence of the amino acid tryptophan did not produce any measurable change in fluorescence intensity. The

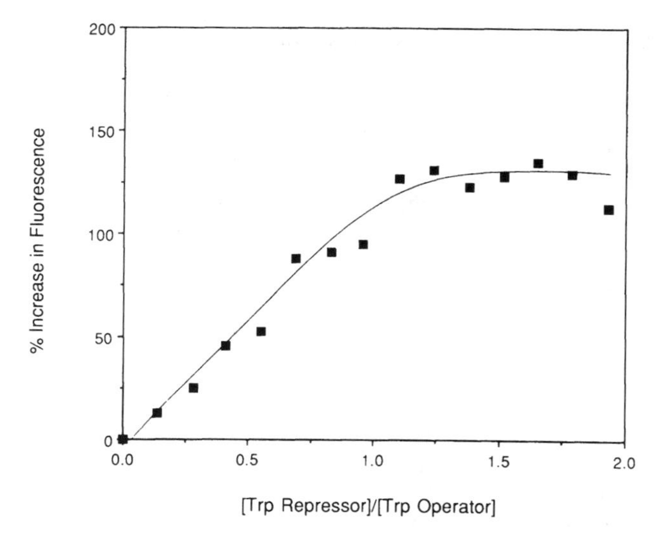

Figure 5. Change in fluorescence of the NBD labelled 20mer with the addition of increasing concentration of the tryptophan repressor.

addition of a non-specific protein (BSA) to a concentration of 125 μM did not result in significant changes in the observed fluorescence. Both of these control experiments argue that the observed increased in fluorescence intensity was directly the result of the formation of the sequence-specific repressor–operator complex.

We additionally compared changes in fluorescence for the titration of sequences, 1, 2, and 3. Both sequences 1 and 2 exhibited similar increases in fluorescence (100–125%) during titration by the repressor. However, sequences 1, 2, and 3. Both sequences 1 and 2 exhibited similar increases in increase in fluorescence). This can be explained in terms of the published repressor–operator complex in that as the fluorophore is moved from position 1 to 2 to 3 it is rotated away from the bulk of the protein present in the complex. Sequences 1 and 2, which place the fluorophore nearer the protein, report a more significant change in environment than does sequence 3. This indicates that the site-specific placement of fluorophores on the DNA backbone can be useful to monitor sequence-specific protein–DNA binding, and may in some cases also be used to map those areas of the DNA helix which are approached most closely by the protein upon binding.

5. DNA containing multiple phosphorothioate diesters

Substitution of one of the non-bridging oxygens of the internucleotidic phosphate diester by a sulphur atom is a relatively conservative modification.

In fact, DNA sequences which contain multiple phosphorothioate diesters are essentially native in structure, and often the physical properties, such as T_m values are virtually indistinguishable from those for unmodified DNA. The substitution of phosphate diesters by phosphorothioate diesters creates a series nucleophilic sites in an otherwise native DNA fragment. However, the presence of multiple phosphorothioates now allows for the introduction of multiple labels, ideally one at each phosphorus residue.

5.1 Synthesis of DNA with multiple phosphorothioate diesters

Phosphorothioates can be introduced into nucleic acids at multiple sites using phosphoramidite or H-phosphonate chemistry by the procedures described in Chapter 4. Multiple phosphorothioate diesters can also be introduced into a DNA fragment enzymatically using polymerases when one or more of the substrate α-thio-2′-deoxynucleotide-5′triphosphates are employed as originally developed in the laboratory of Fritz Eckstein. Replacement of one of the non-bridging oxygens at the α-position of a 2′-deoxynucleotide-5′-triphosphate with sulphur creates two diastereoisomers. All polymerizing enzymes examined to date (including *E.coli* DNA polymerase I or the Klenow fragment, phage T4 and T7 DNA polymerases, *Micrococcus luteus* DNA polymerase, reverse transcriptase, and *Taq* polymerase) use only the S_p diastereoisomer (16, 21, 22) as a substrate (in the presence of a Mg^{2+} cofactor). After polymerization the S_p α-thio-2′-deoxynucleoside 5′-triphosphate, the phosphorothioate diester formed is exclusively in the R_p configuration. In theory the polymerizing enzymes will use only the S_p isomer from a racemic mixture of phosphorothioates, however in practice we have found better results are obtained if pure S_p α-thiotriphosphate is used. This observation is likely a function of the higher purity of α-thiotriphosphate which is obtained if the care is taken to isolate a single stereoisomer rather than any inherent inefficiency exhibited by the enzyme in the presence of the racemic mixture. The procedures described in the section employ the pure S_p α-thiotriphosphates which are readily available (Amersham).

Protocol 7. Enzymatic synthesis of the replicative form of M13mp18 DNA with multiple phosphorothioate diesters in the (−) strand.

Materials

- M13mp18 single stranded DNA
- Universal primer
- Buffer (100 mM Tris–HCl, pH 8.0, 100 mM NaCl, 20 mM $MgCl_2$)
- dNTPαS (dATPαS, dGTPαS, dCTPαS, dTTPαS)

Protocol 7. *Continued*

- ATP
- DNA polymerase I (10 units)
- T4 DNA ligase (6 units)

Method

1. Add DNA and primer to an Eppendorf tube in <10 μl and add 25 μl of buffer. Heat the mixture to 56 °C and then cool it to ambient temperature over a 30 min period. Add all four dNTPαS substrates and ATP such that the reaction volume is 50 μl and the final concentration of each S_p α-thiotriphosphate is 500 μM and ATP is 1 mM. Add DNA polymerase I and T_4 DNA ligase and incubate the reaction mixture at 16 °C overnight (18 h).
2. Heat the reaction mixture to 70 °C and maintain this temperature for 10 min to inactivate the enzymes. Add two volumes of ethanol, incubate at −20 °C for 60 min. Centrifuge at 0 °C for 10 min. Discard supernatant and remove residual ethanol under vacuum. Resuspend the DNA pellet in the appropriate buffer.

In the labelling reaction described in section 6 of this chapter, DNA restriction fragments are labelled at multiple sites with the fluorophore monobromobimane. It has been reported that many restriction endonucleases are sensitive to the presence of a phosphorothioate diester at or near the cleavage site (18, 23, 24). If specific DNA fragments will be prepared from the phosphorothioate containing DNA described in *Protocol 7*, then the protocol must be modified such that only two or three α-thiotriphosphates plus the unmodified triphosphate(s) are used as substrates. This will allow the construction of the RF form of the DNA which still contains many phosphorothioate diesters, but maintains an unmodified phosphate diester(s) at or near the site of restriction endonuclease hydrolysis.

6. Fluorescent labelling of phosphorothioate diesters for the detection of DNA

Fluorophores represent an attractive class of reporter molecules for the detection of nucleic acids because they are stable and directly detectable and could potentially replace radioisotopic labelling procedures in some assays. Many procedures have been described for the introduction of a single fluorophore (25), and in some cases a number of fluorophores (26, 27) into nucleic acids. However, fluorophores do not typically allow the detection of small quantities of nucleic acids ($<10^{-15}$M) in the absence of sophisticated detection systems; and the high sensitivity of radioisotopes for detection by autoradiography typically allows the observation of femtomolar quantities of nucleic acid (28).

6.1 Introduction of multiple fluorophores into DNA containing phosphorothioates

Highly sensitive detection of DNA should be expected if multiple fluorophores, ideally one at each phosphorus residue, can be introduced. In principle, multiple fluorophores can be introduced into DNA fragments containing multiple phosphorothioate diesters essentially under the conditions described in section 3 of this chapter. However, the isolation of DNA fragments labelled in this manner is problematic. After fluorescent labelling, a DNA fragment of hundreds of nucleoside residues contains hundreds of covalently bound fluorosphores (*Figure 6*). Since the charged phosphorothioate diesters are converted to neutral phosphorothioate triesters upon labelling, such fragments would have severely altered properties including but not limited to solubility. Because of the inherent problems which accompany the purification of DNA containing multiple labels bound to the phosphorus residues, we have examined a procedure by which the fluorophores are introduced *in situ*, only after the assay of interest. In the present chapter a technique for introducing multiple fluorophores into DNA after resolution of DNA fragments by polyacrylamide gel electrophoresis is described.

6.2 Post-assay labelling for high detection sensitivity

A simple protocol has been developed which allows for the covalent introduction of multiple fluorescent markers into DNA fragments after gel

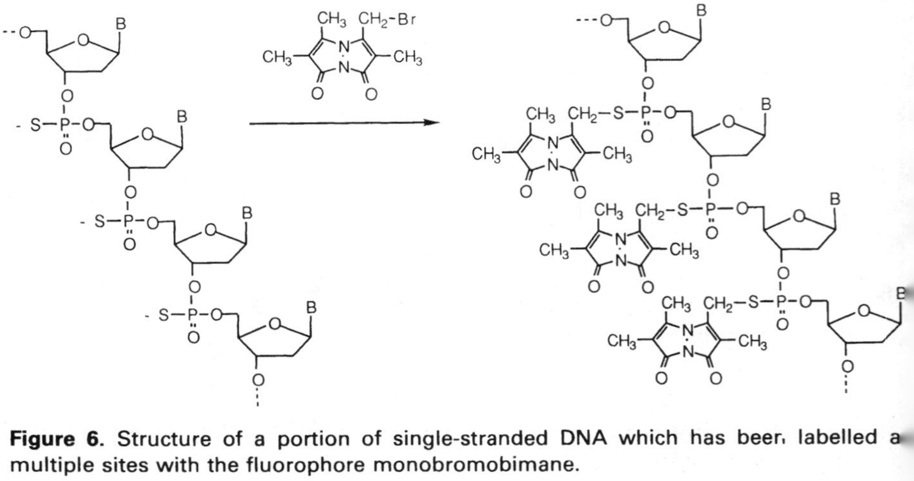

Figure 6. Structure of a portion of single-stranded DNA which has been labelled at multiple sites with the fluorophore monobromobimane.

electrophoresis techniques, that is, while the nucleic acid fragments are still embedded within the polyacrylamide gel matrix. 'Post-assay' fluorescent labelling in this manner employs DNA fragments containing multiple phosphorothioate diesters rather than a single phosphorothioate diester, ideally one at every internucleotidic position. The internucleotidic residues are then alkylated with the fluroescent marker monobromobimane.

Monobromobimane has been chosen as the fluorophore for use in post-assay labelling for a number of reasons. It is expected that diffusion of the bimane into the gel matrix readily occurs as a result of its relatively small size. Monobromobimane exhibits acceptable solubility in aqueous or largely aqueous solutions (the polyacrylamide gel matrix is unstable in the presence of many organic solvents). Monobromobimane is essentially non-fluorescent but becomes highly fluorescent after reaction with a thiol or similar derivative. This latter property is important in that the labelled DNA bands will be highly fluorescent while the background of the gel containing unreacted monobromobimane will be largely non-fluorescent.

Protocol 8. Post-assay labelling of phosphorothioate containing DNA in polyacrylamide gels

Materials

- 10% aqueous acetic acid
- Monobromobimane
- Acetonitrile
- Dimethylformamide

Method

1. Prepare a polyacrylamide gel using standard procedures. Add the desired samples and resolve the fragments by electrophoresis.
2. After completion of the electrophoresis, remove the top plate of the gel. If the gel is denaturing, remove the urea by soaking the gel in 10% aqueous acetic acid for 5–10 min.
3. Prepare the monobromobimane solution by dissolving 0.054 g of monobromobimane in 25 ml of acetonitrile, to the mixture add 25 ml of water (final solution is 4 mM monobromobimane in 50% aqueous acetonitrile).
4. Place the monobromobimane solution in a glass dish or try (slightly larger than the gel) and cover the tray with Saran wrap and then with aluminium foil, or store the solution in the dark.
5. After soaking in 10% aqueous acetic acid, transfer the gel into the 4 mM monobromobimane solution, cover with Saran wrap and aluminum foil, and store at ambient temperature overnight.

Protocol 8. *Continued*

6. Destain the gel briefly by gently shaking it in solutions of 50% aqueous acetonitrile (3 × 15 min). The gel may be somewhat cloudy at this point, a brief treatment of 50% aqueous dimethylformamide (1 × 5 min) clears the gel and enhances the fluorescence.
7. Place the gel on a standard long wavelength UV transilluminator (λ_{max} = 366 nm) and photograph using a sharp cut-off filter.

6.3 Detection sensitivity of DNA labelled with multiple fluorophores

The sensitivity of detection (to the naked eye) of DNA fragments labelled in this manner has been determined by the labelling of differing amounts of DNA of varying lengths (or varying numbers of phosphorothioate diesters) and visualizing the fluorescent bands using a standard transilluminator (λ_{max} = 366 nm; the bimane labelled phosphorothioate exhibits an excitation maximum of 390 nm and an emission maximum of 480 nm). Based upon this analysis, an increasing number of bimane fluorophores attached to the DNA fragments results in a corresponding increase in the detection limit. Fragments containing hundreds of potential labelling sites (phosphorothioate diesters) can be visualized in the low femtomolar range (10^{-15} mol) (*Figure 7*) which is very near the sensitivity commonly achieved with radioisotopes.

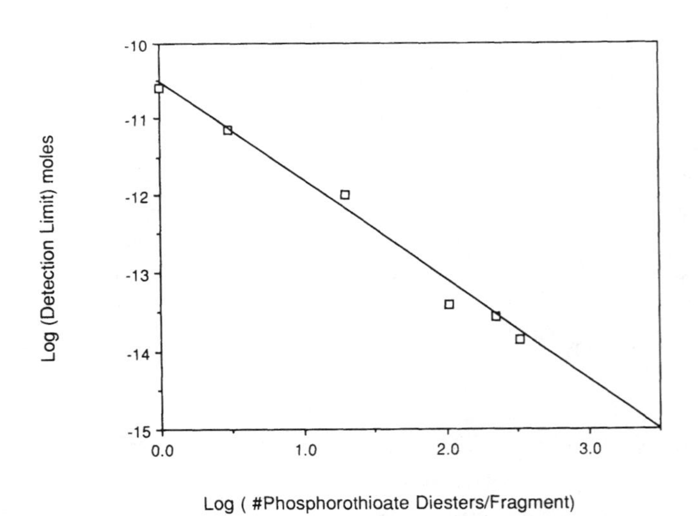

Figure 7. Relationship of detections sensitivity to the number of labelled phosphoro thioate diesters.

7. Fluorescent detection of DNA

The multiple labelling procedures described in this section are most applicable to DNA sequencing and hybridization procedures. At present such procedures are not fully developed but would be expected to occur in the following manner.

7.1 DNA sequencing

DNA sequencing using the Sanger sequencing technology employs a DNA template, a primer, the appropriate polymerizing enzyme, the four 2-deoxynucleoside-5′-triphosphates and one 2′,3′-dideoxynucleoside-triphosphate for each of the four reaction mixtures. DNA fragments amenable to post-assay fluorescent labelling can be prepared if the four triphosphates are substituted by a α-thio-2′-deoxynucleoside-5′-triphosphates. This requires virtually no changes in the standard procedures already developed and generates a series of DNA fragments with phosphorothioate diesters at each and every internucleotidic position. After resolution of these fragments by polyacrylamide gel electrophoresis they can be post-assay labelled with the fluorophore monobromobimane.

7.2 DNA hybridization

In a related approach, a DNA hybridization probe can be prepared chemically or enzymatically such that each internucleotidic residue is a phosphorothioate diester. After transfer of the target DNA sequence to the hybridization membrane the hybridization probe can be added using essentially standard procedures. After removal of the excess probe by washing, the hybridization complex can be labelled with monobromobimane using procedures which are analogous to those described for post-assay labelling in polyacrylamide gels.

8. Conclusions

The alkylation of phosphorothioate diesters provides a simple yet versatile technique for the introduction of labels, or potentially agents of any kind, into the DNA backbone. This procedure does not require the synthesis of modified nucleosides nor the preparation of any unique linkers, but rather relies upon standard DNA synthesis techniques which have been modified in order to incorporate phosphorothioate diesters in place of phosphate diesters. Although the alkylation of an internucleotidic phosphorus residue can introduce unwanted additional chirality into the molecule, specific isomers can be obtained if the corresponding phosphorothioate diastereoisomer is prepared. The triesters formed by the described procedures hydrolyze in alkaline solution but exhibit acceptable stability in acidic, neutral or mildly basic solutions to be valuable for a variety of biochemical and biophysical experiments.

Acknowledgements

We would like to thank Drs D. Boger and S. Munk for samples of the CC-1065 derivative CDPI-Br. This work has been supported by the National Institutes of Health (GM 37065).

References

1. Telser, J., Cruickshank, K. A., Morrison, L. E., and Netzel, T. L. (1989). *J. Am. Chem. Soc.*, **111**, 6966.
2. Telser, J., Cruickshank, K. A., Morrison, L. E., Netzel, T. L., and Chan, C.-K. (1989). *J. Am. Chem. Soc.*, **111**, 7226.
3. Letsinger, R. L., Ahang, G., Sun, D. K., Ikeuchi, T., and Sarin, P. S. (1989). *Proc. Natl. Acad. Sci. USA*, **86**, 6553.
4. Agrawal, S. and Tang, J.-Y. (1990). *Tetrahedron Lett.*, **31**, 1543.
5. Cosstick, R., McLaughlin, L. W., and Eckstein, F. (1984). *Nucleic Acids Res.*, **12**, 1791.
6. Hodges, R., Conway, N. E., and McLaughlin L. W. (1989). *Biochemistry,* **28**, 261.
7. Conway, N. E., Fidanza, J. A., and McLaughlin, L. W. (1989). *Nucleic Acids Res. Symposium Series*, **21**, 43.
8. Fidanza, J. A. and McLaughlin, L. W. (1989). *J. Am. Chem. Soc.*, **111**, 9117.
9. Conway, N. E., Fidanza, J. A., and McLaughlin, L. W. (1990). *Phosphorus, Sulphur and Silicon*, **51/52**, 27.
10. Agrawal, S. and Zamecnik, P. C. (1990). *Nucleic Acids Res.*, **18**, 5419.
11. Froehler, B. C., Ng, P. G., and Matteucci, M. D. (1986). *Nucleic Acids Res.*, **14**, 160–166.
12. Stein, C. A., Subasinghe, C., Shinozuka, K., and Cohen, J. S. (1988). *Nucleic Acids Res.*, **16**, 3209.
13. Connolly, B. A., Potter, B. V. L., Eckstein, F., Pingoud, A., and Grotjahn (1984). *Biochemistry*, **23**, 3443.
14. Stec, W., Zon, G., Egan, V., and Stec, B. (1984). *J. Am. Chem. Soc.*, **106**, 6007.
15. Iyer, R. P., Egan, W., Regan, J., and Beaucage, S. L., (1990). *J. Am. Chem. Soc.*, **112**, 1253.
16. Eckstein, F. (1985). *Annu. Rev. Biochem.*, **54**, 367, and references therein.
17. Iyer, R. P., Phillips, L. R., Egan, W., Regan, J., and Bearcage, S. L., (1990), *J. Org. Chem.*, **55**, 4699.
18. Connolly, B. A., Potter, B. V. L., Eckstein, F., Pingoud, A., and Grotjahn, L. (1984). *Biochemistry*, **23**, 3443.
19. Cosstick, R. and Eckstein, F. (1985). *Biochemistry*, **24**, 3630.
20. Gish, G. and Eckstein, F. (1988). *Science*, **240**, 1520.
21. Eckstein, F. (1983). *Angew. Chem.*, **22**, 423.

22. Eckstein, F. and Gish, G. (1989). *Trends Biochem. Sci.*, **14**, 97.
23. Taylor, J. W., Schmidt, W., Cosstick, R., Okrusezek, A., and Eckstein, F. (1985). *Nucleic Acids Res.*, **13**, 8749.
24. Nakamaye, K. L. and Eckstein, F. (1986). *Nucleic Acids Res.*, **14**, 9679.
25. See references noted in 9.
26. Langer, P. R., Waldrop, A. A., and Ward, D. C. (1981). *Proc. Natl. Acad. Sci. USA*, **78**, 6633.
27. Eshaghpour, H., Söll, D. and Crothers, M. (1979). *Nucleic Acids Res.*, **7**, 1485.
28. Sambrook, J., Fritsch, E. F. and Maniatis, T. (1989). *Molecular Cloning, A Laboratory Manual*, Cold Spring Harbor Laboratory Press, Cold Spring Harbor, NY.

10

Oligodeoxynucleotides for affinity chromatography

ROBERT BLANKS and LARRY W. McLAUGHLIN

1. Introduction

A number of proteins such as polymerases, repressor proteins, transcription factors, and restriction and modification enzymes, bind to their recognition or initiation sites with relatively low dissociation constants (10^{-8}–10^{-12} M) and exhibit a high degree of nucleic acid sequence specificity. Specific binding between a protein and a small molecule is commonly used for the affinity purification of proteins. In these procedures, the specific ligand or substrate analogue is covalently attached to an insoluble chromatographic matrix in such a manner as to facilitate interaction of the ligand with the protein. After introduction of the protein mixture to the matrix, only those proteins which bind effectively to the attached ligand are retained, and thus separated from the mixture introduced to the matrix.

Nucleic acids can be viewed as substrates or ligands for a variety of enzymes and proteins, and could be expected to function as effective ligands for affinity chromatography procedures. The use of sequence-specific nucleic acid ligands was first reported for tRNAs which could be attached to matrices and used for the isolation of specific aminoacyl tRNA-synthetases (1,2). Similar materials containing specific sequences of double-stranded DNA have been recently utilized for protein isolation (3–6). In most cases of nucleic acid affinity chromatography, a naturally occurring nucleic acid fragment or a polynucleotide has been employed as the desired ligand. There are relatively few examples of successful protein isolation with short matrix bound oligonucleotides. (See also Chapter 3 for 2′-O-methyloligoribonucleotides for affinity chromatography)

2. Matrix-bound oligodeoxynucleotide ligands

The use of short matrix bound oligodeoxynucleotides as affinity ligands is advantageous since they can be readily prepared in large quantities by chemical synthesis techniques. However, this approach requires knowledge of the exact protein binding site. Some studies have indicated the potential usefulness of short fragments over longer polymers. For example, oligo(dT)$_{(12-18)}$ cellulose was found to be more efficient than longer cellulose bound polymers in the isolation of murine leukemia virus RNA-dependent DNA polymerase (7). Recently, partially purified transcription factor SpI was further purified on a DNA affinity column composed of a double-stranded oligodeoxynucleotide containing the SpI recognition sequence. Shorter fragments were ligated together to form multimers and covalently bound to cyanogen bromide activated Sepharose CL-25 (4). In a similar approach two components of mammalian transcription factor IIIC were isolated (5). Our own work has described the isolation of a restriction endonuclease employing a matrix bound double-stranded eicosomer (20mer) (6).

2.1 Design of the oligodeoxynucleotide ligand

In order to exploit sequence-specific protein–nucleic acid interactions effectively for the isolation of desired DNA binding proteins or catalytically active enzymes, the sequence of interest must be bound to the column such that the protein can effectively access the binding site. Although DNA polymers have been bound to matrices by physical entrapment, the lack of a covalent bond between the ligand and the matrix leads to loss of the ligand (leaching) over a period of time. With smaller oligodeoxynucleotides this process can be expected to accelerate. UV cross-linking or the use of cyanogen bromide activated matrices have been used to covalently bind nucleic acid ligands (8, 9), but rely upon random covalent attachment of the nucleic ligand to the matrix. This will provide the desired covalent bond but can result in complications if an important functional group within the binding site is lost or, if binding of the nucleic acid to the matrix results in a change in the tertiary structure of the ligand with a corresponding reduction in affinity for the complementary protein (10–12). In the present procedure a more specific reaction between a thiol residue and a matrix-bound epoxide or trifluoromethylsulphonic ester has been employed. The specificity of the attachment of the ligand to the matrix is a result of the small spacer arm with a terminal thiol group bound to the 5′-terminus of one of the oligodeoxynucleotides composing the double-stranded recognition site (see also Chapter 8). By preparing the spacer arm as a phosphoramidite derivative, the spacer arm can be coupled to the growing oligodeoxynucleotide chain during normal chemical synthetic procedures. Coupling of this fragment to a thiol-specific affinity support should occur exclusively though the spacer arm and allow the recognition sequence to be extended into solution away from the matrix surface. Although the use of a thiol-containing spacer arm precludes the possibility of ligating the oligomers to form longer fragments containing multiple recognition sites, the specificity of ligand attachment to the matrix should be advantageous providing that a suitable concentration of ligand can be introduced onto the support.

To control the orientation of the ligand, only one of the strands contains the thiol linker; the second strand is annealed to the covalently bound sequence to provide the double-stranded recognition site as exemplified for the eicosomer matrix described in this chapter (*Figure 1*). This approach can lead to some leaching of the complementary strand from the column, but with radioisotopic labelling of the second strand, such processes can be easily monitored. Radiolabelling of the annealed strand also allows direct monitoring of the concentration of double-stranded DNA within the matrix, which may differ significantly from the amount of single-stranded DNA covalently bound to the matrix.

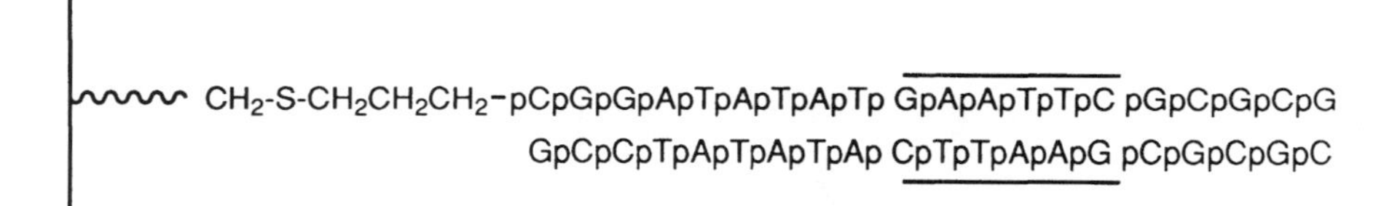

Figure 1. A double-stranded eicosomer (20mer) affinity matrix containing the recognition site for the *Eco*RI restriction endonuclease (underlined).

2.2 Choice of chromatographic matrix

With procedures which involve the chemical synthesis of the nucleic acid ligand, two general approaches are possible. The most commonly used DNA synthesis techniques employ a solid phase or matrix-bound procedures (see Chapter 1). It may be possible to synthesize the DNA fragment directly on a matrix, deprotect the fragment, and directly generate a matrix bound sequence. The second approach involves DNA synthesis and purification followed by a second step to attach the DNA ligand covalently to the matrix. While the first approach appears advantageous for its inherent simplicity, at present the matrices which are effective for chemical synthesis procedures are not typically used for protein isolations, and matrices which function effectively for the isolation of proteins are not compatible with chemical DNA synthesis techniques. The procedures used in this chapter involve the latter approach, in which a DNA fragment is synthesized, purified, and then subsequently attached to the chromatographic matrix.

The procedures described in this chapter produce oligodeoxynucleotides containing a terminal thiol residue attached to a small linker. The chromatographic matrices which function most efficiently to covalently bind these ligands will contain thiol-specific functional groups. Four functionalized matrices fit this requirement: thiol-containing matrices, mercurial matrices, epoxide matrices and trifluoromethylsulphonyl ester (Tresyl) matrices:

Affi-Gel 401 ⊢$OCH_2CONHCH_2CH_2NHCOCH(CH_2CH_2-SH)(NHCOCH_3)$

Affi-Gel 501 ⊢$OCH_2CONHCH_2CH_2NHCOCH_2CH_2CONH-C_6H_4-HgCl$

Epoxy-activated Sepharose 6B ⊢$(O)_2C=NCH(COOH)CH_2CONHCH(OH)CONHCH_2$–epoxide

Tresyl-activated Sepharose 4B ⊢$CH_2-O-SO_2-CF_3$

The first two matrices, those containing thiol or mercurial residues, may result in side reactions during chromatographic procedures if the thiol or mercurial functionalities are not fully saturated with ligand. In either case, the residual functionality would be available to react with other thiol-containing reagents. In theory the residual functionality can be removed by subsequent treatment of the affinity matrix with additional small thiol reagent. Unfortunately these are also the conditions used to remove the matrix-bound ligand from the matrix. The presence of residual functionality may allow for further covalent reactions, for example with cysteine residues of proteins, and result in protein retention by mechanisms other than ligand–protein binding. With these concerns in mind, the procedures described below have employed only the epoxide- or Tresyl-activated matrices. Either of these two functionalities can be removed by subsequent steps to avoid covalent retention of proteins by reaction with the functionalized matrix.

3. Preparation of oligodeoxynucleotide matrices for affinity chromatography

The matrix can be prepared using two procedures which have only slight variations. In the first the oligodeoxynucleotide containing the thiol linker is annealed to its complementary strand and then added to the thiol-specific matrix. In the second, the thiol-containing strand is covalently bound to the matrix and the complementary strand is annealed in a second step. Although in principle both procedures function adequately, we prefer the first procedure since it does not require heating of the chromatographic matrix for the anealling step.

3.1 Preparation of the oligodeoxynucleotide ligand

The oligodeoxynucleotide ligand containing the thiol linker is prepared using phosphoramidite synthesis techniques essentially as described in Chapters 1 and 8. The final coupling step employs either the C6-ThiolModifier (Cruachem, Catalogue no. 20–8405–20) or the corresponding C3 linker (6, 13). The linker contains a trityl-protecting group such that HPLC purification of the oligodeoxynucleotide proceeds as described for DMT-containing oligodeoxynucleotides (see Chapter 1). The trityl protecting group is removed with silver nitrate as described originally by Connolly and Rider (13) to generate the oligodeoxynucleotide carrying a 5′-terminal thiol group (*Figure 2*).

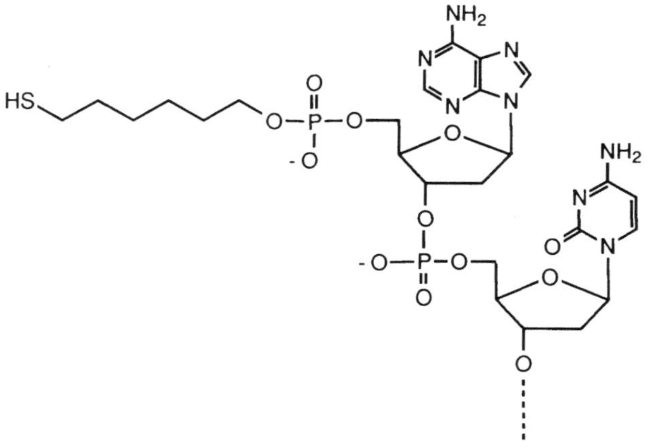

Figure 2. Illustration of the 5′-terminus of an oligodeoxynucleotide carrying a thiol-containing linker.

Protocol 1. Deprotection of the S-trityl oligodeoxynucleotide

Materials

- 40–60 A_{260} units of oligodeoxynucleotide (~ 0.3 μmol)
- 100 mM triethylammonium acetate pH 7.0
- 100 mM aqueous silver nitrate
- 140 mM aqueous dithiothreitol
- 2 × 30 cm Sephadex G-10 column

Protocol 1. *Continued*

Method

1. To the oligodeoxynucleotide in 200 μl of 50 mM triethylammonium acetate pH 7.0 add 15 μl of the silver nitrate solution and incubate the reaction mixture at ambient temperature for 30 min.
2. After 30 min add 15 μl of the dithiothreitol solution and wait an additional 30 min.
3. Centrifuge the resulting suspension, decant the supernatant and desalt it using a column of Sephadex G-10. Lyophilize the collected material to dryness.

The complementary strand (with no thiol ligand) is prepared and purified by standard DNA synthesis techniques (see Chapter 1). In order to monitor directly the extent of ligand loading on the matrix, the unmodified (complementary) DNA strand is labelled at the 5′-terminus with γ-^{32}P]ATP and polynucleotide kinase.

Protocol 2. Radioisotopic labelling of the unmodified oligodeoxynucleotide

Materials

- 0.5 μCi[γ-^{32}P]ATP
- 1.0 A_{260} unit oligodeoxynucleotide
- T4 polynucleotide kinase

Method

1. In a total volume of 200 μl containing 1 A_{260} unit oligodeoxynucleotide, 37.5 μM ATP, 40 mM Tris–HCl (pH 8.7), 10 mM $MgCl_2$, 10 mM DTT, 1.0 μg BSA, 0.5 μCi [γ-^{32}P]ATP, add 10 units of T_4 polynucleotide kinase and incubate for 18 h at 37 °C.
2. Prepare a Sep-Pak (or Nensorb) cartridge by attaching a 10 ml plastic syringe cylinder as a solvent reservoir and wash with 20 ml methanol followed by 20 ml of distilled water.
3. Add the reaction mixture to the cartridge and wash with 15 ml of aqueous 1% methanol to elute any unincorporated ATP and other salts followed by 9 ml of aqueous 20% methanol to elute the 5′-labelled oligodeoxynucleotide.
4. Evaporate the fractions containing the oligomer to dryness and dissolve in 0.4 ml of water. Measure the optical density and determine the incorporated radioactivity in an aliquot of the solution by scintillation

Protocol 2. *Continued*

counting. From these measurements determine the specific activity of the oligodeoxynucleotide (simply expressed as c.p.m./A_{260} unit or c.p.m./nmol).

5. Add this solution to 27 A_{260} units of unlabelled oligodeoxynucleotide and determine the specific activity of the isotopic labelled mixture. Use this solution in the method of *Protocol 3*.

3.2 Attachment of the oligodeoxynucleotide ligand to the Sepharose matrix

The following procedures employ either Tresyl-activated Sepharose 4B (maximum binding capacity of 0.3 μmol/ml of swollen gel) or epoxy-activated Sepharose 6B (maximum binding capacity of 15–20 μmol/ml of swollen gel), both products are available from Pharmacia. The procedures are described for columns of 0.5 ml volume, but larger columns should result with appropriate increases in reagents.

Protocol 3. Attachment of the annealed double-stranded DNA ligand

Materials

- 0.5 ml swollen epoxy-activated Sepharose 6B[a]
- 28 A_{260} units (0.24 μmol) of thiol-containing 20mer (*Figure 1*)
- 32 A_{260} units (0.32 μmol) of complementary 20mer (*Figure 1*)
- 0.2 M Tris–HCl pH 8
- 1.0 M sodium chloride

Method

1. Prepare the epoxy-activated matrix as described by the manufacturer.
2. Dissolve both 20mers (or appropriate sequences) in 0.6 ml of 0.1 M Tris–HCl pH 8.0[b], 0.5 M sodium chloride (oligonucleotide concentration ~0.4 mM) in a 1.0 ml Eppendorf tube and heat the solution to 95 °C, cool slowly to room temperature and finally to 15 °C.
3. Add the washed and filtered matrix to the oligodeoxynucleotide solution at 15 °C and mix the suspension (vortex).
4. Mix the suspension (vortex) at various times during the following 48 h[c].
5. Centrifuge the suspension, remove the supernatant, wash the matrix a number of times with 0.1 M Tris–HCl pH 8.0, 0.5 M sodium chloride.
6. Any remaining unreacted matrix bound epoxide (Tresyl) residues were blocked by incubation of the functionalized matrix in 1.0 M aqueous

Protocol 3. *Continued*

mercaptoethanol, 0.1 M Tris–HCl pH 8.0, 0.5 M sodium chloride overnight at ambient temperature.

7. Wash the matrix three times each with 0.1 M sodium acetate, pH 4.0, 0.5 M sodium chloride followed by 0.1 M Tris–HCl, pH 8.0, 0.5 M sodium chloride to remove any non-covalently bound ligand. Place the matrix into a water jacketted column and equilibrate it with 40 mM Tris–HCl, pH 7.5, 0.2 M NaCl, 5 mM EDTA, 2 mM DTT.
8. To determine the ligand loading, transfer a small portion of the matrix (10–20 mg) to a scintillation vial and lyophilize to dryness. Determine the radioactivity present and calculate the concentration of oligodeoxynucleotide present based upon the specific activity (see *Protocol 2*).

[a] Or Tresyl-activated Sepharose 4B.
[b] For Tresyl-activated Sepharose 4B, sodium carbonate pH 10 is used.
[c] Reaction with the Tresyl-activated Sepharose 4B usually requires only 12–18 h.

Although it is not essential to radiolabel the complementary DNA strand for the isolation of proteins, the concentration of double-stranded recognition sites present can be determined directly from the radioactivity present in a small portion of the lyophilized matrix by scintillation counting (control experiments indicated that the radiolabelled 20mer was not retained on the column when incubated with either matrix in the absence of the thiol-containing strand). With this procedure, the double-stranded recognition sequence (using either matrix) is typically present at concentrations which vary from 6 to 30 μM (6–30 mmol/ml). The concentration of covalently bound thiol-containing oligodeoxynucleotide (single-stranded) may be higher than the values determined in this manner (particularly when some leaching of the complementary strand occurs), but this procedure should accurately reflect the concentration of the *double-stranded* ligand present on the matrix. The ligand loading achieved with these procedures represents only a small number of the total binding sites available on either support. Loading of the matrix increases only minimally when the coupling period for the ligand and the matrix was extended beyond 48 h (a reaction time of one week only raised the matrix bound ligand concentration to 58 μM). Increasing the concentration of the ligand during the coupling reaction to 1.0 mM did not result in a significant increase in the amount of double-stranded ligand bound to the matrix.

4. Protein isolation with oligodeoxynucleotide affinity columns

For effective affinity chromatography, the ligand should have an affinity for the protein ideally in the range 10^{-4}–10^{-8} M. With higher dissociation constants (K_d), the binding interaction will be too weak to effectively retard the flow of the protein. Conversely, with lower K_d values, it may be impossible to desorb bound protein without inactivating the biomolecule or destroying the matrix. The eicosomer matrix described in Section 3.2 was designed for the isolation of the *Eco*RI restriction endonuclease. The dissociation constant for the *Eco*RI restriction endonuclease in the absence of magnesium is dependent upon the salt concentration and has been reported as low as 10^{-11} M (14). However, with short oligodeoxynucleotides and a sodium chloride concentration of 0.2 M, the dissociation constant lies between 10^{-6} and 10^{-7} M (15), a range which is ideal for the present affinity chromatography procedure.

To shift the protein–nucleic acid binding equilibrium to favour the bound complex it is necessary that the concentration of bound ligand is significantly higher than the expected protein dissociation constant with the desired ligand. A matrix containing 30 μM double-stranded nucleic acid provides a ligand concentration which is some 60-fold higher than the dissociation constant for the *Eco*RI restriction endonuclease (about 5×10^{-7} M). Having the matrix-bound ligand concentration significantly higher than the expected K_d is important to retain the protein effectively on the column.

In the present example, it is necessary to isolate the enzyme in the absence of magnesium. Under these conditions the *Eco*RI endonuclease exhibits sequence-specific binding but not the associated hydrolytic properties (which would destroy the affinity matrix). Sequestering the metal ions available in the cell lysate and buffer solutions by using Na_2EDTA appears to be an effective and simple procedure to inhibit the catalytic properties of the enzyme while exploiting sequence-specific recognition.

The endonuclease is obtained directly from the precipitated crude protein fraction without any preliminary purification using other forms of column chromatography. This was partly the result of using an over-expressing cell strain which produced the enzyme in concentrations of 1–3% of the available protein. The isolation of other sequence-specific binding proteins present in lower concentrations may require partial purification by additional chromatographic steps prior to an affinity chromatography procedure.

4.1 Isolation of the *EcoRI* restriction endonuclease

A solution of crude cellular proteins (approximately 20 mg and containing 1–3% endonuclease) is incubated with the matrix in the presence of Na_2EDTA for about one hour to allow for diffusion processes within the matrix and the formation of the enzyme–substrate complex. Elution of non-bound proteins is achieved by washing the column with a buffer containing sodium chloride (0.2 M). This was followed by a single step or multistep increase of the salt concentration to elute the bound protein(s). Fractions of 1 ml were collected, dialysed, and aliquots were analysed by electrophoresis on denaturing polyacrylamide gels. In this case a single step high salt wash (1.5 M) eluted primarily the endonuclease (*Figure 3*). The restriction endonuclease could be obtained in higher purity if a multistep salt gradient was used. The enzyme was then observed to elute from 0.7 to 1.0 M sodium chloride. The endonuclease isolated with the multistep salt gradient was purified to near homogeneity (80–90% pure) as determined by densitometry (*Figure 4*). Typically, 0.1–0.2 mg of enzyme was obtained from a total of 20 mg of the crude protein fraction using a 1.0 ml column. This was typically 30–40% of the restriction endonuclease available in the crude protein precipitate.

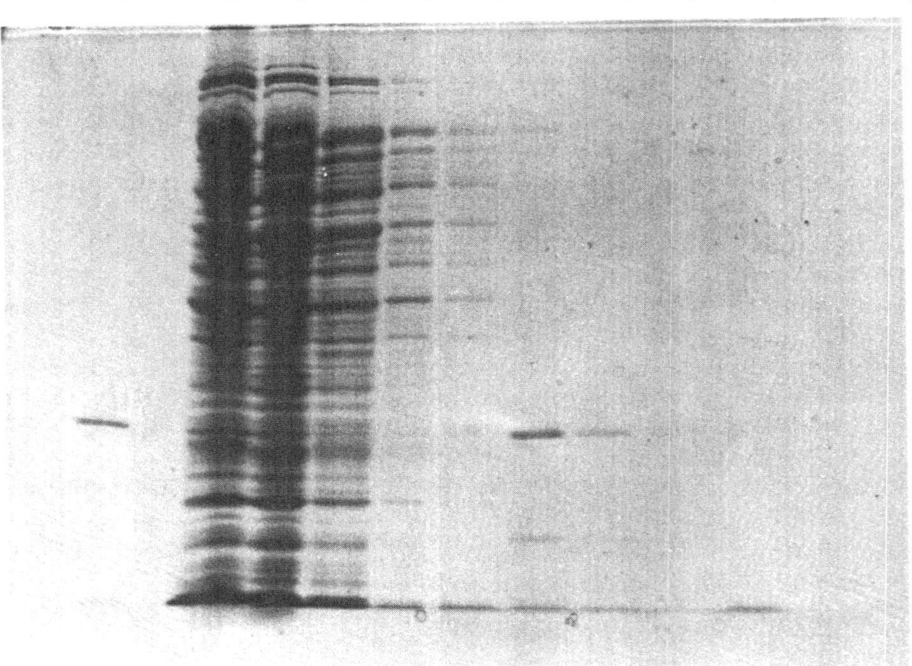

Figure 3. Denaturing 10% polyacrylamide gel electrophoresis of the eluate obtained from the nucleic acid affinity matrix containing the eicosomer. Lane 1: Authentic sample of the *Eco*RI restriction endonuclease. Lane 2: Crude protein mixture obtained from *E.coli* strain M5248 transformed with plasmid pSCC2 after cell lysis and removal of the nucleic acid material by streptomycin sulphate precipitation. The remaining lanes contain material eluted from the 0.5 ml column and collected in 1 ml fractions. Elution was at 15 °C using 0.04 M Tris–HCl, pH 7.5, 5 mM Na_2EDTA, 2 mM dithiothreitol and 0.2 M NaCl. Lane 3: fraction 1; lane 4: fraction 2; lane 5: fraction 3; lane 6: fraction 4. Fractions 5–9 did not contain any detectable protein material. The last four lanes contain fractions eluted with the buffer describe above but contained 1.5 M NaCl. Lane 7: fraction 10; lane 8: fraction 11; lane 9: fraction 12; lane 10: fraction 13.

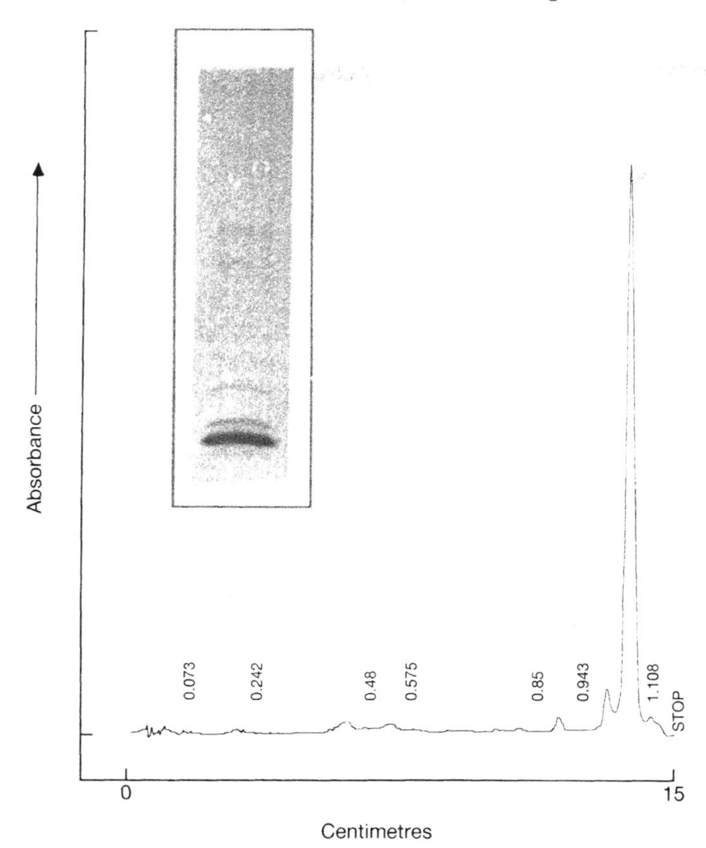

Figure 4. Densitometry scan of the polyacrylamide gel analysis of the isolated endonuclease. Inset: photograph of the stained gel.

Some of the endonuclease was not retained on the column although this was difficult to quantify since the enzyme appeared to be present in a number of column fractions, but at relatively low concentration. This can be accounted for in that during the washes, even at low salt concentrations, dissociation of the bound protein as a function of equilibrium processes can result in its loss from the matrix. Based upon the purity of the protein, the isolated enzyme had a specific activity of 1.87×10^6 u/mg.

Protocol 4. Isolation of the *EcoRI* restriction endonuclease

Materials

- Buffer A: 40 mM Tris–HCl pH 7.5, 0.2 M NaCl, 5 mM Na_2EDTA, 2mM DTT

Protocol 4. *Continued*

- Buffer B: 40 mM Tris–HCl pH 7.5, 1.5 M NaCl, 5 mM Na_2EDTA, 2 mM DTT
- Affinity column (see *Protocol 3*)
- Crude protein fraction (for the *EcoRI* endonuclease see reference 16)

Method

1. Pipette 0.8 ml of crude protein solution (obtained from the work-up of 0.5 g of cell paste containing the pSCC2 plasmid) onto the water jacketted column at 15 °C.
2. After incubation for 1 h, elute the unbound proteins from the column by washing with 20 ml of Buffer A (flow equal to 0.5 ml/min).
3. Elute the bound proteins by washing with 9 ml of buffer containing 40 mM Tris–HCl (pH 7.5), 1.5 M NaCl, 5 mM EDTA, 2 mM DTT (Buffer B).
4. A sodium chloride step gradient can also be employed for the elution of the bound proteins. In this case successive 1.0 ml aliquots of buffer each with a 100 mM increase in NaCl concentration (0.5–1.5 M NaCl) are added to the column. In the present example, the protein elutes between 0.7 and 1.0 M NaCl.
5. Dialyse the eluate containing the protein overnight at 4 °C against 2 mM Tris–HCl (pH 7.5) and 2 mM DTT. Reduce the volume as desired. Dialyse overnight at 4 °C against 20 mM KH_2PO_4 (pH 7.5), 0.2 M NaCl, 1 mM DTT, and 0.1 mM EDTA. Add glycerol to a final concentration of 50% and store the enzyme (0.1–0.2 mg/ml) at −20 °C.

The enzyme purified from the eicosomer matrix was examined for specificity as well as the presence of contaminating non-specific nuclease activity. The restriction digest pattern of bacteriophage λ DNA obtained from an authentic sample of *Eco*RI endonuclease was identical to that of the isolated endonuclease as shown by electrophoresis on 0.8% agarose gels. In order to determine if there was contamination of the enzyme with non-specific nucleases, a ligation–recut assay was done. Digestion, ligation, and redigestion of λ DNA showed no change in the pattern of DNA present. Greater than 90% of the digested fragments could be ligated for redigestion as determined by electrophoresis on 0.8% agarose gels. Incubation of the isolated endonuclease with the decamer, d(CTGAATTCAG), produced the expected hydrolysis products. The chromatogram showed no evidence of nonspecific endonuclease or exonuclease activity present in the enzyme preparation.

The affinity column can be used repeatedly to isolate the endonuclease from crude cell extract with only small decreases in efficiency. Some ligand is lost from the matrix over a period of time as determined from the radioisotopic content. However, the original matrix bound ligand concentration could be regenerated by adding more of the complementary non-thiol-containing oligodeoxynucleotide and heating the matrix to 70 °C followed by slow cooling.

For the present work, epoxy-activated Sepharose 6B was the chromatographic matrix of choice although Tresyl-activated Sepharose 4B can also be employed. The former contains a higher concentration of active ligand binding sites (15–20 μmol/ml for epoxy-activated Sepharose 6B versus 0.3 μmol/ml for Tresyl-activated Sepharose 4B) and a longer spacer arm. However, in our hands, the amount of nucleic acid bound to either support represents only a very small fraction of the sites available for reaction. A higher fraction of the reactive sites could be modified when using Tresyl-activated Sepharose 4B, but with the lower ligand capacity available with this material, we were still unable to prepare a matrix with a ligand concentration significantly higher than 30 μM (modification of 10% of the available sites). This relatively low loading may in part reflect the inherent exclusion properties of the Sepharoses in which case many of the active residues imbedded in the pores would be less available for reaction.

5. Conclusions

At present there are relatively few examples documenting the use of matrix-bound oligodeoxynucleotides for affinity chromatography of proteins. In cases similar to that reported here it may be possible to isolate the protein directly from crude cell lysate in relatively high purity. With proteins present in low concentration, such as some transcription factors, the use of such affinity procedures may function well only as part of a multi-step isolation procedure. The procedures in this chapter must be viewed as only a guide for designing and implementing such affinity techniques for specific proteins.

Acknowledgements

This work was supported by a grant from the National Science Foundation (DMB-8904306) and a Brystol Meyers Company Grant of Research Corporation.

References

1. Remy, P., Birmele, C., and Ebel, J. P. (1972). *FEBS Lett.*, **27**, 134.
2. Joyce, C. M. and Knowles, J. R. (1974). *Biochem. Biophys. Res. Commun.*, **60**, 1278.

3. Rosenfield, P. J. and Kelly, T. J. (1986). *J. Biol. Chem.*, **261**, 1398.
4. Kadonaga, J. T. and Tjian, R. (1986). *Proc. Natl. Acad. Sci. USA*, **83**, 5884.
5. Dean, N. and Berk, A. J. (1987). *Nucleic Acids Res.*, **15**, 9895.
6. Blanks, R. and McLaughlin, L. W. (1988). *Nucleic Acids Res.*, **16**, 10283.
7. Gerwin, B. I., and Milstein, J. B. (1972). *Proc. Natl. Acad. Sci. USA*, **69**, 2599.
8. Litman, R. M. (1968). *J. Biol. Chem.*, **243**, 6222.
9. Alberts, B. M. and Herrick, G. (1971). *Methods Enzymol.*, **21**, 198.
10. Arndt-Jovin, D. J., Jovin, T. M., Bahr, W., Frischauf, A.-M., and Marquardt, M. (1975). *Eur. J. Biochem.*, **54**, 411.
11. Chan, W. W. C. and Takahashi, M. (1969). *Biochem. Biophys. Res. Commun.*, **37**, 272.
12. Harvey, M. J., Lowe, C. R., Craven, D. B., and Dean, P. D. G. (1974). *Eur. J. Biochem.*, **41**, 335.
13. Connolly, B. A. and Rider, P. (1985). *Nucleic Acids Res.*, **13**, 4485.
14. Halford, S. E., Johnson, N. P., and Grinsted, J. (1980). *Biochem. J.*, **191**, 593.
15. Terry, B. J., Jack, W. E., Rubin, R. A., and Modrich, P. (1983). *J. Biol. Chem.*, **258**, 9820.
16. Cheng, S.-C., Kim, R., King, K., Kim, S.-H., and Modrich, P. (1984). *J. Biol. Chem.*, **259**, 11571.

11

Oligodeoxynucleotides with reporter groups attached to the base

JERRY L. RUTH

1. Introduction

The ability to attach reporter groups to nucleic acids has added much to our knowledge of nucleic acid structure and behaviour. Traditional non-isotopic labels developed for cloned DNA fragments used indirect labels such as biotin or haptens. These labels were detected after hybridization by binding to labelled avidin or antibodies. Such indirect reporter groups were used for very practical reasons—most of the methods for preparing cloned probes required enzymatic incorporation of nucleoside polyphosphate precursors. Large hydrophobic groups such as fluorescein-5-isothiocyanate (FITC, sterically-hindered molecules, or biopolymers with activities of their own (enzymes such as alkaline phosphatase) could not be easily incorporated by enzymatic methods.

Significant recent progress in chemical methods has resulted in the ability to synthesize specifically-modified oligodeoxynucleotides. As hybridization probes, such oligodeoxynucleotides have many attractive advantages over cloned DNA fragments. Such advantages include rapid hybridization, lower cost, higher purity, greater selectivity, and better reproducibility. Perhaps more importantly, the number, type, and placement of the modifications such as reporter groups can be precisely controlled.

There are three basic strategies which can be used for labelling single-stranded oligodeoxynucleotides with non-isotopic reporter groups:

- *Enzymatic incorporation* of a modified nucleoside 5′-triphosphate (1, 2) onto the 3′-end of the oligomer,
- *Direct chemical reaction* of an unmodified oligomer by transamination with bisulphite (3), reaction with labelled mutagens (4, 5), or sulphonylation (6), or
- *Incorporation of a modified base during chemical synthesis* of the oligonucleotide (7–24). The base may be already labelled, or the label attached to a 'linker arm' after oligonucleotide synthesis.

For many reasons, the last strategy is the most attractive. This chapter will focus exclusively on the incorporation of linker arm nucleosides into oligodeoxynucleotides during chemical synthesis of the oligomer (see also Chapter 3 for groups attached to the base of 2′-O-methyloligoribonucleotides).

For attachment of non-isotopic reporter groups, the most common approach has been to incorporate a linker arm of some length (usually 3–12 atoms) which is attached to a nucleic acid base on one end, and terminates in a primary reactive group such as an amino, carboxy, or sulphydryl function. Before oligomer synthesis, the reactive group is chemically protected to avoid undesired reaction with active phosphite monomers, and later deprotected to allow attachment of the desired reporter group. For some applications where the reporter group is either protected [such as biotin (20)] or is unreactive (such as dinitrophenyl), the label may be attached to the linker arm nucleoside before oligomer synthesis. Some labelled internal linker arms which lack a base unit (abasic site) have also been described (25, 26). Although many of the techniques are similar, labelling at abasic sites and labelling of RNA are outside the scope of this chapter.

This chapter is broken into three parts. The first part will detail the protocols used to synthesize functionalized 'linker arm' nucleosides, including the 5′-OH protection and 3′-O-phosphitylation, and subsequent oligodeoxynucleotide synthesis. The second part of the chapter will describe some reactions for attaching reporter groups such as ligands (example: biotin), fluorescent or luminescent labels (e.g. fluorescein), and enzymes (e.g. alkaline phosphatase) to the product 'linker arm' oligodeoxynucleotides. The third part of the chapter will briefly summarize hybridization protocols and behaviours of some non-isotopic oligodeoxynucleotide probes.

2. Synthesis of 'linker arm' oligodeoxynucleotides

Pyrimidine nucleosides are the most straightforward to modify. They are more stable to acids and bases than many purine nucleosides, and the amide functions on thymine analogues do not require additional protection before oligomer synthesis. However, the procedures outlined in *Protocols 1–3* are not chemically trivial, particularly for the average molecular biologist. For convenience, the product linker arm phosphoramidites illustrated in *Figures 1* and *2* (R_1 = DMT, R_2 = β-cyanoethyl N,N-diisopropylphosphoramidite) are available commercially from suppliers of oligonucleotide synthesis reagents such as Glen Research and American Bionetics.

Much less work has been done on the modification and attachment of reporter groups to the purine bases of oligonucleotides. From a chemical point of view, the adenine and guanine nucleosides are more difficult to work with due to lessened solubilites, the pH sensitivity of many analogues, and the need for additional blocking groups on the exocyclic amines. However, strategies similar to those for pyrimidine bases are beginning to appear.

Some pyrimidine base analogues which are inherently fluorescent have also been synthesized and incorporated into synthetic oligomers (21, 22); these have been used to study environmental effects in hybrids. Methods have also been developed to synthesize labelled triphoshate analogues of all four nucleic acid bases, which are then enzymatically incorporated into double-stranded DNA (27). The broad application of these last approaches have not yet been demonstrated.

2.1 Synthesis of pyrimidine nucleosides with linker arms at C-5

In a double-stranded DNA helix, the C-5 position of pyrimidines is not involved in hydrogen bonding, and faces outward into the groove of the helix. This allows significant steric tolerance, and makes the C-5 position ideal for attachment of linker arms and reporter groups, with little effect on hybridization. The most common and convenient method of C-5 derivatization involves direct mercuration (28) or halogenation at C-5 of unprotected 2′-deoxyuridine (dUrd) or 2′-deoxycytidine (dCyd), and subsequent alkylation by olefins in the presence of palladium (29). The following protocols use variations of this method to attach either aminoalkyl or carboxyalkyl linker arms to C-5 of pyrimidine bases.

2.1.1 Synthesis of pyrimidine nucleosides with blocked amine linker arms at C-5

An eleven atom spacer arm containing a primary amine can be synthesized easily from the reaction of acryloyl choride with 1,6-diaminohexane. The resulting primary amine is then blocked with trifluoroacetyl using methyl trifluoroacetate to form N-(trifluoroacetylaminohexyl)acrylamide, as described in *Protocol 1*, steps 1–6. 5-Chromomercuri-dUrd (*Protocol 1*, step 8) is then reacted with an excess of the blocked linker arm in the presence of palladium (*Protocol 1*, step 9) to produce a thymidine (dThd) analogue with a trifluoroacetyl-protected eleven atom amine linker arm at C-5 (*Figure 1*, $R_1 = R_2 = H$) (7).

The trifluoroacetyl group is ideal for protection of primary amines during oligonucleotide synthesis (7, 9, 12–17) since it is stable to hydrolysis in the organic acids used to remove 5′-DMT groups during each cycle ($t_{1/2} \geq 300$ min in 10% dichloroacetic acid), yet it is easily removed by ammonia or other amine bases during oligomer deprotection ($t_{1/2} \approx 10$ min at 50 °C); see *Table 1*. A less reactive trichloroacetyl function as a blocking group is too resistant to ammoniolysis during standard oligomer deprotection. Other amine protecting groups such as *tert*-butyloxycarboxyl (*t*-BOC) (10) and phthalimido (11) have also been used successfully.

After the blocked amine linker arm nucleoside is isolated, the 5′-OH of the crude can be protected with a dimethoxytrityl (DMT) group by conventional reaction with DMT-chloride in anhydrous pyridine (*Protocol 1*, step 14). This provides a fully blocked 5′-DMT-5-trifluoroacetylaminoalkyl-dThd analogue (*Figure 1*, R_1 = DMT, R_2 = H) which can be converted to its 3′-O-phophoramidite by standard methods.

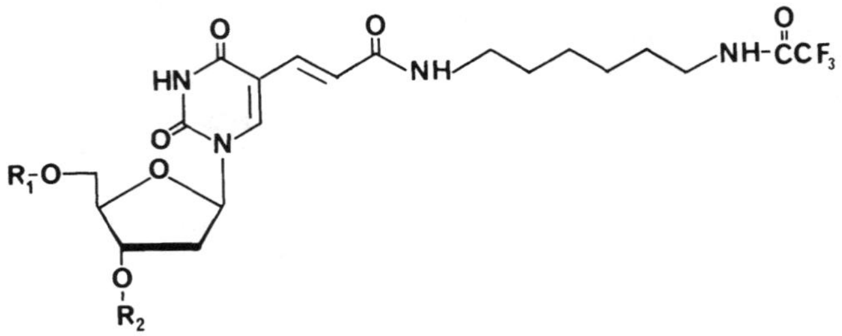

Figure 1. Structure of a blocked C-5 aminoalkyl-dThd analogue (7), 5-[N-(6-trifluoroacetylaminohexyl)-3-(E)acrylamido]-2′-deoxyuridine (R_1 = H- or DMT; R_2 = -H or N,N-diisopropylphosphoramidite).

Protocol 1. Synthesis of 5′-O-dimethoxytrityl-5- trifluoroacetylaminoalkylthymidine (*Figure 1*; R_1 = DMT, R_2 = H) (7)

Synthesize the blocked amine linker arm precursor

1. In a 1 litre round-bottomed flask, dissolve 35 g (0.30 mol) 1,6-diaminohexane in 30 ml chloroform, and stir vigorously. Dissolve 12 ml (0.15 mol) acryloyl chloride in 300 ml chloroform, and add dropwise to the diaminohexane solution. Stir for 15–30 min at ambient temperature.
2. Add 75 ml (0.75 mol) methyl trifluoroacetate dropwise with stirring. Continue stirring for 15 min, then slowly add 78 ml triethylamine. Continue stirring until solution is homogenous.
3. Assay the reaction by TLC on UV-silica plates, eluting with ethyl acetate. Develop the TLC plate by treatment in a chlorine chamber followed by starch–iodine spray. The crude reaction will consist of the desired product (R_f 0.57) and a faster-eluting by-product (R_f 0.7–0.8).
4. Remove solvents by rotary evaporation to a dry cake. Suspend the solid in ethyl acetate, and extract extensively with water.
5. Dry the organic layer containing the crude linker arm over sodium sulphate. Filter off the drying agent. Add approximately 30 g silica gel and rotary evaporate to dryness.
6. Purify by chromatography using a 5 × 50 cm column of silica gel (E. Merck) packed in cyclohexane. Load the crude product in cyclohexane,

Protocol 1. *Continued*

elute with cyclohexane/ethyl acetate (2:1) until the faster-eluting by-product (TLC R_r 0.7–0.8) has eluted (2–4 litres). Elute with cyclohexane/ethyl acetate (1:2) to bring off the desired product (TLC R_f 0.57). Combine appropriate fractions and rotary evaporate to dryness. Dry to constant weight under vacuum. Yield will be 10–12 g (25–30% isolated yield) of blocked linker arm precursor (mol. wt 266) as a white or off-white solid. Store desiccated at 4 °C.

Mercurate 2′-deoxyuridine (dUrd) at C-5

7. Dissolve 20 g (88 mmol) dUrd in 60 ml water. Dissolve 31 g (97 mmol) mercuric acetate in 120 ml water, and add to the dUrd solution. Stir and heat at 55–60 °C for 18–24 h. A thick white precipitate will form.
8. Convert the intermediate 5-mercuri-dUrd to the chloride form by adding 35 ml conc. HCl slowly, and stirring for 1 h. Let cool to ambient temperature. Filter the white precipitate and wash with methanol. Dry under vacuum to constant weight. Yield will be 35–38 g (85–95%) of 5-chloromercuri-2′-deoxyuridine (5-ClHg-dUrd; mol. wt 463) as a white solid. Store at ambient temperature.

Attach the blocked amine linker arm to dUrd

9. In a 500 ml round-bottomed flask, dissolve 7.4 g (28 mmol) blocked linker arm precursor (from step 6 above) in 50 ml acetonitrile with stirring. Add 190 ml 0.2 N $LiPdCl_3$ in acetonitrile. Heat to 50–60 °C for 15–20 min in an oil bath. Add 9.3 g (20 mmol) 5-ClHg-dUrd (step 8) and stir to suspend. Attach a reflux condenser and heat for 6 h at a bath temperature of 110 °C.
10. Allow to cool to ambient temperature. Check a 10 μl aliquot by TLC on UV-silica plates eluting methanol/chloroform (1:4), major product R_f 0.63.
11. While stirring, bubble hydrogen sulphide gas through the reaction to precipitate Hg and Pd metals as sulphides. (Caution: H_2S gas is highly toxic. This step must be done in a vented hood). The reaction will turn very dark.
12. Filter the crude reaction through a Celite cake in a Buchner funnel. To maximize recoveries, resuspend the filter cake in methanol, boil for 2–5 min, and refilter. Combine filtrates and concentrate by rotary evaporation to a foam.

Block the 5′-OH with dimethoxytrityl (DMT)

13. Render the crude material anhydrous by rotary evaporation from 100 ml anhydrous pyridine. Repeat twice.

Protocol 1. *Continued*

14. Dissolve the crude material in 100 ml anhydrous pyridine. Add 7.5 g (22 mmol) dimethoxytrityl (DMT) chloride. Stir for 4 h at ambient temperature.
15. Check for conversion to the 5′-DMT-blocked nucleoside (R_f 0.30) by TLC on UV-silica plates eluting methanol/chloroform (1:9). DMT-containing compounds, including unreacted DMT-Cl, can be visualized by exposure to HCl vapours in a closed chamber.
16. When complete, pour the reaction onto an equal volume of ice-water, mix, then extract twice with 200 ml dichloromethane. Dry the combined organic layer over sodium sulphate, and concentrate by rotary evaporation to a foam.
17. Purify the final compound by column chromatography on a 3 × 30 cm neutralized silica gel column eluting a linear gradient of 2 litres chloroform containing 0.2% triethylamine to 2 litres methanol/chloroform (1:9). Combine appropriate fractions, and concentrate by rotary evaporation to a solid. Yield will be 7–8 g (9–10 mmol) of blocked amine linker arm dThd analogue, 5′-O-dimethoxytrityl-5-[N-(6-trifluoroacetylaminohexyl)-3-(E)acrylamido]-2′-deoxyuridine (*Figure 1*; R_1 = DMT, R_2 = H; mol. wt 794) as a white powder. UV_{max} 302 nm (ε 18 200). The product is stable for more than three years desiccated at 4 °C.

A procedure identical to *Protocol 1* can be used to prepare the corresponding linker arm 2′-deoxycytidine (dCyd); however, after attaching the blocked linker arm, the N^4 exocyclic amine of dCyd must be carefully protected with a suitable group such as benzoyl before adding the 5′-DMT group.

Table 1. Deprotection kinetics of some blocked amine linker arms

Hydrolysis conditions		**Approximate $t_{1/2}$ in min**[a]	
		Amino protecting group	
		Trifluoroacetyl	Trichloracetyl
Water	100 °C	⩾ 300	⩾ 300
10% dichloroacetic acid	25 °C	⩾ 300	–
10% piperidine	25 °C	2	–
	50 °C	1	–
Carbonate, pH 11.4	0.10 M	15	–
	0.25 M	12	–
	0.50 M	6	–
Ammonia (conc.)	25 °C	60	⩾ 300
	50 °C	10	120

[a] Hydrolysis rates of protected aminoalkyl-dThd analogue (*Figure 1*, R_1 = R_2 = hydrogen) were established by TLC and RP-HPLC.

2.1.2 Synthesis of pyrimidine nucleosides with blocked carboxyl linker arms at C-5

An alternative to the incorporation of aminoalkyl linker arms is the incorporation of carboxyalkyl linkers. Since a carboxy linker is more difficult to react selectively in the presence of nucleic acid bases than an amine, care must be taken in attaching reporter groups through a carboxylic acid function. Nonetheless, for some applications, particularly when there is the need to label a single oligomer with two different reporter groups, a carboxylic acid function can be useful. To make a blocked carboxyl linker arm attached to C-5 of dUrd, methyl acrylate is reacted with 5-chloromercuri-2′-deoxyuridine in the presence of palladium to form 5-(methyl 3-acrylyl)-dUrd smoothly and in good yield (7, 8, 29). The 5-(methyl acrylyl)-dUrd can be blocked with DMT at the 5′-OH (*Protocol 1*) and converted to its 3′-phosphoramidite (*Protocol 3*) in a normal manner. After oligomer synthesis, the methyl ester must be removed before ammonia treatment (see section 2.5). Carboxy linker arms in nucleosides or in oligonucleotides can also be produced by reaction of amine linker arms with excess disuccinimidyl suberate (DSS) as described in *Protocol 6*, steps 2–5, or by reaction with glutaric anhydride.

2.2 Pyrimidine nucleosides with linker arms at N^4 of 2′-deoxycytidine

In double-stranded nucleic acids, the exocyclic N^4 of cytosine nucleosides is strongly hydrogen bonded to a complementary guanine base, and does not appear to be sterically available for attachment of reporter groups. N^4-Alkyl-2′-deoxycytidine is also known to be mutagenic, suggesting that proper base pairing during replication is negatively affected by N^4 modification. Several methods are available, however, which allow modification at the exocyclic amine of dCyd, making this an attractive synthetic route.

2.2.1 Synthesis of pyrimidine nucleosides with linker arms at N^4 of 2′-deoxycytidine

Non-isotopic reporter groups have been attached to N^4 of dCyd or 5-methyl-d Cyd (13, 14, 18–20). The three basic approaches for attaching alkyl linkers to N^4 have been:

- *Introduction of a leaving group* such as triazolyl (13, 18) or nitrophenyl (14) at O4 of dThd or dUrd followed by reaction with a diaminoalkane;
- *bisulphite-catalysed transamination* of dCyd with a diaminoalkane (15); or
- *displacement of sulphur* from 4-thiopyrimidines with a diaminoalkane (19).

These reactions with diamines provide a primary amine linker arm attached at N^4 of dCyd which can then be blocked. The nucleoside analogue is protected at its 5′-OH with DMT, and converted to its 3′-O-phosphoramidite in a normal manner. Unlike dCyd in oligomer synthesis, the N^4 positions of the analogues are not further blocked with benzoyl or other protecting groups, making adverse side reactions possible at this site during oligomer synthesis.

The first approach (13) is detailed in *Protocol 2*. The O^4 of dThd or dUrd is first activated by condensation with triazole, then displaced with diaminoethane to initially give a 3-atom amine linker arm. This linker is then extended by reaction with aminocaproate to provide an 11-atom aminoalkyl linker as the final form used in synthesis. *Protocol 2* uses dThd as the initial nucleoside, producing a 5-methyl-N^4-aminoalkyl-dCyd analogue. Substitution of dUrd for dThd should be considered if the increased hybrid stability of 5-methyl-dCyd could be a factor in the use of the product oligonucleotide.

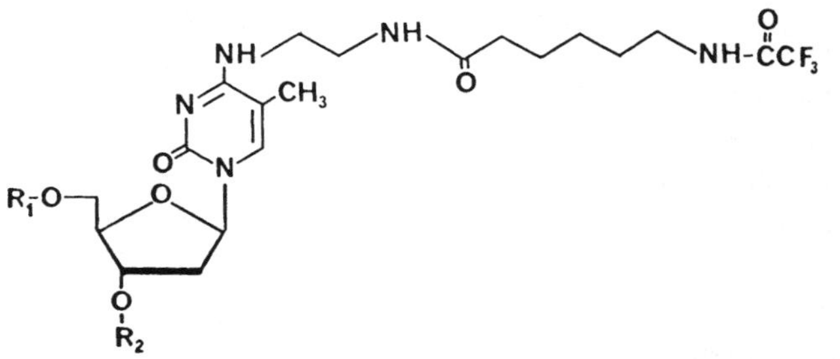

Figure 2. Structure of a blocked N^4-aminoalkyl-dCyd analogue (13) N^4-[N-(6-trifluoroacetylamidocaproyl)-2-aminoethyl]-5-methyl-2′-deoxycytidine (R_1 = H- or DMT; R_2 = -H or N,N-diisopropylphosphoramidite).

Protocol 2. Synthesis of 5′-O-DMT-N^4-aminoalkyl-5-methyl-2′-deoxycytidine-analogues (13)[a]

1. In a 500 ml round-bottomed flask, dissolve 25 g (46 mmol) 5′-O-dimethoxytrityl-dThd in 150 ml acetonitrile. Add 50 ml N,N-dimethylaminotrimethylsilane and stir at ambient temperature for 30 min. Concentrate by rotary evaporation to give 5′-DMT-3′-O-trimethylsilyl-dThd as an oil.
2. In a 2 litre round-bottomed flask, dissolve 51 g 1,2,4-triazole in 300 ml acetonitrile, stir, and add 16 µl trichlorophosphite. Cool in an ice-bath. Add 120 ml triethylamine dropwise with stirring to form a thick slurry. Dilute with 100 ml acetonitrile.
3. Dissolve the crude 5′-DMT-3′-O-trimethylsilyl-dThd oil from step 1 in 100 ml acetonitrile, and add dropwise to the triazole solution from step 2. Stir for 60 min on ice, then allow to warm to ambient temperature for 30 min.

Protocol 2. *Continued*

4. Add 800 ml ethyl acetate and mix. Extract twice with 800 ml 5% aqueous sodium bicarbonate, then with 800 ml 80% saturated aqueous sodium chloride. Dry the organic phase over sodium sulphate, filter, and rotary evaporate to remove solvent. Co-evaporate from toluene then acetonitrile to dry. Crude intermediate TLC R_f 0.65 on UV-silica plates eluting methanol/dichloromethane (1:9).
5. Dissolve the crude 4-triazolo intermediate in 200 ml acetonitrile. Add to 200 ml acetonitrile containing 25 ml 1,2-diaminoethane on ice with stirring. Stir 15 min. Extract and dry over sodium sulphate as in step 4.
6. Dissolve the resulting solid in 200 ml dichloromethane. With stirring add 100 ml of 0.5 M succinimidyl N-trifluoroacetyl-6-aminocaproate (13) in dichloromethane. Stir at ambient temperature for 30 min. Evaporate to dryness, and co-evaporate from 250 ml dry toluene to give a foam.
7. Remove the 3′-O-trimethylsilyl protecting group by dissolving the foam in 400 ml methanol and adding 55 ml 1 M potassium carbonate. Stir 15 min. Add 700 ml ethyl acetate. Extract and dry as in step 4 to give crude 5′-O-DMT-N^4-(trifluoroacetylaminoalkyl)-dCyd.
8. Purify by chromatography on 2 litre silica gel column. Load the crude in 200 ml 0.5% TEA in dichloromethane. Elute column with a linear gradient from 3 litres 2% to 3 litres 6% v/v methanol in dichloromethane/0.5% TEA, collecting fractions of 100 ml. Monitor fractions by TLC on UV-silica plates eluting methanol/dichloromethane (1:9). Combine appropriate fractions (desired compound R_f 0.37, ninhydrin positive) and precipitate fron cold mixed hexanes. Dry by vacuum to constant weight to yield 20–22 g (25–27 mmol) of 5′-DMT-N^4-(trifluoroacetylaminoalkyl)-dCyd (see *Figure 2*). UV_{max} 280 nm (ε 5800); 246nm (ε 4900).

[a] Deoxycytidine analogues lacking the 5-methyl group can also be prepared using this method by starting with 5′DMT-dUrd in place of 5′DMT-dThd (13).

Similar approaches have also been used to synthesize an N^4-hydroxyalkyl-dCyd analogue used to make branched oligomers (18).

2.3 Synthesis of purine nucleosides with linker arms

The strategy for modification of purines is similar to that for pyrimidines. The ideal location for attachment of reporter groups are at sites which have steric tolerance in a double helix, and sites which are not involved in hydrogen bonding. The C-8 position is considered ideal for such attachment in purine nucleosides. An additional consideration is the sensitivity of purines in general to depurination by acid during oligonucleotide synthesis and work-up. Some modifications (such as alkylamino or aminomethyl at C-8 of 2′-deoxyguanosine) increase this tendency dramatically.

Short (10 base) fluorescent oligonucleotides with dansyl attached through C-8 of adenine by an aminoalkylamino linker have been synthesized using phosphotriester chemistry (23). A guanine derivative with anthracene linked through N^2 [N^2-(anthracen-9-methyl)-2′-deoxyguanosine] has been made and incorporated into synthetic oligomers (24) using phosphoramidite chemistry. To date, the broad applications and potentials of such chemistries have not yet been explored.

2.4 Conversion of the blocked nucleosides to 3′-O-ß-cyanoethyl N,N-diisopropylphosphoramidites for oligonucleotide synthesis

To incorporate the linker arm synthons into oligodeoxynucleotides, the blocked nucleosides are converted to reactive monomers containing a 3′-phosphite, phosphate, or phosphonate, depending on the route of oligomer synthesis (for amidite, phosphate triester, or phosphonate chemistries, respectively). Currently, the use of 3′-O-ß-cyanoethyl N,N-diisopropyl-phosphoramidites (CE amidites, or CEA) is the most common synthetic route. *Protocol 3* summarizes the phosphitylation and purification of the phosphoramidite of 5-aminoalkyl-dThd (*Figure 1*). The protocol is quite general, and can be used for most blocked nucleosides. Due to the inherent reactivity/instability of phosphoramidites, only an amount of amidite which can be used in 8–12 months should be made at one time.

Protocol 3. Conversion of blocked nucleosides to 3′-ß-cyanoethyl N,N-diisopropylphosporamidites.

1. In a 100 ml round-bottomed flask, dry 5.0 g (6.2 mmol) 5′-O-DMT-5-trifluoroacetylaminoalkyl-dThd[a] (see *Protocol 1* and *Figure 1*) by dissolving in 25 ml anhydrous pyridine and rotary evaporating to dryness. Re-evaporate from 50 ml anhydrous tetrahydroturan/toluene (3:2) to a foam.
2. Dissolve in 30 ml anhydrous tetrahydrofuran. Add 2.1 ml diisopropyl-ethylamine. Evacuate and flush reaction vessel with dry argon gas. Stir and cool in ice bath.
3. Add 1.5 ml chloro-ß-cyanoethyl N,N-diisopropylphosphoramidite[b]. Stir at 4 °C for 20 min.
4. Assay reaction by TLC on UV-silica plates eluting ethylacetate/cyclohexane (2:1). Desired nucleoside phosphoramidite has R_f 0.30, 0.45 (diastereomers); starting material R_f 0.05–0.15. If reaction is not complete, add additional 0.2 ml aliquots of chlorophosphoramidite.

Protocol 3. *Continued*

5. When the reaction is complete, filter off any insolubles and concentrate the filtrate to a foam. Dissolve in 50 ml ethyl acetate, extract twice with 50 ml water, and dry over sodium sulphate. Filter and concentrate the filtrate to a solid.
6. Purify by chromatography on a 3 × 30 cm silica gel column:
   (a) Pack the column in 2:3 ethyl acetate/cyclohexane (EtOAc/CH) containing 0.2% triethylamine.
   (b) Dissolve the solid from step 5 in minimum 2:3 EtOAc/CH, and load on to column.
   (c) Elute a linear gradient of 500 ml 2:3 EtOAc/CH containing 0.2% (v/v) triethylamine to 500 ml 9:1 EtOAc/CH, collecting 20 ml fractions. Assay by TLC (see step 4).
   (d) Combine appropriate fractions.
   (e) Evaporate to dryness under argon.
7. Dissolve product in 20 ml anhydrous benzene. Microfilter through 0.2 μm nylon. Lyophilize to a white solid. Yield will be about 4.5 g (4.5 mmol) of 5′-O-dimethoxytrityl-5-[N-(6-trifluoroacetylaminohexyl)-3-(E)acryl-amido]-2′-dUrd 3′-β-cyanoethyl N,N-diisopropylphosphoramidite (*Figure 1*), mol. wt 995, as a white powder.

[a] Other 5′-O-DMT blocked nucleosides may be used; chromatographic behaviour (R_f and solvent mixes for elution from silica) will vary accordingly.

[b] The methyl phosphoramidites may be prepared using the same procedure by adding 1.15 ml chloromethyl-N,N-diisopropylphosphoramidite in place of the β-cyanoethyl amidite.

2.5 Oligodeoxynucleotide synthesis using linker arm nucleoside amidites

All of the linker arm analogues discussed above can be used directly in automated DNA synthesizers with no modifications to the normal procedures (see Chapter 1).

Preparation

Before oligomer synthesis, the linker arm amidites are weighed directly into dry amidite bottles for the synthesizer (usually 25–150 mg, depending on the molecular weight of the analogue and the number of additions desired). The amidites are then dissolved in 0.5–1 ml anhydrous benzene, and thoroughly dried by overnight lyophilization. When dry and properly sealed, the amidites are generally stable for 1–2 years desiccated at −20 °C. [Note: linker arm phosphoramidites illustrated in *Figures 1* and *2* (R_1 = DMT; R_2 = β-cyanoethyl N,N-diisopropylphosphoramidite) are available commercially from suppliers of oligonucleotide synthesis reagents such as Glen Research and American Bionetics and can be used directly.]

Oligonucleotide synthesis

For oligomer synthesis, the amidites are warmed to ambient temperature and dissolved to a final concentration of 100 mM (about 90–100 mg/ml) in anhydrous acetonitrile, then put directly onto the amidite port of the synthesizer. The sequence is programmed as usual, inserting the analogue into the desired positions, and the oligomer synthesis completed with no changes in program. Any standard ending program (**DMT-on**, **DMT-off**, **Auto cleave**, or **Manual cleave**) may be used, except when using carboxy ester linker arms (see below).

Cleavage from the CPG support and deblocking of the linker arm

When incorporating trifluoroacetyl-blocked amine linker arms, standard ammonia cleavage and base deblocking by heating in concentrated ammonia will fully deprotect the amine by removing the trifluoroacetyl group as well; see *Table 1*.

When incorporating carboxy linker arms blocked as methyl esters (section 2.1.2), the standard cleavage and deprotection must be modified since the methyl ester must be removed before the ammonia cleavage. The methyl ester is hydrolysed by treating the support-bound oligomer with triethylamine/methanol/water (1:1:8 v/v) overnight at 50–55 °C. This treatment will hydrolyse the oligomer from the support, remove β-cyanoethyl protecting groups from the phosphates, and hydrolyse the methyl ester to leave a 3-atom carboxylic acid linker arm. After evaporation of the solvent, the remaining base-protecting groups can be removed by an additional heating for 6–12 h at 50–55 °C in concentrated ammonia.

Purification and storage of linker arm oligodeoxynucleotides

The crude linker arm oligomers may be purified by any standard method, including electrophoretic gels, reversed phase HPLC, or ion exchange HPLC. The size and charge of the linker arm will have a significant effect on mobility and retention time, particularly for short oligomers (10–30 bases). Generally, each amine linker will slow mobility in electrophoretic gels equivalent to 1–1.5 base units.

After purification, *all interfering components must be removed from the oligomer solution before attempting to label with the reporter group.* Amine-modified oligomers must be separated from common amine buffers such as Tris or ammonium salts. This separation is easily accomplished by spun G-25 columns or ethanol precipitation. For convenience, storage of amine-oligomers at 1 mM (1 mmol/µl) concentration in 2 mM EDTA, pH 7–9, allows direct labelling. Assuming 37 µg per A_{260} unit, concentration of the purified oliogmer can be calculated from the absorbance at 260 nm [concentration (in mM) = 0.114 $A_{260}n^{-1}$, where n is the number of bases in the oligomer]. The amine oligomers can be stored for several years at −20 or −70 °C before labelling. In a similar manner, carboxy-modified oligomers must be separated from common organic acid components such as EDTA or acetate salts before labelling.

3. Attaching reporter groups to linker arm oligomers

The linker arms are reactive functionalities (such as primary amines) which do not normally exist in DNA. As a result, any reporter groups which contains groups reactive towards nucleophilic amines can be used to label the modified oligomers selectively. Such amine-reactive groups are succinimidyl (NHS) esters of carboxylic acids, isothiocyanates (ITC), sulphonyl chloride, and others. The best amine-reactive functions to use are the NHS esters, or ITC derivatives, both of which react cleanly and nearly quantitatively without apparent side reactions at other sites on the DNA. Acid anhydrides will also react cleanly. More reactive functions such as sulphonyl chlorides will react with unmodified sites on the oligomer and give poorer yields. Because of this, there may be some advantages in reacting active sulphonyl chlorides such as Texas Red with amino caproic acid, followed by activation with NHS to form a milder and more selective NHS ester.

3.1 Attaching biotin to amine linker arm oligonucleotides

For many applications biotin is a useful reporter group. It is especially preferred for removing the labelled oligomers or hybrids from solution with avidin or streptavidin. NHS esters of biotin or its aminocaproyl derivatives which are commercially available from several companies react with amine oligomers rapidly and quantitatively, and the biotinylated products are easily purified by reversed phase HPLC. *Protocol 4* describes a general method of biotinylation and purification. Chelators (8, 16) and structural labels such as azidonitrobenzoate (11) may be coupled by similar methods. Biotin labelling by incorporation of a protected biotinylated linker arm phosphoramidite (20) will also provide biotin-labelled oligomers.

Protocol 4. Labelling of amine-modified oligodeoxynucleotides with biotin

1. Dissolve 50 nmol amine-oligomer[a] in 50 μl 200 mM sodium bicarbonate pH 9–10, in a 1.5 ml Eppendorf tube.
2. Dissolve succinimidyl esters of biotin (NHS-biotin) or biotinamidocaproic acid (NHS-X-biotin) in dimethyl sulphoxide to a final concentration of approximately 100 mM (3.5 or 4.5 mg in 100 μl DMSO).

Protocol 4. *Continued*

3. Add 50 μl biotin solution to the amino-oligomer and mix. Incubate for 30–60 min at ambient temperature.
4. To remove the excess biotinylating agent, add μl 250 mM sodium acetate, pH 6.5–7.0 and mix. Precipitate by adding 900 μl ethanol, mixing, and cooling at −70 °C (−20 °C is acceptable) for ≥ 10 min. Pellet the white oligomer precipitate, remove the supernatant. Dry the oligomer pellet briefly under vacuum. Redissolve in 50–100 μl water. [Note: for many applications, the biotin oligomer can be used directly at this point; extent of biotinylation can be determined by 20% PAGE using Stains-all development.]
5. Analyse by reversed phase HPLC on C_8 columns (0.4 × 15 cm) eluting a linear gradient of 23–35% (v/v) acetonitrile in 100 mM triethylammonium acetate, pH 6–7, over 20 min at 1 ml/min. Detect at 254 or 260 nm. This system is used for oligomers of 15–40 bases in length with 1–3 biotins attached; unlabelled probe will elute near the void volume. A single major peak should be observed at 5–10 min for an oligomer with a single biotin label. Maximum column capacity is 200–300 nmols of labelled oligonucleotide.
6. Purify by preparative reversed phase HPLC, combine appropriate fractions, determine total absorbance at 260 nm, and concentrate to dryness. Dissolve in 400 μl 200 mM sodium acetate, pH 6–7, mix, and precipiate by addition of 1.2 ml ethanol. Cool, pellet the oligomer, remove the supernatant (A_{260} should be ≤0.3), and briefly dry the pellet under vacuum. Redissolve in desired storage butter such as 10 mM Tris–HCl, 1 mM EDTA.
7. Determine product purity by electrophorsis on 20% acrylamide/8 M area gel, loading 50–100 pmol per lane (equivalent to 2 μl of A_{260} = 7 of 20–30mer or about 500 ng of DNA). Visualize by Stains-all development. Oligomer with one biotin will electrophoresis slower that unlabelled oligomer by the equivalent of about one base unit. Additional biotins further slow migration.
8. Store the biotinylated oligomer at −20 or −70 °C.

3.2 Attaching fluorophores and lumiphores to amine linker arm oligomers

The strategy for attaching fluorescent or luminescent groups to amine oligomers is similar to biotinylation in *Protocol 4*. However, the separation of unreacted label and the work-up of the labelled oligomer must be altered to account for the differing solubilities of the fluorophores and of the labelled oligomer. Oligonucleotides of 20–30 units in length with only one group such as FITC or rhodamine attached will still dissolve in aqueous solution easily, and will ethanol precipitate normally. However, if two or more fluorophores are attached, the increase in overall hydrophobicity makes ethanol precipitation very inefficient. This will vary greatly depending on the character of the label itself. *Protocol 5* details the attachment of fluoroscein to amine oligomers using FITC. Other amine-reactive fluorophores or lumiphores such as Texas Red (7), succidimidyl 1-pyrenebutyrate (15), pyrenesulphonyl chloride (15), rhodamine isothiocyanates (7), NHS-esters of acridines or isoluminols (7, 13), or other labels with isothiocyanate, sulphonyl chloride, N-hydroxysuccinimide, or anhydride structures may be attached in a manner similar to FITC.

Protocol 5. Labelling of amine-modified oligodeoxynucleotides with fluorescent labels

1. Dissolve 50 nmol amine oligomer[a] in 50 μl 0.5–1 M sodium carbonate, pH 9–10.
2. Prepare a 50 mM solution of fluorescein-5-isothiocyanate (FITC; from Molecular Probes, catalogue no. F-1906) by dissolving 2 mg (5 mmol) FITC in 100 μl DMF.
3. Add 100 μl of the 50 mM FITC in DMF to the 50 μl oligomer solution. Mix well and let stand at ambient temperature for $\geqslant$ 4 h. (Allow sulphonyl choride-activated fluorescers such as Texas Red to react for only ~ 1 h). Quench unreacted fluorophore by addition of 10 μl concentrated ammonia. Let stand for 30 min. min.
4. Add 100 μl 5 mM sodium acetate, pH 6–7, to the crude reaction.
5. Remove excess unreacted fluorescer by using 0.44 × 5 cm Sephadex G-25 spun columns (Isolab, catalogue no. QS-2B):
   (a) Equilibrate column with 3 × 250 μl 5 mM sodium acetate, pH 6–7, by spinning for 5 min at 600–800 *g* (clinical centrifuge).
   (b) Pipette the 250 μl crude reaction onto the column, and spin for 5 min at 600–800 *g*, collecting the eluate in a 1.5 ml Eppendorf tube.
   (c) Wash the column with 2 × 150 μl 5 mM sodium acetate, spinning each for 5 min at 600–800 *g*.
   (d) Determine the approximate probe concentration by absorbance at 260 nm.
6. Analyse by reversed phase HPLC (see *Protocol 4*, step 5). Detect at 260 nm and at the maximum absobance wavelength of the fluorophore (495 nm for FITC), if possible. Oligomers of 20–40 bases in length with one

Protocol 5. *Continued*

FITC will elute at 10–14 min; oligomers labelled with one Texas Red will elute at 14–16 min.

7. Purify by reversed phase HPLC and ethanol precipitate (see *Protocol 4*, step 6), if possible. (Oligomers with three or more fluorophores will precipitate poorly.) Determine final concentration by absorbance at 260 nm. The contribution of the fluorophore at 260 nm, if any, must be considered; for example, FITC has $\varepsilon_{260} \approx 24\ 000$ at pH $\geqslant 7.5$.
8. Determine purity by gel electrophoresis (see *Protocol 4*, step 8) and UV-VIS scanning.

[a] Before attempting to attach reporter groups, remove all sources of ammonium ions or primary and secondary amines (such as Tris) by gel filtration or ethanol precipitation.

3.3 Conjugation of enzymes to linker arm oligodeoxynucleotides

Methods for detecting biotin- or hapten-labelled probes almost invariably use enzyme activity to detect the label, by conjugation of alkaline phosphatase (AP) or horseradish peroxidase (HRP) to avidin or the corresponding antibody. Such methods require significant time and effort for post-hybridization detection. Attaching enzymes directly to cloned probes would be useful (30). However, it has proven difficult to attach enzymes to cloned DNA fragments directly in a reproducible and stable manner. Cross-linking enzymes to cloned DNA (30) results in each enzyme being modified more than once, and at more than one site on the DNA. This 'over-modification' coupled with the inability of most enzymes to survive stringent hybridization conditions (high temperatures, organic solvents, extended hybridization times) results in poor signal and often high backgrounds.

The attachment of enzymes to synthetic oligonucleotides, however, overcomes many of these problems (9). Due to their well-defined nature, oligomer probes can be conjugated cleanly in a 1:1 ratio with the enzyme, thus avoiding excess modification of the enzyme or probe. Hybridizations of oligonucleotides can be done rapidly, at more moderate temperatures, and in simple buffers, preserving enzyme activity and therefore maximizing signal. Surprisingly, the conjugation of enzymes to oligonucleotides has little affect on the behaviour of the oligonucleotide itself (9, 13, 35; see Section 4).

The approach to synthesis of alkaline phosphatase–oligodeoxynucleotide conjugates in *Protocol 6* (9) is the following:

(a) a single internal amine linker arm base is incorporated into the desired oligodeoxynucleotide sequence, and the oligomer is purified away from competing ammonium or amine salts;

(b) the alkaline phosphatase (AP) is attached through the cross-linker; and linkers;

(c) the alkaline phosphatase (AP) is attached through the cross-linker; and

(d) the conjugate is purified by anion exchange chromatography.

The presumed site of attachment to the enzyme is lysine residues. Conjugates prepared by this approach are 1:1 oligomer:enzyme conjugates, with only one site modified on the enzyme. Calf intestinal AP works significantly better as a conjugate than bacterial AP. Since the product conjugates are more highly charged than free enzyme and much larger than free oligonucleotide, the conjugates can easily be purified to homogeneity (although the presence of free enzyme will not generally affect the use of the conjugate). Variations of this approach have been tried which activate the enzyme rather than the oligomer, but this results in loss of enzyme activity and correspondingly lower signal (31).

Protocol 6. Conjugation of alkaline phosphatase to amine-modified oligodeoxynucleotides[a]

Solutions used in the procedure:

- Alkaline phosphatase (AP) reaction buffer: sterile 3 M NaCl, 100 mM $NaHCO_3$, 1 mM $MgCl_2$, 0.05% azide, pH 8.2–8.3.
- G-25 chromatography buffer: 1 mM NaOAc, adjusted to pH 4 with HOAc, filter sterilized.
- P-100 chromatography buffer: 100 mM NaCl, 50 mM Tris–HCl, pH 8.5, filter sterilized.
- Anion exchange buffers: (for Mono Q columns; others will vary somewhat)
  A: 20 mM Tris–HCl, pH 8
  B: 20 mM Tris–HCl, 1 M NaCl, pH 8. Filter before use.
- High salt buffer (HSB): (for storage of alkaline phosphatase conjugates) Sterile 3 M NaCl, 30 mM Tris, 1 mM $MgCl_2$, 0.1 mM $ZnCl_2$, 0.05% azide, pH 7.6.

Conjugation to alkaline phosphatase: (50 nmol oligonucleotide scale) (9)

1. Prepare calf intestinal alkaline phosphatase (AP; mol. wt 141 000; Boehringer Mannheim catalogue no. 556602, at 10 mg/ml):
   (a) Put 20 mg AP directly into a Centricon-30 (Amicon) with 1 ml AP reaction buffer and concentrate by centrifugation at 5000 *g*, maximum.
   (b) Wash by refilling the Centricon with reaction buffer and reconcentrating to 100–200 mg/ml (0.71–1.4 mM).

Protocol 6. *Continued*

(c) Measure the enzyme's concentration by conventional Bradford assay using a standard curve. Determine specific activity of the enzyme in duplicate using *p*-nitrophenyl phosphate (PNPP) as substrate.

(d) Store the enzyme in a sterile tube at 4 °C until use.

2. Prepare a gel filtration column of Sephadex G-25:

(a) Pre-swell and wash 5 g Sephadex G-25 in degassed 100 mM NaOAc, pH 4.

(b) Pack a 0.7 × 45 cm column of G-25. Equilibrate the column at 4 °C with 20–30 ml 1 mM NaOAc, pH 4[b].

3. Prepare a fresh 10 mg/ml solution of DSS cross-linker (di-succinimidyl suberate; Pierce catalogue no. 21555) in dry DMSO. (Di-isothiocyanate cross-linkers such as DITC can also be used (13).)

4. Activate the oligonucleotide[a] by:

(a) Mixing 50 μl 1 mM oligonucleotide solution (50 nmol of oligonucleotide) into a clean Eppendorf tube.

(b) Add 15 μl fresh 1 M $NaHCO_3$, pH 8.5–9.

(c) Add 50 μl of the 10 mg/ml DSS solution to the oligomer. Pipette the solution up and down in the pipette to mix well. **Immediately** go on to step 5.

5. **Immediately** (0.5–2 min) after mixing DSS with the oligomer, separate activated oligomer from excess DSS by:

(a) loading the reaction onto the cold G-25 column from step 2.

(b) Wash on and elute with cold (4 °C) 1 mM NaOAc, pH 4.[b]

(c) Collect fractions of 1–2 ml/fraction. Monitor fractions by UV absorbance (260 nm wavelength is best). The first peak eluting will be the DSS-activated oligonucleotide, which will elute at approximately 8–11 ml.

(d) **Immediately** combine appropriate fractions and go on to step 6.

6. Concentrate the combined fractions containing the activated oligomer to 50–100 μl volume using a Centricon-10 concentrator (Amicon) in a refrigerated centrifuge at 5000 *g*. (This will take 1–1.5 h). Keep cold and **immediately** go on to step 7.

7. React the activated oligomer with alkaline phosphatase:

(a) Add a 2–3 fold molar excess of concentrated (100–200 mg/ml) AP from step 1 to the tube containing the concentrated activated oligomer from Step 6 [i.e., use 14–20 mg AP per 50 nmol, or 150–200 μl of 100 mg/ml AP (100–140 nmol)].

Protocol 6. *Continued*

(b) Mix well by shaking stirring with the pipette tip. Do *not* vortex the mix.

(c) Allow the reaction to sit in the dark for 4–18 h.

8. Remove unconjugated oligomer:

(a) Purify the reaction mix from step 7 on a 1 × 50 cm P-100 column eluting with 100 mM NaCl, 50 mM Tris–HCl, pH 8.5 buffer, at 4 °C.

(b) Monitor by UV absorbance (260 nm). Collect fractions of 50 drops for about 30 fractions. The first UV-absorbing peak is a mixture of oligonucleotide conjugate and free enzyme, and should have an $A_{260/280}$ of 0.8–1.2. Unconjugated oligomer will be in later fractions, with an $A_{260/280}$ ratio of about 1.5–1.7.

(c) Plot absorbance and combine appropriate AP-containing fractions, keeping the product cold.

9. Concentrate the conjugate/enzyme mix to 60–100 μl using a Centricon 30 (Amicon) microconcentrator, refilling Centricon twice with 20 mM Tris–HCl, pH 8, and centrifuging in a refrigerated centrifuge.

10. Purify the desired oligonucleotide–alkaline phosphatase conjugate from free enzyme by anion exchange chromatography[c] on a Pharmacia Mono Q-10 (0.5 × 5 cm) FPLC column:

(a) Load the cold conjugate solution on the column, washing on with 2 × 100 μl 20 mM Tris–HCl, pH 8.

(b) Elute a linear gradient of 0–100% B (0–1 M NaCl in 20 mM Tris–HCl, pH 8) over 60 min at 1 ml/min[c].

(c) Pool the desired fractions and keep on ice.

11. Concentrate the product on a Centicon-30 microconcentrator to ~ 50 μl volume.

12. Dilute the product with sterile High Salt Buffer (HSB) to 500 μl final volume. Measure protein concentration by Coumassie to determine concentration of the conjugate (generally between 1–10 μM).

13. Measure enzyme activity; the specific activity should be within 10% of original, but can be lower if the enzyme has been mishandled.

14. Dilute to desired final volume in HSB (for long-term stability, final should be no more dilute than 0.5 μM, ideally 5–10 μM. (A 5 μM solution is 0.7 mg/ml AP).

15. Determine conjugate purity by electrophoresis on non-denaturing 7.5% polyacrylamide gels. Detect bands by staining for enzyme activity using standard NBT/BCIP reagents for alkaline phosphatase; this allows strong bands with as little as 10 fmol (equivalent to 1.4 ng) enzyme.

Protocol 6. *Continued*

16. Store the conjugate in sterile tubes at 2–8 °C. Expected useful life should be 1.5–3 years if handled carefully.

[a] Before attempting to attach enzymes, remove all sources of ammonium ions or primary and secondary amines (such as Tris) by gel filtration or ethanol precipitation.
[b] The low pH and low temperature are necessary to maximize yields by minimizing the hydrolysis of the activated oligonucleotide. The intermediate is an NHS-activated carboxylic acid ester which hydrolyses rapidly ($t_{1/2}$ of a few minutes) in water at alkaline pH. This intermediate must be kept intact until step 7.
[c] The free enzyme is only slightly negatively charged and elutes at about 0.15 M NaCl, with an $A_{260/280}$ ratio of 0.5–0.9. The conjugate is much more negatively charged, and elutes at about 0.45 M NaCl (usually over 3–5 fractions between 15–30 mins); the desired conjugate has an $A_{260/280}$ ratio of 1.0–1.3, depending on oligomer length and composition. Any residual unconjugated oligonucleotide ($A_{260/280}$ about 1.5–1.7) will elute after conjugate, near the gradient end.

Other enzymes can be conjugated to amine linker arm oligomers using similar methods, including HRP (see *Protocol 7*), β-galactosidase, luciferase, oxidase, and others. HRP conjugates can also be used successfully as probes (9, 13), although HRP in all applications tested is less sensitive than AP; see *Table 2*. For use as hybridization probes, however, most other enzymes are wholly or partially inactivated by the minimum temperatures, salts, and/or detergents necessary even for hybridization of oligomer probes. As a result, enzymes promising maximum signal (such as β-galactosidase) have not been successfully used as probe conjugates. An overview of synthesis and properties of modified oligodeoxynucleotides has been recently published (33).

Protocol 7. Conjugation of amine-modified oligonucleotides to horseradish peroxidase (HRP)

Follow *Protocol 6* with the following modifications:

1. Step 1. Use EIA grade HRP (Boehringer Mannheim); use 4 mg HRP (mol. wt ≈ 40 000) per 50 nmol oligonucleotide; concentrate HRP in 0.1 M $NaHCO_3$ in step 1, in place of the AP reaction buffer. HRP activity does not need to be assayed before reaction;

2. Step 11. Use 100 mM potassium phosphate, pH 6.3, 0.05% azide, as storage buffer in place of HSB.

---

4. Use and behaviour of oligodeoxynucleotides with non-isotopic reporter groups

4.1 Hybridization of oligomer probes

Oligodeoxynucleotide probes can be hybridized to complementary targets in a number of formats such as *in situ*, in solution, or to target immobilized on membranes (Southern blot, Northern blot, dot blot). Hybridization and detection of non-isotopic oligonucleotide probes will vary somewhat depending on format, reporter group, and modality of detection method. The most common hybridization is to target immobilized on membranes. Regardless of reporter group, hybridization to membranes is very similar for most oligomer probes, and only detection steps may vary significantly. A general protocol for hybridization of oligomer probes (including ^{32}P-labelled) is outlined in *Protocol 8*.

Protocol 8. Hybridization of labelled oligonucleotides to complementary sequences on membranes

This general protocol is useful for Southern blots, Northern blots, and dot blots. It has been optimized for alkaline phosphatase-labelled oligomers of 20–30 bases in length using nylon membranes. The protocol can be applied to probes labelled with biotin, haptens, fluorescers, etc. by varying the detection method after hybridization and washing.

Solutions needed: (1 × SSC = 150 mM sodium chloride, 15 mM sodium citrate, pH 7.0)

- Hybridization solution: 5 × SSC, 1% (w/v) SDS, 0.5 w/v% BSA
- Wash 1 solution: 1 × SSC, 1% (w/v) SDS
- Wash 2 solution: 1 × SSC, 1% (w/v) Triton X-100
- Wash 3 solution: 1 × SSC

1. Wet the membrane with hybridization solution[a] (50–100 μl per cm^2 membrane) for 10–15 min at hybridization temperature[b] in a sealable bag or tube using a water bath as a heat source.
2. Remove the solution. Mix the probe to a final concentration of 1–5 nM (1 nM ≈ 7 ng per ml) in fresh hybridization solution and add 50–100 μl per cm^2 to the membrane. Incubate for 30 min[c] at hybridization temperature[b] with agitation.
3. Transfer the membrane to a wash tray or bottle containing 1.5 ml preheated Wash 1 Solution per cm^2 of membrane. Agitate at wash temperature[b] for 10 min.

Protocol 8. *Continued*

4. Remove the Wash 1 solution, and add Wash 2 solution at 1.5 ml per cm^2 membrane. Agitate at wash temperature[b] for 10 min.
5. Remove the Wash 2 solution, and add Wash 3 solution at 1.5 ml per cm^2 membrane. Agitate at ambient temperature[b] for 10 min.
6. Remove Wash 3, and detect by appropriate methods.

[a] Some detection methods require additional blocking of the membrane before hybridization. For example, if alkaline phosphatase-labelled oligomers are to be detected with luminescent substrates such as AMPPD (Tropix) or Lumiphos (Lumogen), the wetting step should include 0.1–0.2% (w/v) casein to prevent background due to the substrate itself.

[b] For most oligomers of 18–30 bases in length, appropriate hybridization temperatures will be 50–60 °C using this protocol, and optimum wash temperatures will be 10 °C lower.

[c] Hybridizations with 5 nM probe for 10 minutes or 1 nM probe for 1 h are ≥ 90% complete; see *Figure 5*. Alkaline phosphatase-labelled probes should not be hybridized for longer than an hour at 60 °C or three hours at 50 °C due to inactivation of the enzyme.

4.2 Effect of modification and reporter group on hybridization

The choice of reporter group and type of oligomer modification depend on the intended use of the oligonucleotide. For most applications, the two most important factors are:

- *The sensitivity of the signal* after hybridization. Better sensitivity allows wider application.
- *Any effect of the oligomer modification or reporter group on hybridization behaviour.* The probe modification should not adversely affect hybridization strength or selectivity.

For general use as hybridization probes, the intent is not to improve behaviour compared with that of an unmodified probe, but to have as little or no effect on its behaviour as possible. Due to the highly ordered nature of nucleic acid structure, helix formation, and resulting hydrogen bonding, any modification made randomly is more likely to have an adverse effect than no effect at all. This is particularly true if direct chemical modifications are made on intact oligonucleotides.

The effect of most oligomer modifications and reporter groups has not been studied by physiochemical methods in detail. Reaction of an unmodified 9-base oligomer with N-acetoxy-2-acetylaminofluorence at a single site reduces the T_m of the 9mer hybrid from 53 ± 3 °C to 35 ± 3 °C, and also decreases the observed hypochromicity (5). Labelled oligomers containing a single internal N^4-dCyd linker arm (see *Figure 2* and *Protocol 2*) show anomalous two-phase melting behaviour, and lack a single helix-to-coil hypochromic transition (15). Anthracene attached to N^2 of an internal guanine base in a

13mer decreases the co-operative helix-to-coil transition from 56.6 °C to 50.6 °C for the unmodified oligomer (24). These type of effects will limit some applications of non-isotopic probes.

Table 2. Comparison of sensitivities for reporter groups attached to oligonucleotides

Reporter	Detection method	Amount of target detected		Relative signal
		moles	molecules	
Acridine*	luminescene	8×10^{-17}	50 000 000	1
FITC	fluorescence	5×10^{-17}	30 000 000	2
HRP	colour (DAB)	1×10^{-17}	8 000 000	6
Biotin	SA-AP; NBT/BCIP	1×10^{-17}	6 000 000	8
5′-^{32}P*	autoradiography (24 h)	6×10^{-18}	3 000 000	17
AP	colour (NBT/BCIP)	2×10^{-18}	1 000 000	50
AP	fluorescence (MUBP)	5×10^{-19}	300 000	166
AP	luminescence (AMPPD)	1×10^{-19}	60 000	833

Amount of target detected considered to be accurate to ± 2-fold. All non-isotopic labels were attached to the same sequence oligonucleotide 22mer (except*) through a single linker arm dThd analogue positioned internally. Hybridization was to plasmid target immobilized on membranes. SA = streptavidin; AP = alkaline phosphatase; HRP = horseradish peroxidase; DAB = diaminobenzidine; NBT = nitro blue tetrazolium; BCIP = 5-bromo-4-chloro-3-indolyl phosphate; MUBP = 4-methylumbelliferone phosphate; AMPPD = '3-adamantyl 4-methoxy 4-(2-phospho)phenyl 1,2-dioxetane' (systematic name is disodium 3-[4-methoxyspiro(1,2-dioxetane-3,2′-tricyclo{3.3.1.1$^{3.7}$}decan)-4-yl]phenyl phosphate; AMPPD is from Tropix Inc.; also called 'lumiphos' from Lumigen, Inc; FITC = fluorescein isothiocyanate.

In contrast, modification of oligonucleotides by attachment of reporter groups at C5 of pyrimidines does not appear to have a measurable effect on hybridization (7–12, 15–17, 34), although short (20–30 base) oligomers with more than three linker arms have not been studied at length. To compare the effect of internal C5 pyrimidine modification and labelling in oligonucleotides, amine linker arm 22mers of the same sequence were synthesized (5′-dCCCGAGCCGATGACTT*ACTGGC and 5′-dCCCGAGCCGAT*GACTTACT*GGC, where T* indicates the sites of the 5-aminoalkyl-dThd analogue; see *Figure 1*. The unmodified 22mer was also synthesized. Standard comparison of melting curves (see *Figure 3*) indicated no significant effect on T_m with either one or two T* analogues. The amine liner arms were labelled with either biotin (see *Protocol 4*) or FITC (see *Protocol 5*). After 5′-^{32}P-labelling, hybridization of the unmodified, linker arm and biotin- or FITC-labelled oligomers against known amounts of plasmid target immobilized on membranes indicated no apparent effect on hybridization signal or background after 24 h of autoradiography (see *Figure 4*). The 22mer containing one T* was conjugated to alkaline phosphatase and purified as in *Protocol 6*. Subsequent hybridization at known probe concentrations to membrane-bound target (see *Figure 5*) indicated that even for oligonucleotides

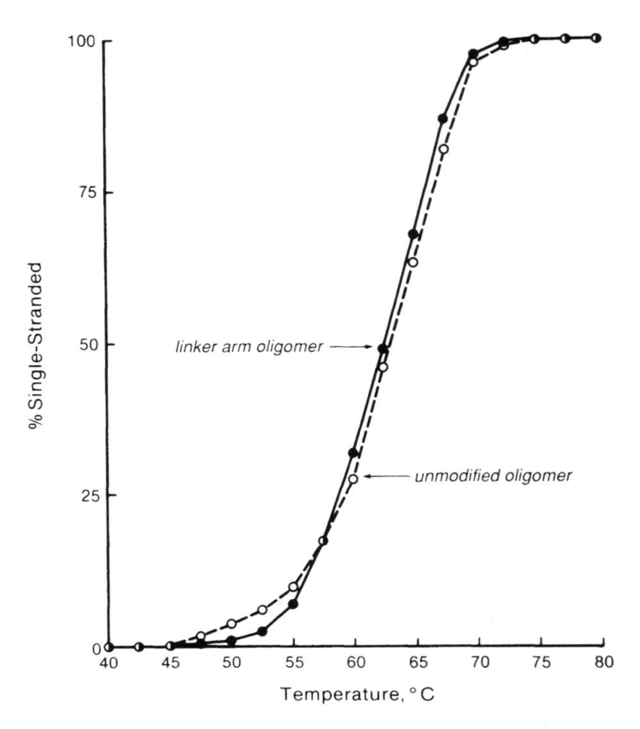

Figure 3. Melting curve of 5′-^{32}P-labelled 22mers. Sequence dCCCGAGCCCGATGACTT* ACTGGC where T* = is either thymidine (○ = unmodified oligomer) or aminoalkyl-dThd (Figure 1) (● = linker arm oligomer). 5′-^{32}P-kinased oligomers were stringently hybridized in 5 × SSC, 1% SDS, to plasmid target on membrane dots, then removed by successive 5 min washes 1 × SSC, 1% SDS, at increasing temperature.

conjugated to relatively large reporter groups (AP mol. wt ≈ 141 000), hybridization kinetics were about the same as unmodified oligomers in solution, with a rate constant $k \approx 4 \times 10^5$ and mol^{-1} s^{-1} (35). The efficiency of hybridization of the AP-labelled or unmodified 22mer probe to either plasmids in solution or fixed on membranes was found to be ⩾ 85 ± 5% of the input target (35). Subsequent results with more than 100 oligonucleotide probes have supported these initial results. This strongly suggests that modification and attachment of reporter groups at C-5 has very little impact on the hybridization strength, selectivity, kinetics, or efficiency of hybridization.

Choice of the reporter group is highly dependent on application. However, for standard hybridizations to targets immobilized on membranes, there is a wide variation of label sensitivities (see *Table 2*), with over 800-fold difference between the least sensitive and most sensitive label. In general, fluorescent or chemiluminescent labels require small volumes and sophistic-

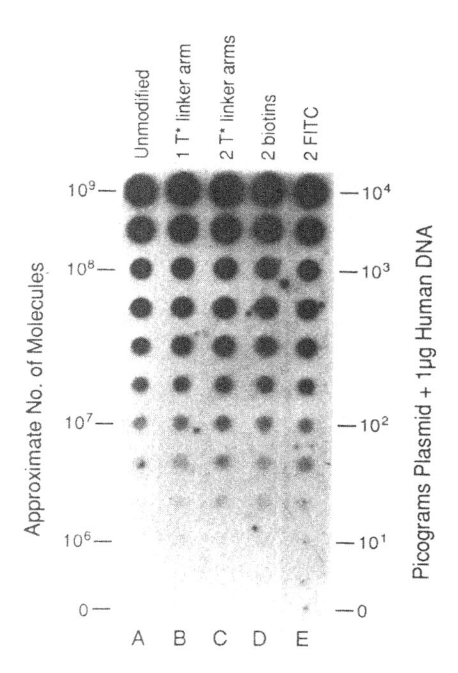

Figure 4. Effect of C-5 linker arms and reporter groups on hybridization. Oligonucleotides 22 bases in length (as in Figure 3) were constructed with either no modified bases, or with one or two aminoalkyl-dThd bases (Figure 1, Figure 3). The oligonucleotides containing two amino-dThd analogues were labelled with either FITC or biotin. The oligomers were labelled with ^{32}P by 5′-end labelling, and hybridized to plasmid target on identical membrane strips. After hybridization and washes (see *Protocol 8*), results were obtained by autoradiography for 18 h at −70 °C with two Q3 screens. No affect of the modifications on sensitivity or specificity were observed.

ated instrumentation for high sensitivity, and therefore tend to be least sensitive in normal use. Of all the reporter groups tested, only alkaline phosphatase is more sensitive than 24 h autoradiography with ^{32}P. The most sensitive reporter group known for oligodeoxnucleotides is alkaline phosphatase using a luminescent dioxetane derivative (AMPPD from Tropix, Inc.), with sensitivies almost 50-fold better than ^{32}P. In applications such as Southern blots, this allows the hybridization and detection of specific bands to be done in one working day, compared with 5–7 days needed to use a ^{32}P-cloned DNA probe for detection of the same bands.

5. Concluding remarks

This chapter has summarized some of the methods which have been developed for attachment of non-isotopic reporter groups through 'linker

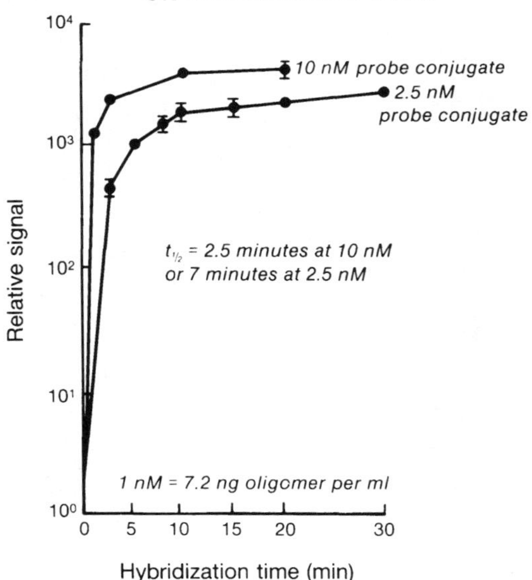

Figure 5. Hybridization kinetics of alkaline phosphatase-labelled oligodeoxynucleotides. AP-labelled oligomer probes were hybridized at either 2.5 nM or 10 nM concentrations to plasmid target immobilized on identical nylon membrane dots in triplicate. At the time points indicated, the dots were removed and washed (see *Protocol 8*). Signal was developed by incubation in buffer containing 4-methylumbelliferone phosphate (MUBP) as a substrate for AP. Signal was quantified by fluorescence emission at 447 nm.

arms' to the bases of single-stranded synthetic oligodeoxynucleotides. Protocols for the synthesis of some particularly useful linker arm nucleosides were presented. Although the syntheses and purifications of the blocked linker arm nucleosides are not trivial, a single synthesis can provide enough linker arm nucleoside monomer to last for several years, even if used daily. As a further convenience, the linker arm nucleoside phosphoramidites illustrated in *Figures 1* and *2* of this chapter are available commercially from suppliers of oligonucleotide synthesis reagents (Glen Research, American Bionetics, and others). Incorporation of the linker arm monomers into oligodeoxynucleotides using standard phosphoramidite chemistry and an automated DNA synthesizer is straightforward. The attachment of non-isotopic reporter groups to the linker arm oligomers is chemically simple, and a large variety of reporter groups can be attached using similar protocols. The methods are flexible, and several small- and large-scale purification techniques are known.

The use of oligomers labelled with non-isotopic reporter groups in hybridizations is relatively simple, and few changes are required due to label. For most work, attachment of linker arms and reporter groups to C-5 of pyrimidines may have the least effect on hybridization. Sensitivities of some non-isotopic reporter groups have now equalled or bettered the sensitivity of ^{32}P label. These advances will greatly increase the use of non-isotopic oligonucleotides as probes in the next few years.

References

1. Murasugi, A. and Wallace, R. B. (1984). *DNA*, **3**, 269.
2. Trainor, G. L. and Jensen, M. A. (1988). *Nucleic Acids Res.*, **16**, 11846.
3. Viscidi, R. P., Connelly, C. J., and Yolken, R. H. (1986). *J. Clin. Microbiol.*, **23**, 311.
4. Forster, A. C., McInnes, J. L., Skingle, D. C., and Symons, R. H. (1985). *Nucleic Acids Res.*, **13**, 745–761.
5. Sanford, D. G. and Krugh, T. R. (1985). *Nucleic Acids Res.*, **13**, 5907.
6. Lebacq, P., Squalli, D., Duchenne, M., Pouletty, P., and Joannes, M. (1988). *J. Biochem. Biophys. Methods.*, 255.
7. a) Ruth, J. L. (1984). *DNA*, **3**, 123.
   b) Ruth, J. L. (1985). *DNA*, **4**, 93.
   c) Ruth, J. L. (1990). US Patent no. 4 948 882 '*Single-Stranded Labelled Oligonucleotides, Reactive Monomers, and Methods of Synthesis*'.
8. Dreyer, G. B. and Dervan, P. B. (1985). *Proc. Natl. Acad. Sci. USA*, **82**, 968.
9. Jablonski, E., Moomaw, E. W., Tullis, R. H., and Ruth, J. L. (1986). *Nucleic Acids Res.*, **14**, 6115.
10. Haralambidis, J., Chai, M., and Tregear, G. W. (1987). *Nucleic Acids Res.*, **15**, 4857.
11. Gibson, K. J. and Benkovic, S. J. (1987). *Nucleic Acids Res.*, **15**, 6455.
12. Cook, A. F., Vuocolo, E., and Brakel, C. L. (1988). *Nucleic Acids Res.*, **16**, 4077.
13. Urdea, M. S., Warner, B. D., Running, J. A., Stempien, M., Clyne, J., and Horn, T. (1988). *Nucleic Acids Res.*, **16**, 4937.
14. Sproat, B. S., Lamond, A. I., Beijer, B., Neuner, P., and Ryder, U. (1989). *Nucleic Acids Res.*, **17**, 3373.
15. Telser, J., Cruickshank, K. A., Morrison, L. E., and Netzel, T. L. (1989). *J. Am. Chem. Soc.*, **111**, 6966.
16. Telser, J., Cruickshank, K. A., Schanze, K. S., and Netzel, T. L. (1989). *J. Am. Chem. Soc.*, **111**, 7221.
17. Meyer, R. B. Jr. Tabone, J. C., Hurst, G. D., Smith, T. M. and Gamper, H. (1989). *J. Am. Chem. Soc.*, **111**, 8517.
18. Horn, T. and Urdea, M. S. (1989). *Nucleic Acids Res.*, **17**, 6959.
19. Roget, A., Bazin, H., and Teoule, R. (1989). *Nucleic Acids Res.*, **17**, 7643.
20. Pieles, U., Sproat, B. S., and Lamm, G. M. (1990). *Nucleic Acids Res.*, **18**, 4355.
21. Inoue, H., Imura, A., and Ohtsuka, E. (1985). *Nucleic Acids Res.*, **13**, 7119.
22. Bischofberger, N. and Matteucci, M. D. (1989). *J. Am. Chem. Soc.*, **111**, 3041.
23. Singh, D., Kumar, V., and Ganesh, K. N. (1990). *Nucleic Acids Res.*, **18**, 3339.

24. Casale, R. and McLaughlin, L. W. (1990). *J. Am. Chem. Soc.*, **112**, 5264.
25. Nelson, P. S., Sherman-Gold, R., and Leon, R. (1989). *Nucleic Acids Res.*, **17**, 7179.
26. Misiura, K., Durrant, I., Evans, M. R., and Gait, M. J. (1990). *Nucleic Acids Res.*, **18**, 4345.
27. Hobbs, F. W., Jr (1989). *J. Org. Chem.*, **15**, 3420.
28. Bergstrom, D. E. and Ruth, J. L. (1977). *J. Carbohydt. Nucleosides, Nucleotides*, **4**, 257.
29. Bergstrom, D. E. and Ruth, J. L. (1976). *J. Am. Chem. Soc.*, **98**, 1587.
30. Renz, M., and Kurtz, C. (1984). *Nucleic Acids Res.*, **12**, 3435.
31. Li, P. P., Medon, P. P., Skingle, D. C., Lanser, J. A., and Symons, R. H. (1987). *Nucleic Acids Res.*, **15**, 5275.
32. Bradford, M. M. (1972). *Anal. Biochem.*, **72**, 248.
33. Leary, J. J. and Ruth, J. L. (1989). In *Nucleic Acid and Monoclonal Antibody Probes*, (ed. B.Swaminathan and G.Prakash), pp. 33–57. Marcel Dekker, Inc., New York.
34. Goodchild, J. (1990). *Bioconjugate Chemistry*, **1**, 165.
35. Podell, S., Maske, W., Jbamez, E., and Jablonski, E. (1991). *Mol. Cell. Probes*, **5**, 117.

12

Oligonucleotides attached to intercalators, photoreactive and cleavage agents

N. T. THUONG and U. ASSELINE

1. Introduction

During the past few years there has been a growing interest in the use of modified oligonucleotides as research tools in molecular biology and as specific inhibitors of gene expression. The potential therapeutic application of oligonucleotides as artificial gene control agents depends on the specificity of their binding to the target nucleic sequence under physiological conditions, their resistance to nuclease, and their uptake by intact cells. The calculation based on the statistical distribution of the nucleic bases showed that short oligonucleotides (11–15 nucleotides in length) could be used to recognize specifically a single target in the mRNA population of the human cell (1). This should facilitate the penetration rate across cell membranes and increase the probability of finding the target sequence in an accessible region. However, a short oligonucleotide might not have a strong enough affinity towards its target sequence. This affinity can be strongly increased by covalently linking an intercalating ring at the end of the oligonucleotide. These molecules can also be used as carriers for active chemical substances to induce irreversible reactions in the target sequences. To achieve this goal, oligonucleotides have to be substituted by a chemical active or photoreactive species. To obtain substances that are stable with respect to nuclease activity, the natural deoxynucleosides having the β-anomeric configuration can be replaced by the synthetic α-anomers (for reviews, see references 1–4).

Two general methods can be used to attach intercalating and active substances to the ends of the oligodeoxyribonucleotides (3, 4). The direct incorporation of the chemical groups during the assembly of the oligomer involves the following:

(a) Introduction via a linker of a functional group (generally a hydroxyl group) to the ligand and appropriate protection of the other functional groups of the two moieties not engaged in the coupling reaction;

(b) coupling reaction to afford the protected intermediate;

(c) unblocking conditions not affecting the conjugated oligomer.

The second strategy involves the introduction of appropriated functional groups in the two unprotected moieties and their specific coupling to provide the conjugated oligodeoxyribonucleotide. Many systems can be used by the intermediate of various functional groups involved in the oligodeoxyribonucleotide such as the following: phosphate, phosphorothioate, amino, thiol, carboxyl and *cis*-diol groups. In this report we will develop only the principal method used in our laboratory, namely the S-alkylation reaction of the phosphorothioate group by the halogenoalkyl group of the ligand.

The chapter first describes the method and a few detailed procedures which allow the incorporation of the appropriate functional group in the ligands. Their coupling with the oligodeoxyribonucleotide chain via either the phosphotriester approach or the phosphorothioate method followed by the general methods used for their purification and their characterization are then outlined. The coupling methods developed in this report can be used to derivatize oligodeoxyribonucleotides built with natural β-D-nucleosides or unnatural deoxynucleosides (α-D, α-L, or β-L). Lastly, some properties and potential applications of these modified oligodeoxyribonucleotides are briefly summarized. An additional part lists commercially available reagents. The attachment of other groups to the 5′-terminus of oligonucleotides is described in Chapter 8.

2. Synthesis of intercalating and active derivatives with a hydroxyl or halogenoalkyl group via a linker

In our model studies, the most commonly used substituent groups consist of an acridine derivative (2-methoxy-6-chloro-9-amino-acridine) as the intercalating agent and the following two sets of molecules as the active groups (some of them also possess intercalating properties): metal chelates such as porphyrin–Fe, phenanthroline–Cu and EDTA–Fe; and photoreactive groups such as azidophenacyl, azidoproflavine, proflavine, psoralen, porphyrin, and ellipticine (*Figure 1*). Among the derivatives of these substances involved in the coupling with the oligodeoxyribonucleotides, only 4-azidophenacylbromide (N_3ØBr) is commercially available, while others such as proflavine [(Pf) $(CH_2)_nBr$] (5), azidoproflavine [(N_3Pf) $(CH_2)_nBr$] (6), and 1-aminoellipticine [(Elli) $(CH_2)_3NH(CO)$ $(CH_2)_6Br$] (7) derivatives involve a multistep synthesis which will not be outlined in this chapter. We describe here (*Figures 1* and *2*) the synthesis of compounds such as the acridine derivatives (1b), [(Acr)$NH(CH_2)_5OH$]); (1c), [(Acr)$NH(CH_2)_5Br$]; the methylpyrroporphyrin xx-derivative (2b) [(MPPo)C(O)$NH(CH_2)_6OH$]; the orthophenanthroline (3c), [(OP)NHC(O)$(CH_2)_5Br$]; the ethylenediaminetetraacetic acid (4c), [Me_3 (EDTA)C(O)NH $(CH_2)_6OH$], and the psoralen

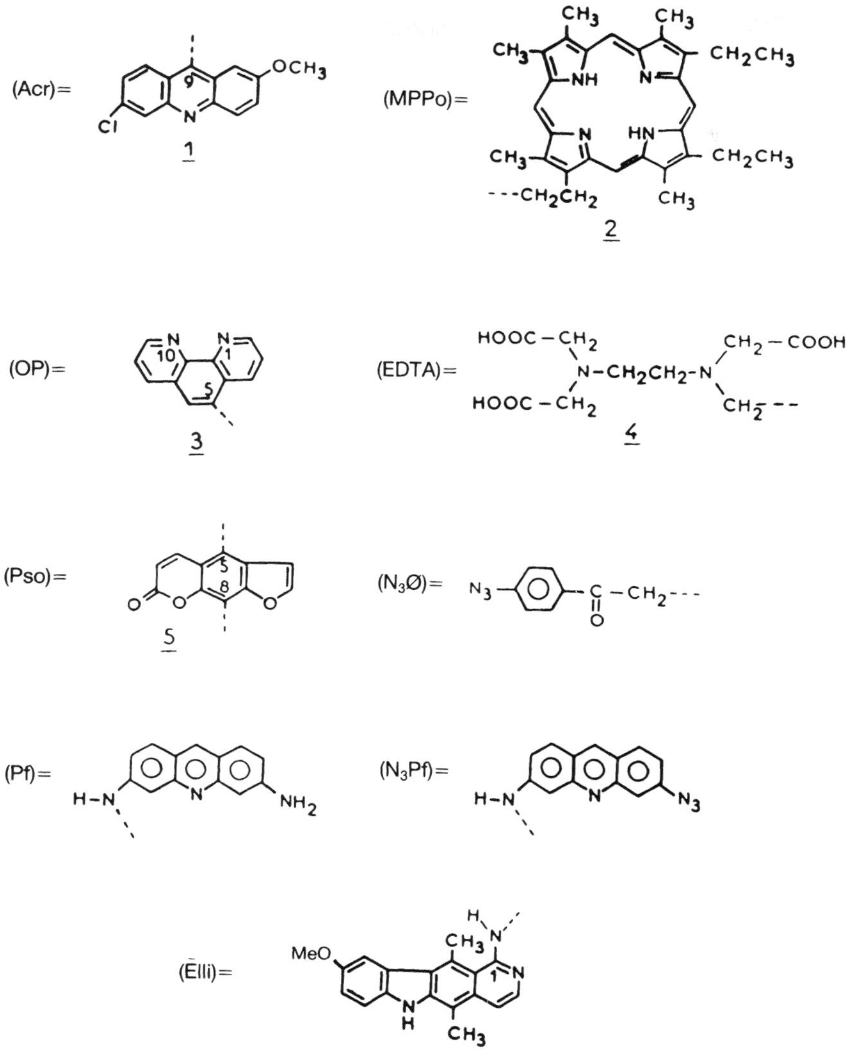

(Acr)= 1

(MPPo)= 2

(OP)= 3

(EDTA)= 4

(Pso)= 5

($N_3Ø$)=

(Pf)=

(N_3Pf)=

(Elli)=

Figure 1. Basic structures of intercalating and reactive groups.

derivatives (5c), 8-[Br$(CH_2)_5$O](Pso); (5d), 8-[HO$(CH_2)_6$O](Pso) and (5g), 5-[I$(CH_2)_6$O](Pso) which can easily be synthesized from commercially available reagents. The synthesis involves the introduction into the ligand structure of a functional group such as a hydroxyl or halogenoalkyl group via a linker to allow the substituent group to adopt a favourable configuration. Thus the best efficiency is assured for the oligodeoxyribonucleotide-conjugate for intercalation, cleavage, or cross-linking.

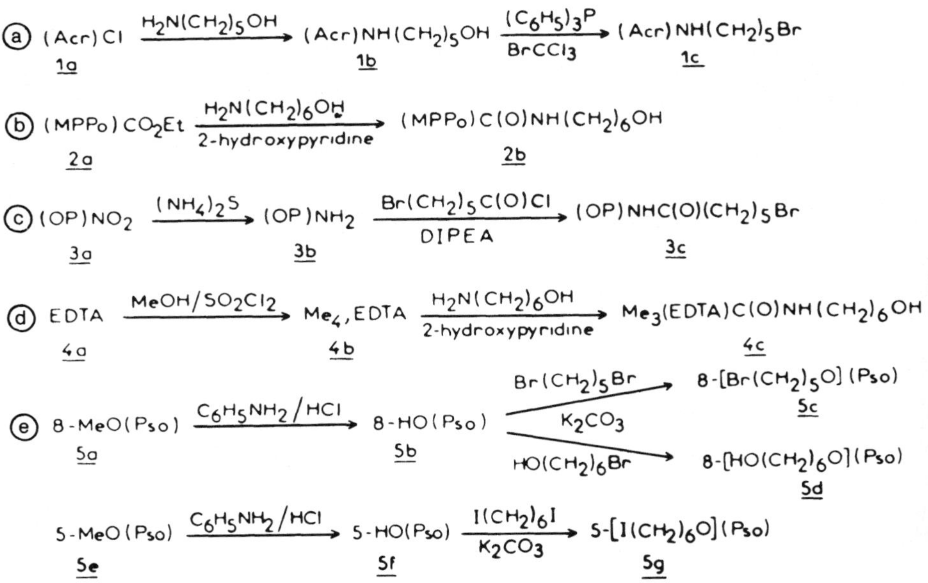

(a) (Acr)Cl **1a** $\xrightarrow{H_2N(CH_2)_5OH}$ (Acr)NH$(CH_2)_5$OH **1b** $\xrightarrow[BrCCl_3]{(C_6H_5)_3P}$ (Acr)NH$(CH_2)_5$Br **1c**

(b) (MPPo)CO_2Et **2a** $\xrightarrow[\text{2-hydroxypyridine}]{H_2N(CH_2)_6OH}$ (MPPo)C(O)NH$(CH_2)_6$OH **2b**

(c) (OP)NO_2 **3a** $\xrightarrow{(NH_4)_2S}$ (OP)NH_2 **3b** $\xrightarrow[DIPEA]{Br(CH_2)_5C(O)Cl}$ (OP)NHC(O)$(CH_2)_5$Br **3c**

(d) EDTA **4a** $\xrightarrow{MeOH/SO_2Cl_2}$ Me_4,EDTA **4b** $\xrightarrow[\text{2-hydroxypyridine}]{H_2N(CH_2)_6OH}$ Me_3(EDTA)C(O)NH$(CH_2)_6$OH **4c**

(e) 8-MeO(Pso) **5a** $\xrightarrow{C_6H_5NH_2/HCl}$ 8-HO(Pso) **5b** $\xrightarrow[K_2CO_3]{Br(CH_2)_5Br}$ 8-[Br$(CH_2)_5$O](Pso) **5c**

8-HO(Pso) **5b** $\xrightarrow{HO(CH_2)_6Br}$ 8-[HO$(CH_2)_6$O](Pso) **5d**

5-MeO(Pso) **5e** $\xrightarrow{C_6H_5NH_2/HCl}$ 5-HO(Pso) **5f** $\xrightarrow[K_2CO_3]{I(CH_2)_6I}$ 5-[I$(CH_2)_6$O](Pso) **5g**

Figure 2. Synthesis of intercalating and reactive derivatives containing a hydroxyl or a halogenoalkyl group separated by a linker.

2.1 Synthesis of the acridine derivatives 1b and 1c (*Figure 2*)

The preparation of the 2-methoxy-6-chloro-9-(ω-hydroxypentylamino)-acridine (**1b**) was achieved by reaction of 5-amino-1-pentanol with the 6,9-dichloro-2-methoxyacridine (**1a**) using *Protocol 1*. The synthesis of 2-methoxy-6-chloro-9-(ω-bromopentylamino)-acridine (**1c**) was carried out by replacing the hydroxyl group of compound (**1b**) by a bromine atom following *Protocol 2*.

Protocol 1. Preparation of the 2-methoxy-6-chloro-9-(ω-hydroxypentyl-amino)-acridine (**1b**)

1. Place **1a** (1 g, 3.59 mmol) and phenol (2.5 g) in a round-bottomed flask equipped with a magnetic stirring bar.
2. Add 5-amino-1-pentanol (740 mg, 7.18 mmol), stopper the flask and heat to 100 °C for 90 min.
3. TLC analysis on a silica gel plate using CH_2Cl_2/MeOH (90:10, v/v) as eluant shows the formation of compound **1b**; R_f (1a) = 0.9, R_f (1b) = 0.15.
4. When the reaction is completed, dilute the mixture with methanol (3 ml) and pour the solution dropwise into a magnetically stirred 2 M sodium hydroxide solution (20 ml) and maintain the stirring for 10 min.

Protocol 1. *Continued*

5. Filter the yellow precipitate of compound **1b** and wash with water until neutral. Crystallize twice from H_2O/MeOH (20:80, v/v). (Yield 80%)[a].
6. Check the absence of sodium phenolate by TLC analysis (same system as described above) followed by spraying with a solution of 2,6-dibromo-4-benzoquinone-N-chloroimine (DBPNC) (50 mg/20 ml) and heating. (Phenol gives a blue-coloured spot.)

[a] TLC and HPLC (reversed phase) analyses show a purity of about 95%. Analysis shows three side-products identified as starting material (1a), 2-methoxy-6-chloro-9-acridone and 2-methoxy-6-chloro-9-aminoacridine (≈ 1%).

Protocol 2. Preparation of 2-methoxy-6-chloro-9-(ω-bromopentylamino)-acridine (**1c**) (Figure 1)

1. Dry compound **1b** (100 mg, 0.29 mmol) by co-evaporation with acetonitrile (three times) and add anhydrous DMF (5 ml).
2. Add triphenylphosphine (76 mg, 0.29 mmol) and $BrCCl_3$ (90 μl, 0.9 mmol) and stir the mixture overnight at room temperature.
3. Concentrate the solution to dryness *in vacuo* and purify the residue on preparative glass-backed plates of silica gel using CH_2Cl_2/MeOH (85:15, v/v) as eluant. TLC analyses on silica gel plates are performed using the same eluant. R_f (**1b**) = 0.28; R_f (1c) ≈ 0.31; yield 60%.

2.2 Preparation of the methylpyrroporphyrin XXI derivative (2b) (*Figure 2*)

Compound **2b** was prepared by direct condensation of 6-amino-1-hexanol with the ethyl ester function of porphyrin **2a** in the presence of 2-hydroxypyridine using *Protocol 3*.

Protocol 3. Preparation of methylpyrroporphyrin XXI derivative (**2b**)

1. Heat a mixture of methylpyrroporphyrin XXI ethyl ester (**2a**) (0.053 g, 0.1 mmol), 6-amino-1-hexanol (0.117 g, 1 mmol), 2-hydroxypyridine (0.019 g, 0.2 mmol) and pyridine (1 ml) in the dark at 100 °C under a nitrogen atmosphere with stirring for 5 h in a round-bottomed flask equipped with a reflux condenser and drying tube containing calcium chloride.
2. Remove the pyridine under reduced pressure, add 10 ml of toluene to the residue, and acidify the mixture with 1 N HCl to pH 2. Extract the organic

Protocol 3. *Continued*

solution with water (2 × 10 ml). Combine the water phases and back-extract with $CHCl_3$ (3 × 10 ml). Dry the combined organic layers over anhydrous Na_2SO_4, filter, and evaporate the filtrate under reduced pressure.

3. Dissolve the red residue in CH_2Cl_2 and chromatograph the solution on a silica gel column (12 × 1 cm). Elute the column with CH_2Cl_2/MeOH (98:2, v/v) and then with CH_2Cl_2/MeOH (95:5, v/v). Collect fractions of about 5 ml and monitor them by silica gel TLC in CH_2Cl_2/MeOH (90:10, v/v). R_f (2a) = 0.8, R_f (2b) = 0.4. Pool the fractions containing pure product as determined by TLC and remove the solvent under reduced pressure to leave a red solid which is kept in a desiccator in the dark (yield ≈ 65%).

2.3 Synthesis of orthophenanthroline intermediates 3 (*Figure 2*)

Reduction of 5-nitro-1,10-phenanthroline (**3a**) by ammonium sulphide following *Protocol 4* affords the 5-amino derivative (**3b**) which was acylated with 6-bromohexanoyl chloride to give the bromoalkyl derivative **3c** following *Protocol 5*.

Protocol 4. Preparation of 5-amino-1,10-phenanthroline (**3b**)

1. In an efficient fume hood, add 280 ml (0.82 mol) of ammonium sulphide (20% w/v in water) to a 1000 ml three-necked flask equipped with a nitrogen inlet tube, a reflux condenser, and a dropping funnel. Heat the flask over a steam bath under a nitrogen atmosphere.
2. Dissolve 5 g (0.022 mol) of 5-nitro-1,10-phenanthroline (**3a**) in 200 ml of absolute ethanol by boiling and add this solution dropwise with a dropping funnel for an hour to magnetically stirred ammonium sulphide at 80–85 °C. Add another 125 ml of ammonium sulphide and reflux for one more hour.
3. Allow the stirred solution to cool to room temperature and extract with chloroform (4 × 300 ml). Pool the chloroform extracts and back-extract with water (2 × 60 ml). Dry the organic solution over anhydrous sodium sulphate, filter, and then concentrate the filtrate under reduced pressure.
4. Dissolve the yellow residue in boiling absolute ethanol (100 ml), filter, and add 70 ml of water to the filtrate. Yellow crystals are harvested after three days and dried in a desiccator. Yield, 50–60%; m.p. ≈ 250 °C (with decomposition at 255–260 °C).

Protocol 5. Preparation of 5-(ω-bromohexanoamido)-1,10-phenanthroline (**3c**)

1. Add dropwise a solution of 6-bromohexanoyl chloride (0.235 g, 1.1 mmol) in anhydrous acetonitrile (2 ml) to a stirred mixture of (3b) (0.195 g, 1 mmol, dried by co-evaporation with anhydrous CH_3CN), diisopropylethylamine (0.148 g, 1.15 mmol), and acetonitrile (12 ml) at room temperature. Then stir the reaction mixture for two hours at room temperature. The mixture becomes homogeneous in the process.
2. Add 8 ml of 5% aqueous $NaHCO_3$ and extracts with dichloromethane (3 × 20 ml) and back-extract with water (3 × 10 ml). Dry the organic phase over anhydrous sodium sulphate, filter, and evaporate under reduced pressure.
3. Dissolve the yellow residue in CH_2Cl_2 and chromatograph the solution on an aluminum oxide 90 (activated, neutral) column. Elute the column with CH_2Cl_2/MeOH (97:3, v/v) and collect the fractions containing pure product as determined by TLC [aluminum oxide $60F_{254}$ neutral type E in CH_2Cl_2/MeOH (97:3, v/v)] R_f (3c) = 0.3, yield = 60–70%[a].

[a] The compound should be stored at −70 °C in a freezer. It decomposes after a few weeks of storage at −20 °C.

2.4 Preparation of EDTA intermediate 4 (*Figure 2*)

The tetramethylester of ethylenediaminetetraacetic acid (**4b**) was obtained by reaction of methanol with EDTA (**4a**) in the presence of thionyl chloride as described in *Protocol 6*. Preparation of the hydroxyalkyl derivative **4c** was achieved by reaction of 6-amino-1-hexanol with the tetramethylester derivative **4b** in the presence of 2-hydroxypyridine following *Protocol 7*.

Protocol 6. Preparation of the tetramethylester of ethylenediaminetetraacetic acid (**4b**)

1. To a stirred suspension of EDTA (50 g, 0.17 mol) **4a** in methanol (170 ml) cooled at 0 °C, add dropwise an excess of thionyl chloride (101 g, 0.85 mol). At the end of the addition maintain the mixture at room temperature with stirring for ten days.
2. Neutralize the reaction mixture with a sodium bicarbonate solution (10% w/v) and extract with diethyloxide, then with CH_2Cl_2. Dry the organic solution over Na_2SO_4 and concentrate to dryness.
3. Fractionate the residue by distillation under reduced pressure, bp **4b** 158–161 °C/0.05 mm Hg. Yield: 80%.

Protocol 7. Preparation of the hydroxyalkyl derivative (**4c**)

1. Heat a mixture of 6-amino-1-hexanol (0.7 g, 6 mmol), 2-hydroxy-pyridine (0.285 g, 3 mmol) and **4b** (2 g, 6 mmol) for 1 h at 100 °C under nitrogen atmosphere.
2. TLC analysis on silica gel plates using CH_2Cl_2/MeOH (90:10, v/v) as eluant followed by spraying with a vanillin solution in concentrated H_2SO_4 (1 g in 100 ml) and heating shows the disappearance of the red-coloured spot corresponding to the 6-amino-1-hexanol ($R_f = 0$) while another dark red-coloured spot corresponding to the expected product **4c** appears, $R_f = 0.4$.
3. Purify compound **4c** on a silica gel column using increasing amounts of MeOH in CH_2Cl_2 (2:98, v/v to 5:95, v/v). Yield 40%.

2.5 Preparation of the psoralen intermediates 5 (*Figure 2*)

The ω-bromoalkyloxy **5c** and the ω-hydroxyalkyloxy **5d** derivatives of psoralen were obtained by demethylation of 8-methoxypsoralen (**5a**) (as described in reference (8) followed by condensation of the hydroxyl group of **5b** with 1,5-dibromopentane or 6-bromo-1-hexanol as described in *Protocol 8*. Starting from the 5-methoxypsoralen (**5e**), the compound **5g** is obtained following the procedure described for the preparation of **5c**.

Protocol 8. Preparation of 8-(5-bromopentyloxy)-psoralen (**5c**)

1. Add successively 1,5-dibromopentane (7.5 ml, 55 mmol) and anhydrous potassium carbonate (1 g, 7.25 mmol) to a solution of 8-hydroxypsoralen (**5b**) (1 g, 4.95 mmol) in anhydrous dimethylformamide (20 ml) contained in a 100 ml round-bottomed flask equipped with a reflux condenser and a calcium chloride drying tube, and heat the flask at 70 °C in the dark under argon atmosphere with stirring for 4 h.
2. Allow the mixture to cool to room temperature and remove the insoluble mineral salts by filtration, then concentrate the filtrate to dryness under reduced pressure.
3. Dissolve the residue in CH_2Cl_2 and chromatograph the solution on a silica gel column. Elute with dichloromethane. Collect the fractions containing the pure product as monitored by silica gel TLC in CH_2Cl_2/MeOH (90:10, v/v). R_f (**5b**) = 0.45, R_f (**5c**) = 0.85. Remove the solvent under reduced pressure to leave a white solid which is washed with pentane. Yield 80%, mp 44.5–45.5 °C.

The synthesis of 8-(6-hydroxyhexyloxy)-psoralen **5d** is carried out as described above for the preparation of **5c** using 6-bromo-1-hexanol instead of 1,5-dibromopentane. After purification on a silica gel column with increasing concentrations of MeOH in CH_2Cl_2 (O–3%), compound **5d** is obtained with a yield of 75%. TLC analysis on silica gel, using CH_2Cl_2/MeOH (90:10, v/v) as eluant, R_f (**5d**) = 0.4.

The preparation of 5-(6-iodohexyloxy)psoralen (**5g**) is carried out as previously described for the preparation of **5c**, starting from 5-methoxy-psoralen (**5e**) and using 1,6-diiodohexane instead of 1,5-dibromopentane. TLC analysis on silica gel CH_2Cl_2/MeOH (90:10, v/v) as eluant gives R_f (**5f**) = 0.50, R_f (**5g**) = 0.87.

3. Attachment of intercalators and cleavage reagents to oligomers

3.1 Synthesis by the phosphotriester method in solution (*Figure 3* example a)

This method needs essentially the preparation of oligodeoxyribonucleotides, fully protected except at their 3′-end by an arylphosphodiester group. These oligodeoxyribonucleotides, which can be synthesized by the phosphotriester procedure (9), are coupled with the hydroxyl group of the linker carrying the intercalating or reactive groups (see section 2). We report here the coupling of an oligodeoxyribonucleotide with 2-methoxy-6-chloro-9-(ω-hydroxypentyl-amino) acridine (**1b**) (see *Protocol 9*). The same method can be used to couple other compounds such as derivatives of porphyrine (**2b**), EDTA (**4c**), and psoralen (**5d**) described in Section 2.

Protocol 9. Coupling of an oligodeoxyribonucleotide-3′-arylphosphodiester (**6**) with (Acr)NH$(CH_2)_5$OH(**1b**) to give **7**

1. Dry the oligodeoxyribonucleotide **6** (prepared according to ref. 9) (20 μmol) and the acridine derivative **1b** (8.4 mg, 24 μmol) by successive co-evaporations with anhydrous pyridine (three times). After the last evaporation, keep the mixture in anhydrous pyridine (3–4 ml).
2. Add 1-(2-mesitylenesulphonyl)-3-nitro-1,2,4-triazole (MSNT) (15 mg, 50 μmol) and allow the reaction mixture to react at room temperature with magnetic stirring for 30–40 min. Monitor the reaction by TLC on silica gel plates using CH_2Cl_2/MeOH (90:10, v/v) or CH_2Cl_2/MeOH (85:15, v/v) as eluant. The reaction product (yellow coloured spot on TLC) shows a higher R_f than the starting acridine derivative (**1b**). Its R_f value depends on the oligodeoxyribonucleotide length. Identification of the expected

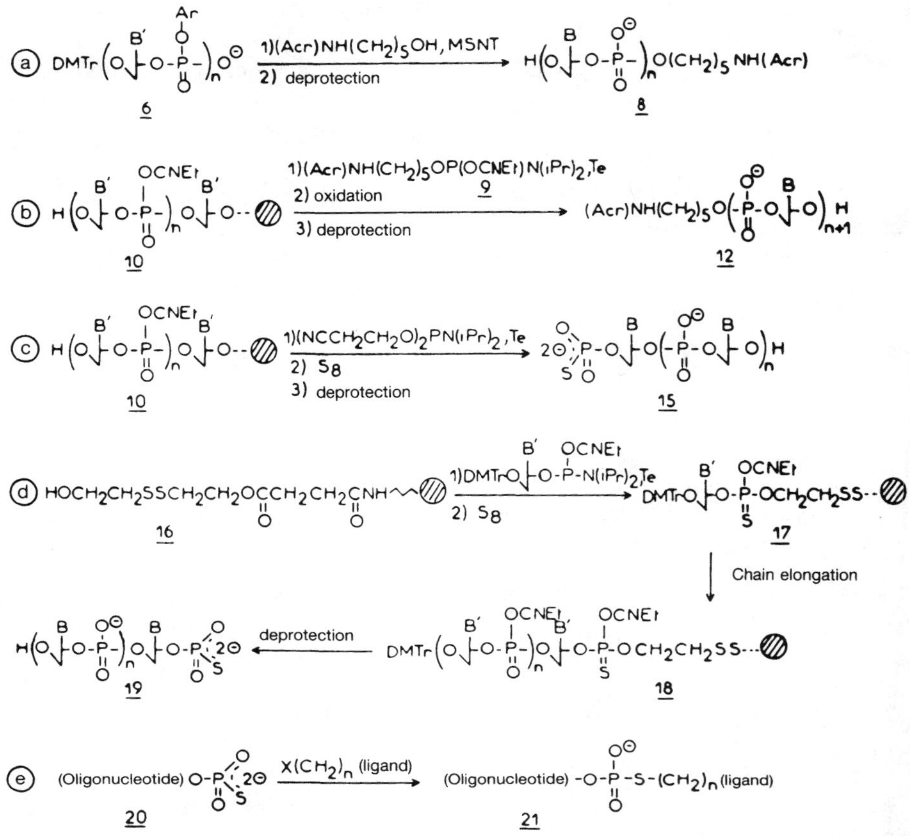

Figure 3. Synthesis of oligodeoxyribonucleotides covalently linked to intercalating and reactive groups. DMTr = dimethoxytrityl; CNEt = 2-cyanoethyl. B = nucleic base; B′ = protected nucleic base; ▨ = solid support.

Protocol 9. *Continued*

product can be confirmed by spraying a 10% perchloric acid solution which reveals the presence of the oligodeoxyribonucleotide. (The yellow coloured spot corresponding to the reaction product becomes orange due to the releasing of the 5′-trityl group of the oligonucleotide.)

3. When the reaction is completed ($\simeq$ 1 h) add ice-water (1 ml) to quench the excess of coupling reagent. Continue stirring for 10 min.
4. After the usual work-up, purify the reaction product by preparative chromatography on glass-backed plates of silica gel 60 PF_{254} with CH_2Cl_2/MeOH mixtures as eluant [usually CH_2Cl_2/MeOH (90:10, v/v) to CH_2Cl_2/MeOH (80:20, v/v) depending on the oligonucleotide length].

Full deprotection of **7** is achieved by a three-step procedure, as described in *Protocol 10*. First, the phosphate groups are dearylated using a mixture of benzohydroxamic acid (BHA) and 1,8-diazabicyclo(5.4.0)-undec-7-ene (DBU). Second, the acyl protective groups are removed from the nucleic bases by an alkali treatment. Last, the product is detritylated in an acidic medium.

Protocol 10. Full deprotection of the oligodeoxyribonucleotide covalently linked to acridine derivative **7** to give **8**

1. Treat the oligodeoxyribonucleotide–acridine conjugate (**7**) (see *Protocol 9*) (previously dried by co-evaporation with pyridine) with a molar solution of BHA and DBU in pyridine (10 equivalents of BHA–DBU for one arylphosphotriester group) for 24 h with magnetic stirring at room temperature.
2. Add a concentrated solution of sodium hydroxide[a] in MeOH/H_2O (50:50, v/v) in order to obtain a final sodium hydroxide concentration of 0.4 N or 0.5 N[b].
3. After 48 h at room temperature, neutralize the reaction mixture by treatment with Dowex 50 (pyridinium form). Remove the resin by filtration and evaporate the solution to dryness.
4. Add an 80% acid acetic solution (3 ml) and keep at room temperature for 30 min. Evaporate the solution to dryness and then co-evaporate three times with ethanol[c].
5. Take up the residue in water (5 ml) and extract with ethylacetate (three times).
6. Purify the crude oligodeoxyribonucleotide contained in the water solution by using the methods described in Section 5[d].

[a] Sodium hydroxide is used instead of the concentrated ammonia commonly employed for the removal of the protective acyl groups from the nucleic bases because of the instability of the bond between the C9 atom of the acridine ring and the N atom of the linker in the presence of ammonia.

[b] For the deprotection of the oligodeoxyribonucleotides covalently linked to substituent groups other than an acridine derivative, concentrated ammonia is used instead of sodium hydroxide for the removal of the acyl groups from the nucleic bases.

[c] Oligodeoxyribonucleotides substituted at their 5′-end do not require detritylation.

[d] TLC analysis of the crude oligodeoxyribonucleotide conjugate on silica gel using iPrOH/NH_4OH/H_2O (65:9:15, v/v) or iPrOH/NH_4OH/H_2O (55:10:25, v/v) gives a good indication of the purity of the expected product. Even though a slight decomposition of the acridine derivative is observed, we use these systems because they are the only ones that give a good separation between oligodeoxyribonucleotides of varying lengths.

3.2 Synthesis on a solid support of oligodeoxyribonucleotides covalently linked to an acridine derivative (*Figure 3*, example b)

The direct incorporation of (Acr)NH$(CH_2)_5$OH (**1b**) either at the 3′-end (via a modified derivatized support) or at the 5′-end (via its phosphoramidite derivative) of the oligonucleotide chain can also be carried out. However the first approach is not very easy to perform, particularly the purification of the deprotected 3′-derivatized oligomers. For these reasons, we will only describe (see *Protocols 11–13*) the synthesis of the 5′-acridine derivatized oligodeoxyribonucleotides in this report. Other compounds which are stable under ammonia treatment required for full deprotection of the oligodeoxyribonucleotide chain can also be prepared by this method.

Protocol 11. Preparation of the 2-methoxy-6-chloro-9-(ω-pentylamino) acridinyl-(2-cyanoethyl)diisopropylamidophosphite (**9**)

1. To an acetonitrile solution (5 ml) of 2-methoxy-6-chloro-9-(ω-hydroxypentylamino) acridine (**1b**) (200 mg, 0.57 mmol, previously dried by co-evaporation with CH_3CN) add, under argon atmosphere, diisopropylethylamine (88.4 mg, 0.12 ml, 0.684 mmol) and 2-cyanoethyl-N-N-diisopropyl-chlorophosphite (150 mg, 0.15 ml, 0.63 mmol) at room temperature with magnetic stirring.
2. Monitor the reaction by TLC on silica gel with CH_2Cl_2/MeOH (80:20, v/v) as eluant. After 30 min the starting material **1b** (R_f = 0.35) is transformed into two compounds (isomers: R_f = 0.55 and R_f = 0.82). When the reaction is completed, CH_2Cl_2 is added (10 ml) and the organic layer is washed with citric acid (10% w/v, 3 ml) then with concentrated NaCl. After being dried over $MgSO_4$, the organic solution is concentrated to dryness.
3. Apply the residue to a silica gel column and elute with a mixture of ethylacetate and NEt_3 (95:5, v/v). Pool the fractions containing the pure product and remove the solvent under reduced pressure. Purified acridine phosphoramidite (9) can be stored at −20 °C for a few months. Prepare an acridine phosphoramidite solution (0.1 M) in anhydrous acetonitrile just prior to use.

Protocol 12. Coupling of the acridine derivative to the 5′-end of the oligonucleotide chain (1 μmol scale)[a] to obtain **11**

1. After the chain elongation using the classical phosphoramidite procedure (10, see also Chapter 1), perform an additional detritylation step.

Protocol 12. *Continued*

2. Allow the acridine phosphoramidite **9** to react with the 5′-terminal hydroxyl group of the oligodeoxyribonucleotide bound to support (*Figure 3*, example 10) using the following modified cycle, and a long reaction time on a DNA synthesizer: a mixture of **9** (0.1 ml of 0.1 M solution in CH_3CN), tetrazole (0.5 ml of 0.5 M solution in CH_3CN) and CH_3CN (0.1 ml) is recycled for 7 min with a flow rate of 1 ml/min on a Pharmacia Gene Assembler. Perform this reaction step twice.
3. Perform the oxidation step using the standard procedure (see Chapter 1).

[a] The incorporation of the acridine derivative can also be performed by manual addition, under argon atmosphere, of a mixture of the acridine phosphoramidite (9) (0.1 ml 0.1 M solution in CH_3CN) and tetrazole (0.3 ml of 0.5 M solution in CH_3CN) to the 5′-detritylated oligonucleotide bound to the support (10) contained in a stoppered short vial. After gently shaking the mixture from time to time by hand for 10 min, the excess solution is removed with a syringe and the oxidation step is performed by addition of an iodine solution (1 ml) (same solution that is used on the synthesizer). The solution is removed after 1 min and the yellow support washed with CH_3CN.

Protocol 13. Deprotection of the solid phase synthesized oligodeoxyribonucleotide–acridine derivative **11** to obtain **12**

1. Treat the solid phase derivatized by the acridine oligodeoxyribonucleotide compound **11** with a sodium hydroxide solution (0.5 M) in $MeOH/H_2O$ 70:30 v/v (3 ml) at room temperature.
2. Remove the solid support by filtration after a few hours and maintain the yellow solution at room temperature for 40 h.
3. Neutralize the oligodeoxyribonucleotide solution by Dowex 50 resin (pyridinium form). Remove the resin by filtration and wash it with water (2 × 2 ml).
4. Concentrate the solution *in vacuo* to remove MeOH and extract the water solution with ethyl acetate (to remove the cleaved protective groups).
5. Purify the oligodeoxyribonucleotide–acridine derivative following the general procedure (see Section 5)

---

4. Synthesis via a phosphorothioate group

4.1 Preparation of oligodeoxyribonucleotides involving a phosphorothioate group at their 5′-end (15) (*Figure 3*) (see Protocols 14–16)

The 5′-phosphorothioate group is incorporated in the oligodeoxyribonucleotide using bis(2-cyanoethyl)-diisopropylamidophosphite followed by a sulphurization step.

Protocol 14. Preparation of bis(2-cyanoethyl)-diisopropylamidophosphite (**13**)

1. To a mechanically stirred mixture of 2-cyanoethanol (21.3 g, 0.30 mol) and diisopropylethylamine (43 ml, 0.33 mol) in dry ether (250 ml), add dropwise under argon atmosphere at 0 °C a solution of commercially available N,N-diisopropylamidodichlorophosphite (30 g, 20.4 ml, 0.15 mol) in dry ether (50 ml).
2. At the end of the addition, allow the stirred mixture to warm up to room temperature for 1 h. Filter the precipitated salt and wash with ether. Concentrate the filtrate by evaporation.
3. Purify the residue by molecular distillation. A falling-film distillation head available from Aldrich (catalogue no. Z 15, 660–4) can be used. ^{1}H-NMR ($CDCl_3$): ς 1.20 [d, 2($\mathbf{CH_3}$)$_2$CH)]; 2.66 [t, 2($OCH_2\mathbf{CH_2}CN$)]; 3.56–3.68 [(m, 2(CH_3)$_2$**CH**)]; 3.78–3.94 [m, 2($OCH_2\mathbf{CH}_C N$)]. The purified product can be stored at −20 °C for one year without degradation.

Protocol 15. Incorporation of the bis(2-cyanoethyl)phosphorothioate group at the 5′-end of an oligonucleotide to obtain **14**

1. After the chain elongation using the classical phosphoramidite procedure (10, see also Chapter 1), perform detritylation step.
2. Allow the bis(2-cyanoethyl)diisopropylamidophosphite (**13**) to react with the 5′-terminal hydroxyl group of the oligodeoxyribonucleotide bound to the support (*Figure 3*, example 10) using the following modified cycle on the DNA synthesizer: a mixture of **13** (0.1 ml of 0.1 M solution in CH_3CN), tetrazole (0.5 ml of 0.5 M solution in CH_3CN) and CH_3CN (0.1 ml) is recycled for 7 min with a flow rate of 1 ml/min on a Pharmacia Gene Assembler. Perform this reaction step twice.
3. Perform the sulphurization reaction as follows: the column is detached from the synthesizer and, using a peristaltic pump, a solution of S_8 (5% by weight) in CS_2/pyridine [50:50 v/v] is passed at 1 ml/min for 90 min, under argon atmosphere, through the support bearing the oligodeoxyribonucleotide (see also Chapter 4).
4. Wash the solid support with a mixture of CS_2/pyridine for 10 min to remove the excess sulphur.

Protocol 16. Full deprotection of the oligodeoxyribonucleotides bearing a 5′-bis(2-cyanoethyl)-phosphorothioate group (**14**) to give **15** (see *Figure 3*).

1. Treat the support carrying **14** with concentrated ammonia (5 ml) 6 h at 55 °C.
2. Evaporate the ammonia solution to dryness.
3. Solubilize the crude material with water and extract with ethylacetate.
4. Purify the oligomer **15** as described in the general protocol (see Section 5).

4.2 Preparation of oligodeoxyribonucleotides carrying a phosphorothioate group at their 3′-end (*Figure 3*, example 19)

This synthesis requires the preparation of a modified derivatized support involving a disulphide bond such as **16** (see *Figure 3*) as described in *Protocol 17*, followed by the addition of the nucleoside 3′-phosphothioate-triester using *Protocol 18*, to give the support **17** (see *Figure 3*). After this, chain elongation and unblocking of the oligonucleotide are performed as described in *Protocol 19*.

Protocol 17. Preparation of 2,2′-dithiodiethyl-derivatized support (**16**) (*Figure 3*).

1. To a magnetically stirred solution of 2,2′-dithiodiethanol (0.6 g, 3.8 mmol) in anhydrous pyridine (15 ml), add dimethoxytritylchloride (1.37 g, 0.4 mmol). The reaction is monitored by TLC analysis on silica gel plates with CH_2Cl_2/MeOH [90:10, (v/v)] as eluant.
2. When the reaction is completed, concentrate the solution under reduced pressure and dissolve the obtained gum in dichloromethane (80 ml). Wash the organic layer once with a 5% aqueous solution of sodium bicarbonate (10 ml), then with H_2O (3 × 10 ml). Dry the organic layer over Na_2SO_4, filter, and evaporate the filtrate to dryness *in vacuo*.
3. Purify the 2-dimethoxytrityloxy-2′-hydroxyethyldisulphide by flash chromatography using increasing concentrations of methanol in dichloromethane [1:99 (v/v) to 3:97 (v/v)] as eluant. Yield 75%. TLC analysis on silica gel plates is performed using dichloromethane/methanol [90:10 (v/v)] as eluant. $R_f \approx 0.66$.

Protocol 17. *Continued*

4. Starting from 2-dimethoxytrityloxy-2′-hydroxyethyldisulphide instead of dimethoxytritylnucleoside using the method described in ref. 11 for derivatizing Fractosil 500, the 2,2′-dithiodiethyl-derivatized Fractosil 500 support is obtained with a loading of approximately 60–70 μmol/g.

Protocol 18. Preparation of the derivatized support with nucleoside-3′-phosphorothioate (**17**) (*Figure 3*)

1. Add a mixture of 5′-O-dimethoxytrityl-N-acyldeoxynucleoside-3′ (2-cyanoethyl-)diisopropylamidophosphite (116 mg, 82 μmol) and tetrazole (11.5 mg, 165 μmol) in anhydrous acetonitrile (1 ml) to the detritylated dithiodiethyl-derivatized Fractosil 500 (16) (300 mg, 20 μmol) and shake by hand for 5 min under an argon atmosphere.
2. Remove the excess phosphoramidite monomer and wash the solid support with acetonitrile.
3. Add 2 ml of an S_8-solution [5% (w/v)] in CS_2/pyridine 50:50 (v/v) mixture to the solid phase and shake from time to time for 30 min. Wash the support with CS_2/pyridine 50:50 (v/v) mixture and then with CH_3CN.
4. Assay the deoxyribonucleotide phosphorothioate loading spectrophotometrically by determining the amount of dimethoxytrityl cation released by the acid treatment of a sample of the support (see ref. 11). The support should have a loading of approximately 40–50 μmol of deoxyribonucleotide per gram[a]. If the support has a lower loading, it can be rederivatized using another sample of the phosphoramidite monomer followed by the sulphurization step. If the loading is satisfactory the unreacted hydroxyl groups are capped as described in ref. 11.

[a] Verifying the derivatized nucleoside-3′-phosphorothioate support **17**. Treat an analytical amount of each derivatized support with dithiothreitol in concentrated NH_4OH (50 mg/ml) for 48 h. TLC analysis on silica gel of the obtained product using iPrOH/NH_4OH/H_2O [65:9:15, (v/v)] as eluant allows to determine the loading of the support with the nucleotide (usually nearly quantitative yield). The 5′-O-dimethoxytrityl-deoxynucleoside-3′-phosphorothioate gives an orange-coloured spot with perchloric acid and a pink-coloured spot with DBPNC.

The R_f values obtained are as follows:

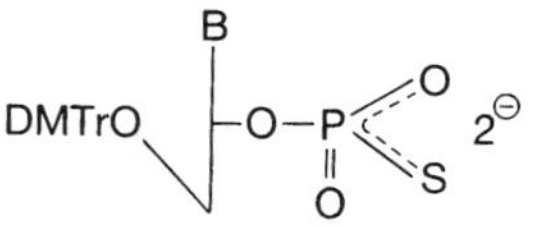

DMTrO–(sugar, B)–O–P(=O)(O)(S) $2^{\ominus}$

B = A $R_f \simeq 0.65$ B = C $R_f \simeq 0.60$
B = G $R_f \simeq 0.60$ B = T $R_f \simeq 0.55$

Following the same procedure, derivatized nucleotide-3′-phosphorothioate supports involving unnatural nucleosides such as α-D-deoxynucleosides, α-L-T and β-L-T can also be obtained.

Protocol 19. Synthesis of oligodeoxyribonucleotides carrying a phosphorothioate group at their 3′-end (**19**) (*Figure 3*)

1. Starting from the support **17**, perform the sequential growth of the oligodeoxyribonucleotide chain **18** by using the phosphoramidite procedure including the oxidation step (see Chapter 1).
2. At the end of the last coupling cycle, add a detritylation step to unblock the 5′-hydroxyl function.
3. Treat the protected oligodeoxyribonucleotide bound to the support with a 0.1 M dithiothreitol solution in concentrated aqueous ammonia at room temperature for 48 h[a].

[a] This one-step process allows both the cleavage of the disulphide bridge followed by elimination of the β-mercaptoethyl group and removal of the classical protective groups (acyl from nucleic bases and cyanoethyl from phosphates). Purification is achieved as described for the 5′-phosphorothioate-modified oligodeoxyribonucleotides.

4.3 Coupling reaction between oligodeoxyribonucleotides containing a phosphorothioate group (20) and the substituent group involving a halogenoalkyl linker to give 21 (*Figure 3*)

A water-soluble substituent group can be coupled to the oligonucleotides in water. Usually the oligodeoxyribonucleotide (10–20 OD units) and the halogenoalkyl derivative (1–2 mg) are solubilized in 500 μl–1 ml of water. For the other compounds, various mixtures of water and organic solvent can be used such as water/DMSO, water/DMF and water/MeOH. Coupling can even be performed in an organic medium[a]. The coupling reaction involving short oligodeoxyribonucleotides can be monitored by TLC analysis on Merck silica gel plates (Merck 5554) using the following eluting systems: first CH_2Cl_2/MeOH 80:20 (v/v), then iPrOH/NH_4OH/H_2O 65:9:15 (v/v) for the oligodeoxyribonucleotide ≤10mer and iPrOH/NH_4OH/H_2O 55:10:25 (v/v) for 10–20mer oligonucleotide length. Usually the excess chromophore is eluted at the top of the solvent and the coupled oligodeoxyribonucleotide is eluted with a slightly higher R_f than that of the starting oligodeoxyribonucleotide containing the phosphorothioate group. The time necessary for the coupling reactions varies. Sometimes the reaction is nearly complete after only a few hours (azidophenacyl bromide). In other cases, the incubation time needed is around 48 h. Usually the yield of the coupling reaction can reach 80–90%.

[a] Lyophilized oligodeoxyribonucleotide (sodium salt) (10 A_{260} units) is dissolved in MeOH (0.6 ml) in the presence of 15-crown-5 (or 18-crown-6 in the case of potassium salt) (0.015 g) to solubilize the oligodeoxyribonucleotide. The halogenoalkyl linker carrying the substituent such as the acridine and psoralen derivatives (1–2 mg) is added to the oligonucleotide solution and the mixture is incubated with stirring at room temperature.

When the reaction is completed or when no further change can be observed[b], the oligodeoxyribonucleotide conjugate is isolated. Various strategies can be used depending on the solubility of the halogenoalkyl-containing compounds. In the case of water-soluble compounds, the excess halogenoalkyl-containing compound can be separated from the oligodeoxyribonucleotide by gel filtration on a G-10 or G-25 column from Pharmacia. The reaction product and the unreacted oligonucleotides containing the phosphorothioate group are eluted at the void volume of the column. In other cases, the organic solvent is evaporated *in vacuo* and the residue is taken up in water (2 ml) and the excess halogenoalkyl-containing compound is extracted (three times) with $CHCl_3$. In all cases, full purification of the reaction product is achieved by ion exchange or reversed phase chromatography following the procedures described in Section 5.

[b] In the case of phenanthroline, we sometimes obtain a low yield for the coupling reaction or even no reaction at all. Liquid chromatography analysis using various systems (see Section 5, *Figure 4*, for conditions) shows two other products besides the starting oligodeoxyribonucleotide containing the phosphorothioate group and the expected oligodeoxyribonucleotide conjugate. These two compounds are identified as the corresponding oligodeoxyribonucleotide containing phosphate group and the oligodeoxyribonucleotide phosphorothioate dimer obtained by disulphide bond formation. Treatment of the latter with dithiothreitol leads to the starting material. The formation of these side-products can be explained by the oxido-reductive process generated by the simultaneous presence of phenanthroline, thiophosphate, oxygen, and divalent cations.

5. Purification, analysis, and identification[a,b]

5.1 General methods

The following description is for standard procedures used for oligodeoxyribonucleotides (10–20mer) derivatized at either the 3′- or the 5′-ends by an intercalating agent or a reactive group. To illustrate this method, results obtained with a 16mer involving various modifications at the 5′-end are summarized in *Figure 4*. In most cases, the crude oligodeoxyribonucleotides obtained from the above protocols are purified by ion exchange chromatography on Mono Q and Mono P columns and checked by reversed phase HPLC on a Lichrospher 100 RP column. In some instances, ion exchange purified materials are sufficiently pure for most purposes (particularly those involving a phosphorothioate group which can be, after desalting, directly used for their subsequent coupling with the chosen ligand). However to obtain very pure modified oligodeoxyribonucleotides, reversed phase HPLC on a Lichrospher 100 RP 18 column is needed as the final purification step.

[a] Prior to chromatography analysis and purification, the crude oligonucleotide solution is filtered through a 0.45 μm disposable filter to remove any solid particles.

[b] *Warning*: In order to avoid chelation of the phosphorothioate, porphyrin, phenanthroline, or EDTA derivative attached to oligodeoxyribonucleotides, all solvents and buffers used for their purification need to be treated with Chelex 100 resin to remove the divalent cations.

Z \ System	a	b	c	d
H	20 min 48 sec	10 min 6 sec	13 min 3 sec	15 min
$(Acr^{\oplus})NH(CH_2)_5O-P(O^{\ominus})(=O)-$	21 min	14 min 15 sec	13 min 36 sec	-
$(OP)NH(CO)(CH_2)_5S-P(O^{\ominus})(=O)-$	21 min 48 sec	-	15 min 50 sec	-
$(SPso)O(CH_2)_6S-P(O^{\ominus})(=O)-$	24 min	18 min 42 sec	17 min 45 sec	-
$2\ominus\,(O)(O)P(=O)-$	21 min 36 sec	9 min 3 sec	14 min 32 sec	8 min 36 sec
$2\ominus\,(O)(S)P(=O)-$	22 min 42 sec	9 min 3 sec	15 min 33 sec	13 min 20 sec
$Oligo-P(O^{\ominus})(=O)-S-S-P(O^{\ominus})(=O)-Oligo$	25 min 48 sec	10 min	23 min 36 sec	14 min 40 sec

Figure 4. Retention times of the 5′-modified 16mer Z-d$^{5'}$-(TTTTCTTTTCCCCCCT) in different chromatographic systems. (**a**) Mono Q HR 5/5 (Pharmacia); linear gradient of NaCl (0–50% 1.5 M over 37.5 min) in NaH_2PO_4 0.01 M, pH 6.8, 20% CH_3CN, 1 ml/min. (**b**) Lichrospher 100 RP 18 5 μm, 4 × 125 mm (Merck); linear gradient of CH_3CN (5–24% by volume over 20 min) in 0.1 M aqueous TEAA, pH 7, 1 ml/min. (**c**) Gen-Pak FAX 4.6 mm × 10 cm (Waters); linear gradient of NaCl (20–50% 1 M over 25 min) in 0.025 M Tris, pH 7.5, 0.5 ml/min. (**d**) Delta-Pak C_4 5μm 100 Å, 3.9 mm × 15 cm (Waters); linear gradient of CH_3/CN (5–16% by volume over 20 min) in 0.1 M aqueous TEAA, pH 7, 0.7 ml/min. Z indicates the group attached to the 5′-end of the oligonucleotide.

This method allows easy separation of the underivatized or the starting oligodeoxyribonucleotide and the oligodeoxyribonucleotide conjugate which is often more lipophilic[c].

Other systems such an ion exchange Gen-Pak FAX column or a reversed phase Delta-Pak C_4 column which allows good separation of the unsubstituted oligonucleotide, phosphate, and phosphorothioate-containing oligomers (for example see *Figure 4*) can be used to complete the analysis and sometimes to achieve separation where some of the above described systems were unsuccessful.

Detection of the oligonucleotide conjugates during analysis and purification steps are performed, in all cases, by measuring the absorption at 254 nm to detect the oligonucleotide part. For all compounds described in this chapter, except for the EDTA and phenanthroline derivatives, two-fold detection can be used at $\lambda = 254$ nm and at a wavelength where only the substituent groups absorb the light.

After purification, the absorption spectra of the oligodeoxyribonucleotide conjugates, in most cases, allow the confirmation of the structure of the synthesized compounds by exhibiting the absorption characteristic for acridine, proflavine, porphyrin, psoralen and ellipticine derivatives. Moreover, nuclease degradation of the oligonucleotide conjugates, followed by the monomer unit analysis, confirms complete removal of productive groups and base composition of the oligonucleotide together with the position of the attachment of the light-absorbing group on the oligodeoxyribonucleotide.

5.2 Systems commonly used for analysis and purification

5.2.1 Ion exchange chromatography

Analysis and purification by ion exchange chromatography are performed on a Mono Q HR 5/5 or HR 10/10 column from Pharmacia. Elution is performed with a linear gradient of NaCl (0–100% of 1 M) in bis-tris 0.01 M, pH 6 and 20% MeOH or a linear gradient of NaCl (0–100% of 1.5 M) in NaH_2PO_4 0.01 M, pH 6.8 and 20% MeOH, using a flow rate of 1 ml/min with an HR 5/5 column or 5 ml/min with an HR 10/10 column. A Mono P HR 5/5 column can also be used with a linear gradient of NaCl (0–100% of 1.5 M) in NaH_2PO_4 0.01 M, pH 6.8 and 20% MeOH with a flow rate of 1 ml/min and Gen-Pak FAX column 4.6 mm × 10 cm (Waters) with a linear gradient of NaCl (0–1 M NaCl) in Tris 0.025 M, pH 7.5 using a flow rate of 0.5 ml/min.

Desalting of the collected fractions, after ion exchange purification, is carried out on a column packed with Lichroprep PR 18 (Art 13900 from Merck).

[c] This does not apply to phenanthroline derivatives which are eluted with poorly resolved peaks having a retention time higher than that expected for oligomer conjugates presenting similar lipophilicity. In most cases, orthophenanthroline-containing oligodeoxyribonucleotides can be separated from the starting oligonucleotide using an ion exchange column (Mono Q and Mono P).

5.2.2 Reversed phase HPLC

The reversed phase column (125 mm × 4 mm) used for analysis is packed with 5 μm Lichrospher RP 18 from Merck. The following elution system is used: a linear gradient of CH_3CN 5–80% in 0.1 M aqueous triethylammonium acetate, pH 7, with a flow rate of 1 ml/min. Reversed phase purifications are carried out on a 250 mm × 10 mm column packed with 10 μm Lichrospher RP 18 (Merck). The elution system described above is used with a flow rate of 4 ml/min. Reversed phase chromatography can also be performed on a Delta-Pak C_4 column 5 μm 100 Å, 3.9 mm × 15 cm (Waters) with a linear gradient of CH_3CN 5–80% in 0.1 M aqueous triethylammonium acetate, pH 7, with a flow rate of 0.7 ml/min.

6. Properties and potential applications of the oligodeoxyribonucleotides covalently linked to intercalating and reactive groups

(For review see references 1–3, 12 and references therein)

6.1 Interactions between the oligodeoxyribonucleotide–intercalator conjugates and their complementary sequences

Oligodeoxyribonucleotides, built from either β-D-deoxyribonucleoside (13) or α-D-deoxyribonucleoside units (resistant to nuclease activity) (14, 15) and covalently linked to an acridine derivative, can recognize specifically their complementary sequence both on RNA and DNA single-stranded targets by a mini-double helix formation. With β-D-deoxyribonucleosides-containing oligodeoxyribonucleotides, the binding of both strands occurs in the antiparallel orientation in DNA/DNA and DNA/RNA hybrids. Contrary to the oligo-β-D-deoxyribonucleotides, α-D-deoxyribonucleosides-containing oligodeoxyribonucleotides bind in a parallel orientation with respect to a complementary sequence contained within a single-strand DNA (15). However, when the oligo-α-deoxyribonucleotides bind to an RNA sequence, the orientation of the two strands might be sequence-dependent (1). Oligo-α-thymidylates bind in the antiparallel orientation whereas multibase containing α-sequences bind in the parallel orientation. In all cases, the intercalating agent provides an additional binding energy that stabilizes the complexes formed with the complementary sequence (13–15). Stabilization depends on the linker length between the oligonucleotide and the acridine derivative as well as its attachment site on the oligonucleotide (13, 16). Homopyrimidine oligo-α-deoxyribonucleotides involving thymidine and cytidine have also been shown to bind to the major groove of the duplex

DNA in the antiparallel orientation with respect to the purine-containing strand, whereas the corresponding β-oligomers bound only when they were synthesized in a parallel orientation (17). Intercalating agents covalently linked to a homopyrimidine oligodeoxyribonucleotide have been shown to stabilize the triple helix (17). Intercalation takes place at the junctions between the double and the triple helix.

6.2 Biological effects of oligodeoxyribonucleotides covalently linked to intercalating agents

Oligo-β-deoxyribonucleotides covalently linked to an acridine derivative have been shown effective for inhibiting gene expression *in vitro* or in cell culture. This has been reported in prokaryotes for gene 32 mRNA in T4 phage (18) and in eukaryotes for rabbit β-globin mRNA (19). Oligodeoxyribonucleotides covalently linked to an acridine derivative have also shown antiviral effects such as inhibition of the cytopathic effect of the influenza virus (20) and inhibition of SV40 DNA replication (21). Oligodeoxyribonucleotide–acridine conjugates can also have an anti-oncogenic effect by inhibiting p21 Ha-ras protein synthesis (22). Oligodeoxyribonucleotide–acridine derivatives can also be efficient in killing parasites in cultured cells as shown for *Trypanosoma brucei* (23). Recently, oligo-α-deoxyribonucleotides covalently linked to an acridine derivative via their 5′-end targeted to the cap site and the initiation region have been shown efficient for inhibiting the translation of mRNA β-globin (24). The mechanism for this inhibition does not affect RNase H activity which mediates the antisense efficiency of the β-oligodeoxyribonucleotides directed against mRNA.

6.3 Irreversible modifications induced by modified oligonucleotides in the target sequences

Two families of oligonucleotides carrying either metal-chelating or photoreactive groups have been synthesized. Of the metal chelate family EDTA–Fe (25), phenanthroline–Cu (26), or porphyrin–Fe (27) have been conjugated to oligodeoxyribonucleotides. These compounds, in the presence of reducing agents such as thiol-containing substances and oxygen, generate $OH^{\cdot}$ radicals or metal-oxo derivatives that attack the deoxyribose or ribose of the target nucleic sequence and lead to the cleavage of the phosphodiester bonds (25–27). These oligonucleotide metal–chelate conjugates have been shown efficient to selectively cleave both single- and double-stranded DNA. As observed for oligonucleotide–intercalator conjugates, efficiency depends on the length of the linker and on the site chosen for attachment of the reactive group on the oligonucleotide (28). The highest yield was obtained with phenanthroline derivatives.

Of the photoreactive group family proflavine (5), azidoproflavine (6), azidophenacyl (29), porphyrin (30), psoralen, and ellipticine (31) derivatives were covalently linked to oligonucleotides. Upon visible or near-UV irradiation, these substances induce the cross-linking of the oligodeoxyribonucleotide to the target nucleic acid. In the case of ellipticine derivatives, irradiation leads to the cleavage of both strands of the target DNA at neutral pH (31). These oligodeoxyribonucleotide–ellipticine conjugates represent the first case reported of a sequence-specific artificial photoendonuclease.

7. Conclusions

The interest in these oligodeoxyribonucleotides covalently linked to intercalating and reactive groups is two-fold. First they are valuable as research tools for genetic analysis, to assign function to genes and to elucidate mechanisms in molecular biology. Secondly, they show an important potential as therapeutic agents.

References

1. Hélène, C. and Toulmé, J. J. (1989). In *Oligodeoxynucleotides: Antisense Inhibitors of Gene Expression* (ed. J. S. Cohen), pp. 137–172. Macmillan Press, London.
2. Hélène, C., Le Doan, T., and Thuong, N. T. (1989). In *Photochemical Probes in Biochemistry* (ed. P. E. Nielsen), pp. 219–229. Kluwer. Academic Publishers, Dordrecht, The Netherlands.
3. Thuong, N. T., Asseline, U., and Montenay-Garestier, T. (1989). In *Oligodeoxynucleotides: Antisense Inhibitors of Gene Expression* (ed. J. S. Cohen), pp. 25–51. Macmillan Press, London.
4. Knorre, D. G., Vlassov, V. V., and Zarytova, V. F. (1989). In *Oligodeoxynucleotides: Antisense Inhibitors of Gene Expression* (ed. J. S. Cohen), pp. 173–196. Macmillan Press, London.
5. Praseuth, D., Le Doan, T., Chassignol, M., Decout, J. L., Habhoub, N., Lhomme, J., Thuong, N. T., and Hélène, C. (1988). *Biochemistry*, **27**, 3031.
6. Le Doan, T., Perrouault, L., Praseuth, D., Habhoub, N., Decout, J. L., Thuong, N. T., Lhomme, J., and Hélène, C. (1987). *Nucleic Acids Res.*, **15**, 7749.
7. Ducrocq, C., Wendling, F., Tourbez-Perrin, M., Rivalille, C., Tambourin, P., Pochon, F., Bisagni, E., and Chermann, J. C. (1980). *J. Med. Chem.*, **23**, 1212.
8. Schönberg, A. and Aziz, G., (1953). *J. Am. Chem. Soc.*, **75**, 3265.
9. Stawinsky, T., Hozumi, T., Narang, S. A., Bahl, C. P., and Wu, R. (1977). *Nucleic Acids Res.*, **4**, 357.
10. Caruthers, M. H. (1987). In *Synthesis and Applications of DNA and RNA* (ed. S. A. Narang), pp. 47–77. Academic Press, London.
11. Atkinson, T. and Smith, M. (1984). In *Oligonucleotide Synthesis: a Practical Approach* (ed. M. Gait), pp. 35–81. IRL Press, Oxford.
12. Hélène, C. and Thuong, N. T. (1988). In *Nucleic Acids and Molecular Biology*. (ed. F. Eckstein and D. M. J. Lilley), vol. 2, pp. 105–123. Springer-Verlag, Heidelberg.
13. Asseline, U., Delarue, M., Lancelot, G., Toulmé, F., Thuong, N. T., Montenay-Garestier, T., and Hélène, C. (1984). *Proc. Natl. Acad. Sci. USA*, **81**, 3297.

14. Thuong, N. T., Asseline, U., Roig, V., Takasugi, M., and Hélène, C. (1987). *Proc. Natl. Acad. Sci. USA*, **84**, 5129.
15. Sun, J. S., Asseline, U., Rouzaud, D., Montenay-Garestier, T., Thuong, N. T., and Hélène, C. (1987). *Nucleic Acids Res.*, **15**, 6149.
16. Asseline, U., Toulmé, F., Thuong, N. T., Delarue, M., Montenay-Garestier, T., and Hélène, C. (1984). *EMBO J.*, 3, 795.
17. Sun, J. S., François, J. C., Montenay-Garestier, T., Saison-Behmoaras, T., Roig, V., Thuong, N. T., and Hélène, C. (1989). *Proc. Natl. Acad. Sci. USA*, **86**, 9198.
18. Toulmé, J. J., Krisch, M. M., Loreau, N., Thuong, N. T., and Hélène, C. (1986). *Procl. Natl. Acad. Sci. USA*, **83**, 1227.
19. Cazenave, C., Loreau, N., Thuong, N. T., and Hélène, C., (1987). *Nucleic Acids Res.*, **15**, 4717.
20. Zerial, A., Thuong, N. T., and Hélène, C. (1987). *Nucleic Acids Res.*, **15**, 9909.
21. Birg, F., Praseuth, D., Zerial, A., Thuong, N. T., Asseline, U., Le Doan, T., and Hélène, C. (1990). *Nucleic Acids Res.*, **18**, 2901.
22. Saison-Behmoaras, T., Tocqué, B., Rey, I., Chassignol, H., Thuong, N. T., and Hélène, C. (1991). *EMBO J.* , **10**, 111.
23. Verspieren, P., Cornelissen, A. W. C. A., Thuong, N. T., Hélène, C., and Toulmé, J. J. (1987). *Gene*, **61**, 307.
24. Boiziau, C., Kurfurst, R., Cazenava, C., Roig, V., Thuong, N. T., and Toulmé, J.-J., (1991), *Nucleic Acids Res.*, **19**, 1113.
25. Boidot-Forget, M., Chassignol, M., Takasugi, M., Thuong, N. T., and Hélène, C. (1988). *Gene*, **72**, 361.
26. François, J. C., Saison-Behmoaras, T., Barbier, C., Chassignol, M., Thuong, N. T., and Hélène, C. (1989). *Proc. Natl. Acad. Sci. USA*, **86**, 9702.
27. Le Doan, T., Perrouault, L., Hélène, C., Chassignol, M., and Thuong, N. T. (1986). *Biochemistry*, **25**, 6736.
28. François, J. C., Saison-Behmoaras, T., Chassignol, M., Thuong, N. T., and Hélène, C. (1989). *J. Biol. Chem.*, **264**, 5891.
29. Praseuth, D., Perrouault, L., Le Doan, T., Chassignol, M., Thuong, N. T., and Hèlène, C. (1988). *Proc. Natl. Acad. Sci. USA*, **85**, 1349.
30. Le Doan, T., Praseuth, D., Perrouault, L., Chassignol, M., Thuong, N. T., and Hélène, C. (1990). *Bioconjugate Chem.*, **1**, 108.
31. Perrouault, L., Asseline, U., Rivaille, C., Thuong, N. T., Bisagni, E., Giovannangeli, C., Le Doan, T., and Hélène, C. (1990). *Nature*, **344**, 358.

Appendix: Chemical suppliers

All reagents and solvents used in solid phase oligodeoxyribonucleotide synthesis should be of the highest purity in order to avoid contaminants which cause undesired by-products and inefficient reactions. HPLC grade solvents are recommended. Chemical sources are listed below.

dimethoxytritylchloride	Merck
2,2′-dithiodiethanol	Aldrich
ammonia	Prolabo

dichloroacetic acid	Merck
1,2-dichloroethane	SDS
acetic anhydride	Merck
2,4,6-trimethylpyridine	Merck
iodine	Merck
anhydrous sodium sulphate	Prolabo
sodium bicarbonate	Prolabo
NaCl	Merck
dithiothreitol	Aldrich
S_8	Prolabo
5-amino-1-pentanol	Aldrich
6,9-dichloro-2-methoxy-acridine	Aldrich
18-Crown-6, 15-Crown-5	Aldrich
phenol	Prolabo
Vanillin	Merck
N,N-diisopropylethylamine	Aldrich
NaH_2PO_4	Merck
acetic acid	Merck
bis-tris	Aldrich
benzohydroxamic acid (BHA)	Aldrich
1,8-diazabicyclo(5.4.0)-undec-7-ene (DBU)	Aldrich
2,6-dibromoparabenzoquinone-N-chloroimine (DBPNC)	Prolabo
2-cyanoethanol	Aldrich
phosphorus trichloride	Prolabo
perchloric acid	Prolabo
triphenylphosphine	Janssen
bromotrichloromethane	Janssen
Chelex 100 resin	Biorad
Dowex 50 W	Aldrich
Lichroprep RP18	Merck
methylpyrroporphyrin XXI ethyl ester	Aldrich
5-nitro-1,10-phenanthroline	Aldrich
G-10, G-25	Pharmacia
EDTA	Aldrich
$SOCl_2$	Prolabo
Kieselgel F. 254 TLC plates (used for all analysis mentioned in this work)	Merck Art 5554
Kieselgel 60 PF 254 glass backed plates	Merck Art 9385
Merck Silica gel (70–230 mesh)	Merck Art 7734
2-hydroxypyridine	Aldrich
6-amino-1-hexanol	Aldrich
calcium chloride	Prolabo
HCl	Prolabo

ammonium sulphide	Aldrich
6-bromohexanoyl chloride	Aldrich
aluminum oxide 90	Merck
aluminum oxide 60 F 254 neutral (type E)	Merck
8-methoxypsoralen	Aldrich
5-methoxypsoralen	Aldrich
1,5-dibromopentane	Aldrich
1,6-diiodohexane	Aldrich
argon	Prolabo
silica gel	Prolabo
anhydrous potassium bicarbonate	Prolabo
triethylamine	Prolabo
1-(2-mesitylene sulphonyl)-3-nitro-1,2,4-triazole	Aldrich
N,N-diisopropyldichlorophosphite	Aldrich

Acknowledgements

This work was supported by the CNRS, the Ligue Nationale Française contre le Cancer and Rhône-Poulenc-Santé. We wish to thank Professor D. Sigman (California) for communication of a detailed procedure for the 5-amino-1-10-phenanthroline preparation. We thank M. Chassignol and V. Roig for their contribution to this work.

A1

Suppliers of specialist items

Aldrich Chemical Company, Milwaukee, WI 53201, USA; The Old Brickyard, New Road, Gillingham, Dorset SP8 4JL, UK

Alltech Assoc. Inc., 2051 Waukegan Road, Deerfield, IL 60015, USA

American Bionetics Inc., 21377 Cabot Blvd., Hayward, CA 94545, USA

Amersham International, Amersham Place, Little Chalfont, Buckinghamshire HP7 9NA, UK; 2636 S. Clearbrook Drive, Arlington Heights, IL 60005, USA

Applied Biosystems, 850 Lincoln Centre Drive, Foster City, CA 94404, USA

BDH Ltd, Broom Road, Poole, Dorset BH12 4NN, UK

Beckman Instruments, 2500 Harbor Boulevard, PO Box 3100, Fullerton, CA 92643, USA

Berghof Labortechnik GmbH, Harretstrasse 1, D-7412 Eningen a.A., FRG

Biosyntech, Biochemische Synthesetechnik GmbH, Stresemannstrasse 268–280, D-2000 Hamburg 50, FRG

Boehringer Mannheim GmbH, Postfach 31 01 20, D-6800 Mannheim 31, FRG; PO Box 50816, Indianapolis, IN 46250, USA

ChemGenes Corp., 925 Webster Street, Needham, MA 02192, USA

Clonetech Laboratories, 4030 Fabian Way, Palo Alto, CA 94303, USA

CPG Inc., 32 Pier Lane West, Fairfield, NJ 07006, USA

Cruachem Ltd, Todd Campus, West of Scotland Sciene Park, Acre Road, Glasgow G20 0UA, UK

Diagen, Institut für molekulare Diagnostik GmbH, Niederheider Strasse 3, PO Box 130247, D-4000 Düsseldorf, FRG (distributer in the UK: Hybaid Ltd., 111–113 Waldegrave Rd., Teddington, Middlesex)

Eastman Kodak Co., Kodak Laboratory and Speciality Chemicals, 2400 Mt. Road Blvd., Rochester, NY 14650, USA

Eppendorf, PO Box 65 06 70, D-2000 Hamburg, FRG; distributed in USA by Brinkmann Instruments

Fisons Scientific Apparatus, Bishop Meadow Road, Loughborough, Leicestershire LE11 0RG, UK

Fluka Chemie AG, Industriestrasse 25, CH-9470 Buchs, Switzerland

Gelman Sciences, 600 S. Wagner Rd., Ann Arbor, Michigan 48106, USA

Glen Research Inc., 44901 Falcon Place, Sterling, VA 22170, USA

Hichrom Ltd, The Markham Centre, Station Rd., Theale, Reading, Berkshire, UK

Hybrid Ltd, see Diagen
J. T. Baker Inc., 222 Red School Lane, Phillipsburg, NJ 08865, USA
Janssen Life Sciences Products, Lammerdries 55, B-2430 Olen, Belgium
LKB, see Pharmacia LKB Biotechnology
Lumigen Inc., 1921 Pembridge Place, Detroit, MI 48207, USA
Medicell International Ltd., 239 Liverpool Rd., London N1 1LX, UK
E. Merck, Frankfurter Strasse 250, Postfach 41 19, D-6100 Darmstadt 1, FRG
Milligen-Biosearch, 81 Digital Drive, Novota, CA 94949, USA
Millipore Corp., 80 Ashby Road, Bedford, MA 01730, USA: B.P. 307, Saint-Quentin, 78054, France
Molecular Probes, PO Box 22010, 4849 Pitchford Ave., Eugene, OR 97402, USA
Nunclone Inc., Nunc Inc., Naperville, IL 60556, USA
Penisula Laboratories, 611 Taylor Way, Belmont, CA 94002, USA
Pharma-Waldhof GmbH, Hansa-Allee 159, Postfach 11 07 32, D-4000 Düsseldorf, FRG
Pharmacia LKB Biotechnology AB, S-75182 Uppsala, Sweden; 800 Centennial Avenue, Piscataway, NJ 08854, USA
Pierce Chemical Company, Rockford, IL 61105, USA; PO Box 1512, 3260 BA Oud-Beijerland, The Netherlands
Prolabo, 12 rue Pelée, B.P. 369, F-76526 Paris Cedex 11, France
Romil Chemicals Ltd., 63 Ashby Road, Loughborough, Leicestershire LE12 9BS, UK
Sigma Chemical Co., PO Box 14508, St Louis, MS 63178, USA (distributer in the UK: Fancy Road, Poole, Dorset BH17 7NH)
SDS (Solvent, Documentation, Syntheses), 15, Quai Jules Guesde, F-94400 Vitry, France
Tropix Inc., 47 Wiggins Ave., Bedford, MA 01730, USA
United States Biochemical Corp., PO Box 22400, Cleveland, OH 44122, USA
Waters/Millipore, Maple Street 34, Milford, MA 01757, USA

Index

acridine derivatives
  for coupling to oligodeoxynucleotides 286
affinity chromatography
  chromatographic matrix 243
  oligodeoxynucleotides 241
affinity depletion 79
affinity selection 83
alkaline phosphatase
  conjugation to oligonucleotides 271
2-aminodeoxyribofuranoside 162
anti-oncogenic effect 304
anti-parasitic effect 304
antisense effect 79, 90, 150
antiviral effect
  derivatized oligodeoxynucleotides 304
  methylphosphonate oligonucleotides 150
  phosphorothioate oligonucleotides 89
application of backbone modified DNA
  sequence specific protein–DNA binding 228

base-modified oligonucleotides 155
  characterization 179
  3′-phosphoramidites 174
  probes for protein–DNA interaction 181
  purification 176
  reactive probes 182
  spectral probes 181
  synthesis 157, 176
Beaucage reagent 97, 214
3H-1,2-benzodithiole-3-one 1,1-dioxide 97
biological application
  phosphorodithioate oligonucleotides 132
  phosphorothioate oligonucleotides 89
biotin 200
  linked to base 267
biotinylation 72
bis (2-cyanoethyl)-diisopropylamido-
    phosphite 296
bis (2-cyanoethyl)-phosphorothioate 296

CDPI
  attachment to phosphorothioate
    backbone 221
cleavage reagents
  attached to oligodeoxynucleotides 283
coumarin 200

deoxynucleoside 3′-O-pyrrolidino-S-
    (2,4-dichlorobenzyl) phosphoro-
    thioamidites 114
detection of DNA
  by backbone labelled fluorescent
    olignucleotides 233

diagnostic probes
  oligonucleotides with non-isotopic reporter
    groups 205
DNA–RNA oligonucleotides 45

EcoRI
  affinity chromatography 249
EDTA derivatives
  for coupling to oligodeoxynucleotides 289
enzyme
  linked to base 270
eosin 200

fluorescein 200
  attachment to oligonucleotides 269
  attachment to phosphorothioate
    backbone 220
fluorescent detection of DNA
  DNA hybridization 237
  DNA sequencing 237
fluorophore
  linked to base 268

gene expression
  inhibition by oligodeoxynucleotides 304

H-phosphonate chemistry 98
H-phosphonate linkers 194
hybridization
  oligomers with reporter group
    base attached 275

intercalators
  attached to oligodeoxynucleotides 283
irreversible modification of oligonucleotides
  by oligonucleotides 304

linker arms
  attached to base 256
  attached to terminus 283
lumiphore
  linked to base 268

M13mp18
  enzymatic incorporation of
    phosphorothioates 232
mercurate 2′-deoxyuridine 259
methylphosphonate oligonucleotides
  antisense activity 150
  antiviral activity 150
  characterization 143
  diastereomers 137
  nuclease resistance 149
  synthesis 137

methylpyrroporphyrin XXI derivatives
  for coupling to oligodeoxynucleotides 287
5-methyl-2-pyrimidinone
    deoxyribofuranoside 168
2′-O-methylribonuleotides 49
  2′-O-methyl-2,6-diaminopurine
      riboside 68
  2′-O-methyladenosine 60
  2′-O-methylcytidine 56
  2′-O-methylguanosine 64
  2′-O-methyluridine 50
  application
    affinity depletion 79
    affinity selection 83
  biotinylation 72
  controlled pore glass
    loading 71
monobromo bimane 200

N-trifluoracetyl-6-aminohexanol 189
nitrophenyltetrazole 70
non-isotopic reporter groups
  attached to backbone 211
  biotin 200
  coumarin 200
  eosin 200
  fluorescein 200
  linked to base 267
  monobromo bimane 200
  oligonucleotides with 185, 200
  rhodamine 200
  Texas Red 200
nuclease resistance
  methylphosphonate oligonucleotides 149
  phosphorodithioate oligonucleotides 132
  phosphorothioate oligonucleotides 87
nucleoside 3′-phosphoramidites 174

oligodeoxynucleotides
  5′-terminal modifying agents 191
oligodeoxynucleotides
  affinity chromatography 241
    chromatographic matrix 243
    EcoRI 249
  anti-oncogenic effect 304
  anti-parasitic effect 304
  anti-viral effect 89, 150, 304
  attachment to Sepharose 247
  automated synthesis 1
    change of chemistry 15
    coupling efficiency 13
    stability of reagents
  automated synthesis limitations 19
  base modified 155
  biotin linked 267
  fluorophore linked 268
  hybridization
    base attached reporter groups 275
  intercalator linked 283
    biological effects 304
    interaction with complementary
      sequences 303
  irreversible modification 304
  linked to cleavage reagents 283
  linked to photoreactive groups 283
  linker arm synthesis 256
  lumiphore linked 268
  non-isotopic reporter groups
    linked to base 267
    sensitivity comparison 274
  phosphorodithioates 109
  phosphorothioates
    attachment of intercalators 295
    for stability 87
  phosphorylation 246
  phosphotriester synthesis
    attachment of intercalators 291
  reporter groups
    attached to base 255
    at backbone 211
    at 5′-terminus 185
  solid-phase synthesis
    phosphoramidite method 1
oligonucleotide
  enzyme linked 270
oligonucleotide methylphosphonates 137
oligonucleotide phosphorodithioates 109
oligonucleotide phosphorothioates
  chemical synthesis 87, 211
  enzymatic synthesis 232
oligonucleotides
  base modification 155, 255
  reporter groups at 5′-terminus
    via amino link 200
    via sulphydryl link 200
oligoribonucleotides 25
  enzymatic analysis 45
  gel purification 42
  HPLC purification 39
  materials/reagents 32
  mixed DNA/RNA 45
  phosphorothioate-containing 46
  purification 36
orthophenanthroline derivatives
  for coupling to oligodeoxynucleotides
    288

pentylnitrite 159
phosphorodithioate oligodeoxynucleotides
  analysis, purification 125
  biological applications 132
  resistance to nucleases 132
  synthesis 109
phosphorothioate diesters
  attachment of intercalators 295, 297
  reactivity 219

phosphorothioate oligonucleotides 87
application 88
antisense 90, 93
antiviral agents 89
autolytic processing 89
enzyme biochemistry 89
interaction with proteins 89
mutagenesis 89
attachment of reporter groups at backbone 211
enzymatic synthesis 232
with fluorescent label detection of DNA 232
oligoribonucleotides 46
product isolation 100
^{35}S-labelling 102
single diastereomer 216
synthesis
Beaucage reagent 97
3H-1,2-benzodithiole-3-one 1, 1-dioxide 97
H-phosphonate chemistry 98
phosphoramidite chemistry 98
stereoselectivity 92
sulfurization 97
tetraethylthiuram disulfide 47
at terminus 295
phosphorothioate triesters
stability 227
phosphorylation
oligonucleotides 246
phosphotriester method
attachment of intercalators 291
photoreactive groups
attached to oligodeoxynucleotides 283
post-assay labelling of DNA 234
PROXYL spin label
attachment to phosphorothioate backbone 220
psoralen derivatives
for coupling to oligodeoxynucleotides 290
purine deoxyribofuranoside
synthesis 157
purine nucleosides
with linker arm 263
pyrimidine nucleosides
with linker arm 257

reporter groups
attached to backbone 211
attached to base 255
attached to 5′-terminus 185
sensitivity comparison 274
rhodamine 200

Sepharose
activation 244
solid-phase oligodeoxynucleotide synthesis
phosphoramidite method 1

tetraethylthiuram disulfide 47
Texas Red 200
6-thiodeoxyguanosine 162
4-thiothymidine 168
tris (pyrrolidino) phosphine 113
tryptrophan repressor
detection of binding
backbone modified DNA 229

vesicular stomatitis virus 150

Printed in the USA/Agawam, MA
September 22, 2015

623858.027